THE OXFORD HISTORY OF ENGLISH LITERATURE

General Editors

†JOHN BUXTON and †NORMAN DAVIS

XIII

THE OXFORD HISTORY OF ENGLISH LITERATURE

Certain volumes originally appeared under different titles (see title-page versos). Their original volume-numbers are given below.

I. Middle English Literature 1100–1400
J. A. W. Bennett; edited and completed by Douglas Gray
(originally numbered I part 2)

II. Chaucer and Fifteenth-Century Verse and Prose
H. S. Bennett *(originally numbered II part 1)*

III. Malory and Fifteenth-Century Drama, Lyrics, and Ballads
E. K. Chambers *(originally numbered II part 2)*

IV. Poetry and Prose in the Sixteenth Century
C. S. Lewis *(originally numbered III)*

V. English Drama 1485–1585
F. P. Wilson; edited with a bibliography by G. K. Hunter
(originally numbered IV part 1)

VI. English Drama 1586–1642: Shakespeare and his Age
(forthcoming; originally numbered IV part 2)

VII. The Early Seventeenth Century 1600–1660: Jonson, Donne, and Milton
Douglas Bush *(originally numbered V)*

VIII. Restoration Literature 1660–1700: Dryden, Bunyan, and Pepys
James Sutherland *(originally numbered VI)*

IX. The Early Eighteenth Century 1700–1740: Swift, Defoe, and Pope
Bonamy Dobrée *(originally numbered VII)*

X. The Age of Johnson 1740–1789
John Butt; edited and completed by Geoffrey Carnall
(originally numbered VIII)

XI. The Rise of the Romantics 1789–1815: Wordsworth, Coleridge, and Jane Austen
W. L. Renwick *(originally numbered IX)*

XII. English Literature 1815–1832: Scott, Byron, and Keats
Ian Jack *(originally numbered X)*

XIII. The Victorian Novel
Alan Horsman *(originally numbered XI part 1)*

XIV. Victorian Poetry, Drama, and Miscellaneous Prose 1832–1890
Paul Turner *(originally numbered XI part 2)*

XV. Writers of the Early Twentieth Century: Hardy to Lawrence
J. I. M. Stewart *(originally numbered XII)*

THE VICTORIAN NOVEL

ALAN HORSMAN

CLARENDON PRESS · OXFORD
1990

Oxford University Press, Walton Street, Oxford OX2 6DP
Oxford New York Toronto
Delhi Bombay Calcutta Madras Karachi
Petaling Jaya Singapore Hong Kong Tokyo
Nairobi Dar es Salaam Cape Town
Melbourne Auckland
and associated companies in
Berlin Ibadan

Oxford is a trade mark of Oxford University Press

Published in the United States
by Oxford University Press, New York

This volume was originally to have been numbered XI part 1

British Library Cataloguing in Publication Data
Horsman, Alan 1918–
The Victorian novel. — (The Oxford history of English literature; 13)
1. Fiction in English, 1837–1900. Critical studies
I. Title 823.809
ISBN 0–19–812216–0

Library of Congress Cataloging in Publication Data
Horsman, E. A. (Ernest Alan), 1918–
The Victorian novel / Alan Horsman.
p. cm. — (The Oxford history of English literature; 13)
Includes bibliographical references.
1. English fiction—19th century—History and criticism.
I. Title. II. Series.
PR871H67 1990
823'.809—dc20 90–34108
ISBN 0–19–812216–0

Translated from the author's disks by
Latimer Trend & Company Ltd, Plymouth
Printed and bound in Great Britain by
Courier International Ltd,
Tiptree, Essex

Preface

My debt is great to the University of Otago, to my colleagues and students for innumerable discussions and disagreements, to the University Council and Research Committee for periods of leave and for travel grants, without which it would have been impossible to complete this book, and to the staff of the Central Library. I have to thank Professor Ian Donaldson for transcripts of Trollope's marginalia, and my wife, members of my family, and Freda von Lindheim for help with other materials from distant places. Professor Kathleen Tillotson and Professor D. W. Harding have generously read and commented on some chapters in draft, as the General Editors have on all; their encouragement has meant much to the brief chronicler of authors not noted for brevity. The chronicle depends on a great deal more bibliographical work than was admitted to the Select Bibliography and the assistance of Briar Gordon and Brian Jenkins, research assistant to Geoffrey Tillotson, is gratefully acknowledged. I am grateful too for secretarial help, over the long period of the volume's gestation, from, in succession, Elsie Moodie, Anaig Fenby, and Mary Sullivan. The final form of the typescript owes everything to the interest and the word-processing skills of Jane Jones.

NOTE References to earlier volumes use the new numbering (p. ii above).

A. H.

Contents

1. Introduction

It is tempting to say of the word Victorian what Henry James said of Venice: 'It is a great pleasure to write the word but I am not sure there is not a certain impudence in pretending to add anything to it.' When the first thing to be added is the word fiction, it would be pleasant to follow him in the disclaimer: 'I do not pretend to enlighten the reader; I pretend only to give a fillip to his memory . . .' At any rate, the reader can expect to be enlightened only if he has fresh memories of the fiction or is prepared to acquire them. The actual novels are the heart of the discussion, which is concerned with the difference made to their description and interpretation by considering the chronology and circumstances of their composition. The circumstances in any one case include the books preceding or accompanying it, by this author or by others, in this genre or in others; in order to select from the vast evidence, while offering a connected account, the end-products must be grouped according to author or sub-genre. But whatever the grouping—and complete consistency in this could be attained only in a study of far fewer works than those of some fifty years—the aim is the better understanding of particular novels, some of them among the most remarkable ever written.

The subject, as well as the sequence of the present series, demands a beginning not with the accession of Victoria but with the death of Scott in 1832. The plan of the series required this volume to end with the death of George Eliot, that is, with the novelists whose most important work was accomplished by then, even though they continued to write after 1880. (So Hardy and a group of minor authors, including Rhoda Broughton and Mrs Oliphant, whose best novels belong to the eighties and later years, are excluded.) This too is appropriate to the subject, since most of the writers included are those for whom Scott was the great exemplar, both in the subjects he chose and in his very career, which had advanced prose fiction to a position rivalling that of poetry.

Although dividing the continuum of literary history into periods may be held to resemble outmoded methods of explaining the geological record by reference to a series of catastrophes, a division at 1832 has the justification that that date did actually seem to many readers, writers, and publishers of fiction to signal a decisive overturning of an older order. Novelists and publishers had shared with the general public in the apprehensions of the 1830s arising from civil disorder in the countryside in the south (in support of a living wage), from strikes in the industrial north, and from popular agitation for the reform of parliament. They had also had their own professional catastrophe. Scott's death in September 1832 followed years of over-work to pay off the debts of Ballantyne and Co., printers with whom his affairs were involved; his publishers, Constable and Co., had gone bankrupt in 1826, and they had not only commanded a uniquely high 31*s*. 6*d*. for each of his novels, from 1821, but had also planned, by a lowering of the price of reprints to levels hitherto unexampled in the century, to make books more widely available. The failure of such a firm added to other publishers' anxieties about their own survival.

In 1832 the bibliographer Thomas Dibdin, on the title page of his *Remarks on the Present Languid and Depressed State of Literature and the Book Trade*, capitalized FEAR not only 'of bailiffs and tax-gatherers,' but 'of *Reform*, of *Cholera*, and of BOOKS'. The state of popular feeling made many people fear the subversive potentiality of print—a stamp duty on newspapers, though lowered in 1837 to 1*d*. per sheet, was not removed until 1855; a tax on paper remained until 1861. In 1832, writers, from Wordsworth down, found that anxiety about the state of the realm limited public interest in imaginative literature. One novelist, Bulwer, in his *England and the English* (1833), believed that the only novels likely to succeed in the early thirties were those which 'either developed the errors of the social system, or the vices of the legislative'. Other kinds certainly achieved publication, but there is no doubt publishers felt the risks with them to be greater than they had previously been. Although the fears of a revolution to match the French one of 1830, and perhaps even of 1789, were not realized, they persisted into the period of growing prosperity following the second wave of railway building, 1844–7. But publishers were hardly reassured by the crises in each decade—when the movement for a People's Charter failed in 1848, when the Crimean War broke out in 1854, and when further agitation for extension of the franchise came to a head in 1867, after the death of its most powerful opponent, Lord Palmerston.

Of Constable's two price strategies, the higher one was obviously the less risky. Publication in three volumes at 10*s*. 6*d*. a volume allowed costs to be covered fairly quickly from the sale of a relatively small edition (usually about 1,000). The writer could be offered, usually by the outright sale of the copyright, enough to make writing worth while—£100, Trollope believed, when negotiating for one of his first successes, *Barchester Towers*, in 1857: 'if a three-volume novel be worth anything it must be worth that'. Most of the population lived on less than that per year.

These conditions were favourable to the publication of a large number of new titles, though not, except for the most popular, of a large number of copies of each. As the population increased and as a greater proportion of this number came to be living in cities and to belong to the professional classes, the demand for new fiction of a certain quality, or a certain appearance of quality, grew also, to be met not by buying but by borrowing from subscription libraries. For one guinea, only two-thirds of the price of a single new novel, it was possible to borrow as many as could be read in a year. Of course there were waiting lists for the most popular, but multiple copies were bought by the circulating libraries, the chief of which after 1852 was Charles Edward Mudie's. From the middle of the century Mudie was able to bid for a proportion of the copies of any new novel he thought would interest his readers. Although this proportion was often large enough for him to drive down the figure he paid, the chance of more quickly clearing an edition was so attractive that the question whether a book would be taken by Mudie became a regular factor in a publisher's calculations. Acceptability of course referred not only to a novel's chance of proving popular but also to its content for family reading. Meredith's *The Ordeal of Richard Feverel* failed the test in 1859; Meredith believed it took ten years for his reputation to recover. But Mudie only spoke for his customers and most novelists accepted the right of the paying public to call the tune. Thackeray might protest in the Preface to *Pendennis* that 'since the writer of *Tom Jones* was buried, no writer of fiction among us has been permitted to depict to his utmost power a MAN', but once he was editor of the *Cornhill* magazine and responsible for the size of its circulation, he was as careful as any publisher about what he allowed into it.

Some of the constraints arising out of publication in three volumes, in particular the limited *sale* to the public of the first edition, could be avoided. Almost by chance, with *Pickwick Papers* in 1836, Dickens rediscovered the advantages of publication in monthly numbers. This was an older method of enabling readers to buy a long novel without having to pay for it all at once. But now each number was specially written for reading within the month. Running for nineteen months, at 1*s*. per number (of 32 pages) and 2*s*. for the double concluding number, a novel could be bought complete for less than two-thirds of the price of a three-decker—and it was longer, and illustrated. Its commercial success in Dickens's hands was crucial to the turn in the fortunes of his first publishers, Chapman and Hall, and important to their successors when he transferred to Bradbury and Evans. The capital which they could then amass enabled them to make full use of new technology in printing and binding and to advertise each new work widely, at the same time as the developing railways facilitated its distribution.

Other novelists followed Dickens in a method of publishing which gave them not only payment for their work while it was going on but a sense of immediate contact with readers whose pleasure or displeasure showed in the figures of monthly sales. Thackeray, recognized as Dickens's nearest rival, used the method for five of his seven big novels. Trollope used it for only a minority of his, but these included such outstanding work as *Orley Farm* (1861–2) and *The Way We Live Now* (1874–5).

Publication in periodicals offered similar advantages. This too was a method which had eighteenth-century precedents. The *Metropolitan Magazine* revived it in the 1830s for what proved the four most popular of Marryat's novels. Most of Surtees appeared first in the *New Sporting Magazine*, from 1831 to 1851, and most of Lever in the *Dublin University Magazine*, from 1837 to 1859, though both authors also combined this method with issue in monthly parts, as Dickens did in the case of *Oliver Twist*. A number of magazines which counted on monthly serial fiction were for a time edited by novelists, *Bentley's Miscellany* by Dickens from its foundation in 1837 until early 1839, when Ainsworth followed him, only to leave to found his own *Ains-*

worth's Magazine (1842–54). After 1860 magazine serials tended to displace novels published separately in monthly parts, since technological advances and the removal of the tax on paper now enabled magazines to compete by lowering the price to 1*s.* per number, substantial in quantity and containing other kinds of reading as well as fiction. Some of these too had novelist editors, Thackeray for the *Cornhill*, Trollope for *St Paul's Magazine*, M. E. Braddon for *Belgravia*, Mrs Henry Wood for the *Argosy*.

Most serial fiction was published in volume form when completed—in some cases just before the serial ended, in order to stimulate sales. After the first interest had waned, both serials and three-deckers might be reprinted in a cheaper single volume, such as those of Bentley's 'Standard Novels', offered at 6*s.* from 1831. In the 1840s travel by railway, smoother (on the whole) and more frequent than most travel hitherto, stimulated the sale of cheap reprints, the price of which had been brought down by the immensely successful 'Parlour Library' of Simms M'Intyre, at 1*s.* from 1847, followed by Routledge's 'Railway Library' in 1848 and Bentley's 'Shilling Series' and 'Railway Library' in 1852–3. Samuel Phillips, writing in *The Times* in 1851 on 'The Literature of the Rail', attacked such series for debasing public taste, even though for many of their titles they relied upon reprints of novels which had first appeared under more respectable auspices. The prevalence by 1850 of publication in single volumes shows it was not *necessary* for new novels to make their first appearance in three; that system was kept flourishing by its financial advantage to writers, publishers, and lending libraries.

Serial publication not only lowered the purchaser's immediate outlay and the price of the completed novel; it also offered a greater quantity for this price. Its 300,000 words, half as long again as the usual three-decker, not only allowed for subjects of larger compass—extending in the chief examples to a vision of a whole society in process of change—but also raised problems of construction peculiar to itself. Dickens was quick to see the advantages of engaging his readers' long-term interest and of making each number contribute to it. His publishers would have been content with mere episodes in *Pickwick Papers*; he was determined to offer a novel. Increasingly in the novels that followed he took pains that each number should mark a definite stage in a developing action, a stage worth thinking over in the month before the next. The form had the advantage that it set limits to how much the reader might absorb at a time and required him to assimilate each section more thoroughly, even if unconsciously, than in a consecutive reading of the whole. There is some evidence that Thackeray, who appeared to plan less carefully, revised (so far as time allowed) in order to increase the point of each instalment and prompt the reader's subtler reconsideration. Trollope appears to have found, after *Framley Parsonage* (1860–1), that serial publication stimulated the interweaving of plots, each of which threw light on the others. Eventually even George Eliot saw the advantage of putting out a novel in parts, though longer and at two-monthly intervals. This was at first a consequence of the length caused by the multiple plots in *Middlemarch* and of the publishers' resistance to a novel in four volumes. But the author, while the parts were appearing, recognized that 'the slow plan of publication has been of immense advantage to the book in deepening the impression it produces'.

Readers of three-volume novels would similarly have to wait before borrowing the next volume. That too was expected to have, in some degree, a unity of its own and to lead to a specific stage of elucidation. The control needed to achieve this was perhaps not quite so unremitting as that of the serial writer—something, in fact, like that which Dickens tended to exert additionally, in order to reach a decisive halfway point with the ninth or tenth number. And, for any novel of even the minimum standard length, it was accepted that reading would be broken into parts of some kind. So novelists became adept at developing methods of emphasis to assist continuity of impression across the breaks. Repetition of idiosyncrasies of character, or of descriptive settings of a certain type or tonality, led to suggestive rhythms; contrasts stimulated reflection, however simple they might be—town life versus country life, or the habits characteristic of a past and of a present generation, both of them matters made topical by the actual pace of change, especially in the second and third quarters of the century.

While the history of the novel is in part the history of such methods, it also itself forms part of a larger history of human awareness, in so far as it throws into relief the subjects which were thought interesting enough to be clarified by these methods. For example, Scott's awareness, in the postscript to *Waverley* (1814), that 'There is no European nation, which, within the course of half a century, or little more, has undergone so complete a change as this kingdom of Scotland' was developed, for the mid-century and the regions to the south, with prodigal force and invention, and a subtle interest in the diversity of possible attitudes toward change itself. Within this dominant subject others took their place, the changes, for instance, in attitudes to children and adolescents, their relations with parents, and their position when left without them.

The appeal to an audience aware of these changes, even if it was an appeal more to their curiosity than to their powers of reflection, required the subject to be presented in the form of a sequence of events that could have occurred; whether it was concerned with external incident or the life of the mind and feelings, the connections between one event and another had to be those which 'experience teaches us to look for', to quote the *Westminster Review* for 1853. The problems of doing this might be eased in so far as a plot could be made to depend on the connections found in some common human occupation like sailing or hunting, or in some sequence that had actually occurred. Harriet Martineau was exceptional in finding the need for such connections inhibiting: too complicated for their proper action to be seen anywhere but in life itself, they were, she believed, beyond the novelist's power to invent—echoing Carlyle, for whom history, if it were to be written by the tracking of chains of cause and effect, would find no 'writer fitted to compose it', for he would 'require All-knowledge'. Most novelists were content that the starting-point should be given and unquestioned, so long as the connection of events thereafter offered a sequence which could prompt the reader to reflect on it as if it had actually occurred.

Fiction like this, in which, to quote Scott in his 'Essay on Romance' (1824), 'the events are accommodated to the ordinary train of human events', stood in contrast to that in which, he said, 'the interest turns upon marvellous and uncommon incidents'. In Victorian practice this second mode tended to appear in surroundings of realistic fiction and to be excused as conforming to 'real life'. If, for example in *Oliver Twist*, chapter 17, sudden melodrama made the connections of the narrative appear strained, Dickens claimed to be justified by referring to 'the transitions of real life from well-spread boards to death-beds'. A more elaborate justification, in the preface to *Bleak House*, finished with an attempt to reconcile novel and romance: 'I have purposely dwelt on the romantic side of familiar things.' It was out of such an attempt that the Brontës developed the best of their published fiction, conscious as they were of the claims both of common life and of their own alternative world of insatiable dream. Emily's intricate structure of actuality and dream in *Wuthering Heights* stimulated Charlotte to go beyond the simple realism of her first mature novel, until eventually, in *Villette*, she found a way of playing the one method off against the other, to give the reader the pleasures not only of surprise but of comparison and assessment. In her own way, George Eliot followed her in *Daniel Deronda*. The sub-genre of the sensation novel, to the origins of which both Dickens and the Brontës contributed, grew out of a similar attempt to reconcile the pleasures of the strange and unexpected with those which came from seeing exactly how certain events had occurred. It led eventually to the forcing of the connections of the sequence in the interests of a thesis, in the later novels of Wilkie Collins, and to the closed form, if not the dead end, of the detective novel—the riddle, the more strange the better, to which the answer was to be discovered in the commonplaces of crime.

The controversies carried on about such matters, both within the novels and outside them, in order to safeguard the artist's freedom while meeting the demand that he attend to the ordinary train of events, were, if the *Westminster Review* was to be believed, the controversies characteristic of 'a scientific, and somewhat sceptical age'. But most of them seemed, at least in the first half of the century, to be arguments with the attacking voice of Carlyle, far more widely attended to than the voice of science or scepticism because he spoke the language of religion. He was both the friend of the novelists and their enemy. When his *Critical and Miscellaneous Essays* were collected in 1839, Thackeray spoke of him as the writer about art 'who has worked more than any other to give it its independence'. One essay, 'Characteristics', made 'Literature ... but a branch of Religion', only to elevate it into, 'in our time, ... the only branch that shows any greenness'. Its task, in the essay on Diderot, was 'to understand and record what is *true*'. The instrument of this understanding, in *Sartor Resartus* (1833–4), was the imagination; for Carlyle asserted, like Wordsworth, that 'not our Logical, Mensurative Faculty but our Imaginative one is King over us', and, like Coleridge, that symbol and figure could turn the Finite into 'the embodiment and revelation of the Infinite ... to stand visible and as it were, attainable there'. This book, George Eliot believed, had constituted, even for readers who 'had the least agreement' with Carlyle's opinions, 'an epoch in the history of their minds'.

His opinions were not easy to escape, in touch as they were with what he called 'the realities of everyday existence'. In *Chartism* (1839) he was concerned to impress the way in which 'New Eras ... come quietly, making no proclamation of themselves'. He led his readers to find the evidence of their own New Era in 'the condition-of-England question', in 'Cash Payment as the sole nexus between man and man', in 'English Commerce with its world-wide convulsive fluctuations', making the working man's trade 'of the nature of gambling'. Arguments about Chartism were to be seen as '*our* French Revolution', in the hope that 'we ... may be able to transact it by argument alone'. To this end the means were to be 'universal Education' to improve the quality of the argument, and 'general Emigration' to give the argument a chance, by reducing the pressure of population upon it, as well as upon the means of subsistence, and opening a vision of Europe 'on the verge of an expansion without parallel'. Such matters formed the groundswell in the preoccupations of many of the novelists.

Increasingly they found reason to regard Carlyle as one of themselves. In 1832 he had invited writers to consider as the true Epic of our Time not 'Arms and the Man' but 'Tools and the Man', the tools to include the pen, 'wherewith to do battle against UNREASON', and the Epic to be written in prose. In *The French Revolution* (1837), he produced the epic of the earlier time which gave warning to his own, writing as what his essay 'On History' had called 'the Artist in History'—not 'the Artisan', who was merely one of those who 'labour mechanically in a department, without eye for the Whole'. Such an artist, it turned out, could not frame a vision of the Whole without using methods like those of a novelist, selecting details for the swift characterization of a multitude of actors and repeating the details to emphasize a pattern in the actions, impressing the present reality by, as he said in a letter, 'flame-picture', in order to show how 'a Nation ... becomes *trans*cendental', a spectacle beyond the experience of the senses on which the account at first depended. Quoting with approval, 'the eye sees . . . what the eye brings means of seeing', he undertook to help his readers to the means of seeing. The novelists were among them. Thackeray reviewed the book; Dickens read and reread it; to Trollope it was 'Homer in Prose'; Charles Kingsley had his working-class hero praise it as a 'great prose poem', when speaking with gratitude of the effect of all Carlyle's writing 'on thousands of my class and of every other'. To George Eliot, in 1855, his 'greatest power' was 'in concrete presentation. No novelist has made his creations live for us more thoroughly than Carlyle has made . . . the men of the French Revolution . . .'

Yet Carlyle seemed to have little time for fiction. From 1832 in the essay 'Biography' to 1867 in 'Shooting Niagara', he denounced fiction as 'of the nature of *lying*'. That his history avoided lying, however, could be for most readers only an assumption when, despite the references in the footnotes, the authority of the account came from the manner of its presentation. *Sartor Resartus*, which even resembled a novel (by Sterne) in its form, had asked, of the documents on which it was supposedly based, 'what if many a so-called Fact were little better than a Fiction; if here we had no direct Camera-obscura Picture . . . ; but only some more or less fantastic Adumbration, symbolically, perhaps significantly enough, shadowing-forth the same!' The novelists may not all have gone so far as Meredith, who by 1895 admired him 'as a great romancer but not as an historian', but, on Carlyle's own showing, the boundary between his work and theirs was often hard to see. In *Past and Present* (1843) it was clear that his attacks were only upon the kind of 'Fiction, "Imagination", "Imaginative Poetry" &c., &c.,' which could not be 'the vehicle for truth, or *fact* of some sort'. The Waverley Romances which he spoke of as 'worthless' had 'taught all men this truth . . . that the bygone ages of the world were actually filled by living men'—a very large concession, and one of the bases of his own success in history. The *Iliad* was 'nothing of a fiction'; Dante had nothing to do with it either; and even the 'mimic Biographies' of the meanest novel could deliver some part of 'the significance of Man's Life'. Writers reading the 1837 exhortations to 'Study Reality', to 'search out . . . *its* quite endless mystery' in the 'glimpses of Romance' furnished by 'our own poor Nineteenth Century', naturally took it that they need not, as he had urged, 'sweep their Novel-fabric into the dust-cart' if they were to follow his further advice and 'betake them with such faculty as they have to understand and record what is *true*'. Such paradoxical denunciations constituted not only a spur but a programme. It was only by overlooking them that Henry James could claim, in 1884, that until 'a short time ago' the English novel 'had no air of having a theory, a conviction, a consciousness of itself behind it—of being the expression of an artistic faith . . .'. It was a faith like his own in the refusal to believe that 'the novelist is less occupied in looking for truth . . . than the historian'.

More than twenty years before, when Carlyle urged him to 'write heestory', Meredith had been able (if correctly reported) to reply 'that novel writing was his way of writing history'. A host of novelists, better and worse, in the years between Scott's death and George Eliot's could have said the same. Common judgement has distinguished those who entertained by enquiring into the truth of human motive and action, from those who were satisfied with an appeal to curiosity about what happened rather than about why. The methods they evolved for the control of the long narratives demanded by the contemporary market gave increasing scope, after major examples of the mid-century like *Vanity Fair*, *Wuthering Heights*, and *Bleak House*, for the reader's powers of consideration and comparison. Multiple narrators of the story stirred the pleasure of seeing the difference they made to the kind of human truth that could be revealed. In a few special cases, like *Villette* and *The Newcomes*, the reader was allowed to choose the ending he thought best fitted, in view of his experience up to that point. More commonly an ending was offered for the reader's own judgement of it, based on the cumulative effect of the comparisons he had made between one major movement of the action and another, often with only their implications to guide him. The scornful phrase of one of the historians, 'the drowsy spell of narrative', hardly applied. A spell was exerted, undeniably, but it stirred to an activity which was a major part of the zest of reading.

2. Earlier Minor Novelists

Most of the novelists who flourished in the first half of the century would now be called minor. They have their place in the development of fiction not only because of one or two unusual books still worth reading, but also because their work was read by or discussed in the presence of major authors to whom it suggested interests and aims which might be extended and refined. This is especially true of Thackeray, but also even of the most original of the major novelists, Dickens. The multitude of authors who are no longer read established fashions, indeed what may be called a language of evaluation, which authors of infinitely greater range and technical accomplishment might then use to give their readers entry into regions of the imagination more remote—just how remote it has taken a century to recognize fully. While it is characteristic of major authors to push back the boundaries of human awareness like this, in the nineteenth century they had also to make their readers enjoy being tempted into such new regions, even though with fear. There was security in the modes of awareness established by minor literature. After all, since genius only rarely entered into the matter, these modes merged into those common in politics and the pulpit, journalism and conversation. Some of the minor fiction, indeed, like that of Disraeli or Charles Kingsley, only just detaches itself, if at all, from these other activities and so gains a dubious double title to stand as essential background to the major writing.

I

Although Napoleon had been finally defeated as long ago as 1815, novels concerned with service in the Great War, as it came to be sometimes called, did not appear in any quantity until after Captain Marryat[1] took over the *Metropolitan Magazine* in 1832.

[1] Frederick Marryat, 1792–1848, son of a banker-merchant, joined the Royal Navy in 1806 and was a commander when he began to write novels in 1829. After retirement from the navy he edited the *Metropolitan Magazine*, 1832–5, associating with it other naval

When he retired from the Royal Navy in 1830, he already had one novel to his name, *Adventures of a Naval Officer, . . . Frank Mildmay* (1829). It would have been hard to predict from this book his future, at first as a humorist and eventually as a charmer of children. Mildmay's amorous affairs, carried on in a language deriving from Restoration comedy, heavy with faded sentiment and routine mockery of 'the sex', compete unsuccessfully for attention against the macabre and humiliating facts of the life he encounters at sea: the 'very curious . . . muscular action' of a disembowelled youth (in chapter 11), or the buoyancy of a corpse bloated with its own gases until they escape 'with a loud whizz' (chapter 8). Marryat was still at sea himself when the novel appeared, with an epigraph from *Don Juan*:

> My muse by no means deals in fiction;
> She gathers a repertory of facts,
> Of course with some reserve and slight restriction,
> But mostly traits of human things and acts.

It has rather less 'reserve' and 'restriction' than the novels which follow, but is nevertheless making a careful literary selection—as is shown by the contrast with the tasteless horrors of Michael Scott (in chapter 3 of *The Cruise of the Midge*, for instance) and with the freedom of Marryat's own conversation as reported by Dickens in a letter of 1846. Nevertheless, that Marryat is selecting among facts gives the novel its strength. In this it is representative of all his work.

Marryat needs the discipline of facts because he has difficulty in imagining consistently and coherently without them. The constraints upon the freedom of fantasy, which in a major novelist are part of the thing imagined (because of the complex inter-relations among its parts), are clear to Marryat mainly in situations which bring to bear his own practical experience of complicated facts as constraints upon human action—the weather, the working of a ship, the lie of the land, above all the necessity to earn a living. His people are most clearly identified by their ability or inability to make use of such constraints, after recognizing that submission is

---

writers like Edward Howard (?1793–1841), who succeeded him as editor, and Frederick Chamier (1796–1870). From 1835 Marryat was a gentleman farmer and a traveller in America and on the Continent (where his wife remained).

unavoidable. The Spanish steersman under fire, in chapter 5, 'a patient, meek, and undaunted old man, who uttered no complaint', is a moving early instance. 'I liked to judge of causes and effects,' says Mildmay in chapter 3 and this is the source of the author's successes, from the fireships to the shark waiting outside his hero's cave.

It is a pity that the development of the hero's character by means of such experiences is too much a matter of routine moralizing, for Mildmay has perhaps greater potential interest than any other Marryat hero, with his pride, his fascination with energetic study, his 'crying with rage and excitement' in battle, his alternations of depression and 'delirious high spirits'. But all this excellent material is not sufficiently focused upon the exact relation of such a man to such experiences—upon the induction into adult life which is Marryat's main subject.

Increasingly he came to depend on incidental comedy of humours until in *Peter Simple* (1832–4) the formula of *Every Man out of his Humour* enabled him to make some of the adventures into the means by which the hero might lose his original 'simplicity'. Marryat over-reached himself by having this character tell his own story, so that the problem arose of so managing his imperfect apprehension of things as to convey how different it was from the reality. The problem was made all the more noticeable by having the narrator claim in chapter 14 to have written up the adventures, full as they are of seaman-like intelligence, from a journal kept at the time by the simpleton. Writing fast, Marryat had no time to think about such matters; he did not even care that the device which was intended to bind the book together could in fact be used in only a few episodes and for the most part was forgotten. Yet he continued the attempt when moving beyond the subject of naval service in a sequence of books for which the title was a name to which an adjective or adjectival expression was attached. In *Jacob Faithful* (1833–4) and *Japhet in Search of a Father* (1834–6) he began with the hero as an infant: his experience could then be presented as so surprising, from his lack of preparation for it, that no comment on his 'humorous' reaction to it was necessary. Of the two, Thackeray's 'dearly beloved *Jacob Faithful*' showed the greater interest in the hero's introduction to life because a simple didactic intent led to more concern with his nature. An orphan before either Oliver Twist or Jack Rattlin, Jacob invited the reader to sympathize with his probity, not with his predicament, as his innocent eye became informed but not corrupted by the violence of the action. But his intelligence, his independence as a waterman, the distinction between friendship and love which the attachment of his master's daughter forced him to draw, and his mild sensibility to the contemplative pleasures of river-sailing, made him more than a mere cipher in the varied action which Thackeray relished. (He praised the novel beside those of Dumas and Wilkie Collins, all 'read . . . with the most fearful contentment of mind'.)

The famous Japhet, without even the slight individualizing traits of Jacob, depended for his immense appeal at the time simply on his state as an orphan and his wish to become a son. Searching for his father revealed no more about himself and little about the world, because the search was a matter of reckless invention and coincidence. Marryat would have already seen enough of Howard's *Rattlin the Reefer* (which began in the *Metropolitan* two months earlier than *Japhet*) for the subject to be suggested to him, and perhaps for him to determine to avoid sentiment. There is here a distinct watershed between the attitudes of Marryat and of Howard towards the orphan. The hard clarity of the humours which Japhet encounters—including his father, the personification of Anglo-Indian irascibility, who quarrels with him at once, and the genial, card-sharping Major (whom Thackeray surely remembered in *Barry Lyndon*)—shows clearly where Marryat stands.

Mr Midshipman Easy (1836) returns to Marryat's strongest vein, the discipline which a theoretical mind may receive from acquaintance with facts. Easy has too much of his author in him so that, far from being strongly resistant to such discipline, he finds the conditions at sea offer him a much better chance of self-fulfilment than anything he has learned from his father's radicalism. His ignorance is instructed not only by acquaintance with storm and fire at sea, the sort of occasion to which Marryat can always rise, but also by an intelligent ex-slave in a long passage of adventures in which Marryat does not need to press the anti-radical moral. Even the more freely-ranging fantasy of exotic adventure, in a love affair between a midshipman and a Moorish beauty, gains in plausibility and interest when Marryat can make it into part of a chain of cause and effect connected with the provisioning of a ship. He presents Easy as awaking to the process of cause and effect and then running on, in chapter 14, to misapply the notion, so that he feels guilty because of his own slightest contribution to the causes of evil effects. The reader is not surprised, therefore, to find a definition of the novel, in chapter 21, as 'a series of amusing causes and effects'. What the author fails to see is his own tendency to over-emphasize 'amusing' and so to make his causes appear merely those which he has selected in order to entertain. In the passages of sea-going adventure, his native sense of the natural causes we disregard at our peril gives his narrative a substance which it lacks when mere fantasy takes over.

The coarse-grained farce and melodrama of *The Snarleyyow* (1836–7) and *The Phantom Ship* (1837–9) intervene before he can find his proper method of making a whole novel in *Poor Jack* (1840). Here he limits himself to a simpler range of facts and to a setting, in and around Greenwich and its Hospital, which, simply but adequately imagined, suffices to relate them to each other. So Marryat has no need to try to conceal the fundamental simplicity of mind which is his greatest charm. He has adequate reason to bring in characters from high life to keep going a slight romantic intrigue, as well as the pensioners with their yarns and the pilot to whom Jack is apprenticed, or to develop the unsatisfactory marriage of his parents. This latter part of the novel, the nearest Marryat ever got to a mature treatment of a relationship between the sexes, compares favourably with the presentation by Harriet Martineau of Hope's marriage in *Deerbrook* and Mrs Gore's much less specific treatment of the marriage of her *ingénue* in *The Hamiltons.* The uneasy relation of the hero to these parents contrasts with his treatment in apprenticeship to a pilot, in which he finds his place in the world, learning a profession despite temptations to make his way more quickly by nefarious means. Even the happy ending has the backing of the authentic detail of the best of the novel, since it is secured by a specific adventure with a French privateer and by a legacy from a dealer in odds and ends whose shop has been clearly present to the reader, particularly when Jack nurses her there, braving the stuffiness and the fleas. It all puts up a good claim to be Marryat's best novel.

After this his best work is for children, with the exception of *Percival Keene* (1842) which transfers the interest of a search like Japhet's to a milieu like Midshipman Easy's. In all but one of the children's books—and that the least successful and least coherent,

The Mission (1845)—he isolates a group of people including children and shows them surviving by their own efforts. It forms a parable of Victorian self-reliance, no doubt, but primarily it gives opportunity for a whole work to be framed from Marryat's characteristic subject of the constraints which are set upon human action. *Masterman Ready* (1841–2) originated in his dissatisfaction with the way these constraints were presented in *The Swiss Family Robinson*, which, his preface relates, his own children had wished him to continue. (It had been widely read ever since becoming available in English, a year after its first publication in German in 1812–13.) It was characteristic of Marryat that he should object to the careless flora and fauna in Wyss's book, since these are inconsistent with the climate presented but, along with the climate, they constitute the main facts which the castaways have to face. His own castaways are subject to more realistic constraints, conveyed in crisp detail. Ready, the second mate of 64 whose resourcefulness, after a lifetime of such constraints, secures their survival, is naturally the hero. The story of his life, counterpointed against the main action throughout the second volume (chapters 30–44), is the story of his efforts to escape from successive kinds of constraint—of home, of poverty, and of prisons, Dutch and French. Finally, humbled by misfortune, he finds freedom as a respected common seaman and eventually (though first in the order of narrating) understands the status of the individual when he sees the order and discipline on board a man-of-war: there 'every man becomes of individual importance; as any neglect of any one in the duty allotted to him immediately puts everybody else out, and everything goes wrong: besides the blame is always certain to be given to the party who is guilty' (chapter 30). He is the quintessential Marryat character, whose qualities, even his sententiousness, have been earned.

There is a little too much 'theoretical' instruction from Seagrave who, as he does less, is less attended to. It may be doubted whether his somewhat officious information has ever, as Marryat in the preface hoped it would, stimulated the curiosity of many readers. But with Ready it is another matter; it is not only for children that he remains believable and moving, through the connection between what he says and does now and what he has been in the past. The climax, Ready's death, unlike the notorious death of the hero of *The King's Own* (1830), emerges naturally from the whole novel. The didactic intention is plain: it is 'the neglect of . . . duty allotted' which, by wasting the water, 'puts everybody else out' and sends Ready, the most resourceful member of the besieged party, out for more, and to his death. But this situation is less simple than Ready or Marryat would have it; 'blame' is not 'certain to be given to the party who is guilty', since the child who wasted the water simply did not understand and was reasonably taking it from the nearest source, which happened to be the waterbutt filled against the possibility of siege.

The separation of the moral from its crucial illustration by over thirty chapters makes the example much less pat, while leaving it to be seen *as* an example. The intervening treatment of the culprit's repeated naughtiness is no doubt intended to make the reader award blame; but this, like the crudity of the judgement upon the child in the process of learning, is something added to the action and not essential to it: the action has been concerned with the discipline of reality; from it there stand out clearly the self-centred young child at one end and the mature second mate at the other, while the development of this maturity appears in the narrative of Ready's life. Something more than simple blame arises from a climax in which the mature man is destroyed as a result of the inevitable limitations of someone at the other end of the disciplining process.

As an image of life itself it has something of the dignity of tragedy. To do this in a children's book is remarkable. Inevitably, such a book is written consistently from the older standpoint and so forfeits some of the delight of the child's view of things which distinguishes the early parts of *Poor Jack* and *Jacob Faithful*. In his last completed novel, *The Children of the New Forest* (1847), Marryat reduces the didactic weight of the book and allows much more to be discovered by the children for themselves. Survival in temporary isolation from their society again throws into relief the means of survival, in sharp details such as children enjoy and Marryat excels in presenting. It is no mere idyll, though the emphasis does fall on the satisfactions following hard work. The young are allowed, too, to discover the world beyond the forest and out of this emerges the impartiality which the narrative preserves between the two sides in the civil wars; if Roundheads can be boorish, Cavaliers can be quarrelsome, and lose the battle of Worcester in consequence. The happy ending depends plausibly on the children's encountering a man who has the confidence of the commonwealth government but fears Cromwell's 'advances to that absolute authority for which the King has suffered' (chapter 16) and believes that, like himself, 'thousands who took an active part against the King, will, when the opportunity is ripe, retrace their steps'.

In *The Little Savage* (1849), which may have been as far advanced as the second chapter of the second volume when Marryat ceased work on it, the child's viewpoint is adopted entirely, as he narrates his own survival. But what he learns is the code which all the Marryat heroes sooner or later accept. Ill-treated at first by the other survivor on his island, he quickly learns his own power once his tyrant is blinded. So the process of survival becomes moral as well as physical, the clarity of the physical detail making acceptable the simplifications of the moral development. Disciplined by the necessity to appeal to children, Marryat manages all this with an impressive absence of rhetorical fuss.

So in the final count it is quite fair to Marryat that he should be remembered as an author for children. In that capacity he shows all his distinctive gifts and is less often tempted beyond his proper range. It is only strange, in a man of such high spirits in actual life, that he allows himself very little humour when writing for children. But the children's books are all the stronger for this, since Marryat's humour has dated. It was highly contrived. After *Frank Mildmay*, he lost the ability to be humorous by understatement—Mildmay jumping from the fireship 'with all the nimbleness suitable to the occasion', for example, or his final remark about the widow of the cuckolded bosun: 'nor was she ungrateful to the ship's company for their sympathy'. He was too often satisfied with the literary 'humours' and insensitive horseplay which he inherited from the eighteenth century and which his audience obviously enjoyed.

Of Marryat's associates on the *Metropolitan Magazine* between 1832 and 1835, those who produced fiction had already proved themselves as writers, Howard with the miscellaneous journalism which probably included the nautical sketches appearing in the *Literary Magnet* from 1824 to 1827, Chamier with *The Life of a Sailor*, which had appeared in the *Metropolitan* from May 1831. The only one of their novels which is worth attention, Howard's *Rattlin the Reefer*, began in the magazine as *The Life of a*

Sub-Editor, in September 1834. It broke off in February 1836, just as the hero got possession of letters describing his birth. The fiction of a search for a father was by this time displacing the attention to fact upon which much of the earlier part of this novel had depended: it was bound for a conclusion of high melodrama, where the villain (like Sikes two years later in *Oliver Twist*, chapter 50) was to hang himself accidentally while escaping from pursuit. Dickens may have known the book. He certainly knew the *Metropolitan Magazine*; moreover a letter, in which he proposes contributing to it as early as December 1833, mentions 'my proposed Novel', probably *Oliver Twist*. Howard made much more use of the children's viewpoint than Marryat did in either *Jacob Faithful*, which was finishing in the *Metropolitan* when *The Life of a Sub-Editor* began, or *Japhet in Search of a Father*, which ran there at the same time. It is not merely that the first fifteen years of the hero's life occupy one volume of Howard's novel, but also that the reader's sympathy for the feelings of the child, apparently orphaned, is more persistently demanded. This is in tune with the concern for juvenile suffering shown by philanthropists and by contemporary novelists like Caroline Norton, at least with her verse pamphlet of 1836, *A Voice from the Factories*. But when Mrs Norton attempts to deal with the sufferings of orphans in a novel, *Woman's Reward*, in 1835, she has them precociously assuming the responsibilities and irresponsibilities of adults. Howard, on the other hand, partly because he is describing his own life, writes wholly from within the child's experience, when he glimpses his mother, or when religion leads him to haunted despair (chapter 7) or preternatural exaltation (chapter 17). Curiously, the rest of his experience is pure Marryat—from the brutalities of school to those of the navy, from the hero's eventual bodily size and invincibility to his exotic love for a mulatto slave girl, which is sealed in stage rhetoric at the opening of the third volume ('Whilst I breathe, dearest, thou shalt never writhe under the lash . . . '). But feeling is not conveyed only in such clichés: there is often a simple particularity of detail which, as in Dickens, authenticates the emotion, like the piece of wood seen six inches from the eyes in the fever of chapter 7.

Howard's other novels accentuate the cliché at the expense of the particularity and thicken the vein of verbal facetiousness—'The fleet foe which is not always the same as the foe's fleet' (in

The Old Commodore, 1837). *Outward Bound* (1838) goes beyond Marryat's nautical Gothic in *Frank Mildmay* and *The Snarley-yow*—it has a dead steward sewn up with the drunken captain who has murdered him into one shroud and committed to the deep—and it laces this with a passionate attachment between brother and sister à la Byron and Chateaubriand. It hardly needed *Jack Ashore* (1840) to complete Howard's decline from the popularity of *Rattlin the Reefer*, which, in the year following its volume publication, was reissued twice, and again, in abridged form, in 1838, 1850, and later. Marryat's name as editor had something to do with this; indeed the novel appeared throughout the century among collected editions of Marryat's works, where it was not till 1897 that Howard's authorship was acknowledged (even though this was no secret and had appeared in obituary notices in 1842, for instance in the *Gentleman's Magazine*).

The formula of Marryat at his most mechanical—adventures at sea leading to matrimony and a legacy or even a peerage ashore—was repeated by Frederick Chamier in *The Unfortunate Man* (1835) and William Johnson Neale in *Gentleman Jack* (1837). Neither had any distinction of style, but neither was much cumbered with himself and so each could narrate a simple sequence of events from a single viewpoint (the unexpected start of the Spithead mutiny in Neale's chapter 12, for instance) without clouding it with his own feelings. The trite loquacity of Howard's reefer, sentimental at the masthead in volume ii, chapter 12, contrasts, for instance, with Chamier's account of the taking of a ship by pirates, as observed from that position in his ninth chapter. It is fair to add that the reefer can make his exalted feelings, even while he excuses them, add to the comedy of the enveloping action when he removes his hat in his 'silent devotions' and is taken to be dutifully counting the convoy. This is repeatedly his way: by the grave of his mother, he says, tears 'would have been very pretty and pathetic—but I was too hungry' (volume iii, chapter 17). Chamier, Neale, and, it must be added, Michael Scott, are tasteless by comparison. Each of the first two has a long series of novels to his name if not to his credit; Scott has only two, *Tom Cringle's Log* (1831–3) and *The Cruise of the Midge* (1834–6). Although more specific in the detail he presents, Scott shows an equal absence of interest in any but the most routine reactions of the hero to it. Only the nature of the detail of violence, torture, and loathsome practical joking has prevented the second of these from being as much a boy's book as the first. In coarse high spirits and casual realism Scott is superior to Chamier and Neale, but all of them show Marryat as above all literate, in the sense that he knows what he is doing with words and even with the form of the novel. With little originality in either, he has given thought to both; it may be only rudimentary thought, but it shapes the force and fluency of his habitual utterance so as to give us a strong sense of what he admires, not in abstract terms, but in relation to the matters which his language makes most distinct, human character under the forming hand of circumstances out of doors.

The service novel in the hands of Charles Lever,[2] on the other hand, is mere romance. When he sends to war *Charles O'Malley the Irish Dragoon* (1840–1), *Jack Hinton the Guardsman* (1842–3), and *Tom Burke of 'Ours'* (1844), the work has no form but allows the historical events to supply the semblance of it: the defeat of Napoleon furnishes a ready-made climax for O'Malley or Burke, so that the inadequacy of the picaresque romance at the centre passes almost unnoticed. The military reportage, praised throughout the century, has worn badly. Lever's episodic boisterousness still captivates, but the outline of the action owes too much to cliché and commonplace: the simple contrasts, of comradeship with enmity, of 'maddening' (a favourite word) excitement with sudden mourning, or of quiet landscape or calm weather with human violence, the flimsy repetitive rhetoric ('the mighty tournament ... of the two great nations', 'the mighty movement of the day', 'the dread force of some mighty engine'), the standardization of mass emotion ('the impatient, long-restrained burst of unslaked vengeance ... that wild cry, ... from rank to rank'), all of these, from the famed accounts of the Douro and of Waterloo, have the credulous enthusiasm of the romancer unconcerned with precision. It is hard to believe that Stendhal has already given his unheroic account of Fabrice at Waterloo in the third chapter of *La Chartreuse de Parme* in 1839. Small episodes, like O'Malley's encounter with some hospitable French officers on the Azava, are

[2] Charles James Lever, 1806–72, was born in Dublin and trained in medicine at Trinity College. In 1842–5 he edited the *Dublin University Magazine*, gained the friendship of Anthony Trollope, and, on Thackeray's Irish visit, interested him in the fictional possibilities of the battle of Waterloo. After 1845 Lever lived in Italy and at the time of his death he was British Consul at Trieste.

vivid but slight, the participants still defined by the stereotypes of the heroic background—the 'highly-wrought chivalrous enthusiasm' of the French, the guerrillas admired for 'something fine, something heroic in the spirit of their unrelenting vengeance'. A single Irishman's reminiscences of Fontenoy and Culloden in *Maurice Tiernay* (1850–2) are more impressive because an Irishman speaks them, so that, for instance, the finding of a resemblance between fighting and fox-hunting is in character, instead of further cheapening the false heroics, as it does in the worst parts of *O'Malley* .

II

No doubt Lever's service novels of the 1840s were read as historical fiction. Certainly the work of most of the writers who attempted to follow Scott in this vein was more like Lever's than Marryat's. In the first novel of W. H. Ainsworth,[3] *Sir John Chiverton* (1826), written in collaboration with J. P. Aston, the authors' feet were kept on the ground in so far as an actual building, the history of which was developed for the benefit of the armchair tourist, became the centre of the story. This formula (adopted five years before Hugo's *Notre Dame de Paris*) Ainsworth continued in four of the novels which followed: *Rookwood* (1834), *The Tower of London* (1840), *Old St. Paul's* (1841), and *Windsor Castle* (1842–3). The other main ingredient was at first pseudo-historical crime, once Bulwer had shown the way with *Eugene Aram* in 1832. The success of Dick Turpin in *Rookwood* was repeated in *Jack Sheppard* (1839): in both cases the fact that the criminals were given a crude vitality and an individualizing speech entirely denied to other characters was taken to indicate approval of their actions. Ten years later Ainsworth was still disavowing the 'intention of holding up vice to admiration'. His attempt at more careful historical reconstruction in *Crichton* (1837) showed only how the results of a certain amount of research

[3] William Harrison Ainsworth, 1805–82, was born in Manchester and came to London as a law student. His early novels brought unusual social and financial success and the editorship of magazines like *Bentley's Miscellany* and *Ainsworth's Magazine*. Although his returns from historical fiction fell off after 1864 he continued to produce it all his life. John Partington Aston, 1805–82, was a lawyer, a friend of Ainsworth's at Manchester Grammar School.

could be obscured by the animal spirits and simple romantic allegiances of this 'forward talking young man' (as Crabb Robinson called him in 1825). An inability to grow up endeared him to his public during a long career, but his kind of history and his range of emotions, both of them little more than those of a clever schoolboy, have effectively prevented his revival. *Old St. Paul's* has had a rather longer life than the others largely because it caught from Defoe's *Journal of the Plague Year* an interest in the detail of ordinary human dealings and not merely in their picturesque exaggeration.

Ainsworth's rival for most of his career was G. P. R. James.[4] A direct protégé of Scott's, he had pretensions to serious learning and a quite remarkable eye for sequences of history that yielded opportunities for spying, mistaken identity, and disguise. He was given to contrasting romance with the 'true history' he was putting before his readers. What he wrote was interesting, however, for its extraordinary events rather than for its people or the nature of their life. Where he was conscientiously concerned with the historical issues—in *Richelieu* (1829), with the Cardinal's policy, or, in *Henry Masterton* (1832), with neutrality in the civil war versus loyalty to Charles I—they were of importance to his work only as background to the adventures. Even in the cases noted in Section v below, in which the background was contemporary and the events of the story might have been expected to illuminate it, they failed to do so. And where it belonged to earlier centuries, the background did not prevent the emotions from being those of the nineteenth century. A French nobleman, exiled and disguised as a brigand in sixteenth-century Savoy, could speak of himself, in *Corse de Leon, or the Brigand* (1841), as 'lord in the moonlight on the hillside', a 'bird of prey' who is 'brother of the eagle', one of 'the bold, the free, the strong' who 'fall back on the law of God and wage war against the injustice of man'. The language of emotion was that of the contemporary stage—'Viper', 'as false as hell', 'when first I kissed those dear pouting lips'—like the gestures, 'cheek hot, brow frowning', 'biting his lip and straining his eye'. Such things make it impossible that James should again be taken

[4] George Payne Rainsford James, 1799–1860, the son of a London physician, was able in 1843 to secure from the publishers Smith Elder an agreement to pay £600 for each new work; they had eventually to break the agreement to stem the flood.

seriously, even if Thackeray had not accurately expressed in parody his amusement at the lumbering atmospherics of James's chapter openings and the wooden and worthy moral which was all his story could substantiate. In 1888 R. L. Stevenson, by then no novice, could find him capable of 'a good, dull, interesting tale, with a genuine old-fashioned talent in the invention when not strained'. Invention is all too frequently strained, in captures, sorcery, and sudden death, but there is a lucidity in the slow staple of the narrative which keeps expectation alive if it can be patient, as in the gradual discovery of the circumstances of the death of Lord Dewry in *The Gypsey* (1835) or the sequence which leads the hero of *The Convict* (1847) to be wrongly convicted of murder.

The line between historical romance and ordinary melodrama is not always discernible now, though at the time it counted for a great deal. *The Literary Gazette* in 1833 quoted with approval Mrs Bray's[5] opinion that 'Historical romance, when compared to what is wholly fictitious, is as superior in its character as historical painting is to mere academical composition.' Yet the history was often all but invisible and the ingredient of mystery the main thing. Even where the historical backdrop is as elaborate as it is in T. C. Grattan's *The Heiress of Bruges* (1834), only crime and the mystery of a ghost keep the novel going: here and in the same author's *Agnes de Mansfeldt* (1835) or in Leitch Ritchie's *The Magician* (1846), history gives no more than a flavour of the exotic. Mrs Bray herself uses it as a handy intensifier of sentiment when she makes the two longest of her *Trials of the Heart* (1839) historical novellas of La Vendée; civil war renders human relationships apparently more complicated but actually more simple, to reveal what she calls the 'history of the feelings'.

For some writers geography will furnish ready-made romance as easily as history. Mrs Gore (Section III below) can fill out the three volumes of *The Courtier Days of Charles II* (1839) with the alternative attractions of distance in time and in space, supplementing historical tales with others of a time nearer to the present which take place in France, Austria, and Peru. The pseudo-archaic solemnities of her *Agathonia: a Romance* (1844) are even further

[5] Anna Eliza Bray, née Kempe, 1790–1883, who was first married to C. A. Stothard, historical draughtsman to the Society of Antiquaries, wrote popular history and biography as well as historical romances.

from her proper vein, though in this case some of the blame may rest on Beckford, who is reputed to have helped her, and on the Disraeli of the eastern tales for which Beckford had an extravagant admiration. Agathonia's grandfather, a primitive protestant in seventh-century Rhodes and a prophet of the future ascendancy of England, has a touch of Disraelian fantasy at its most deplorable. Novelists seemed no more able than painters (in the decade of Haydon's suicide) to resist the temptations of the grandiose and the inane when the subject was historical.

One who did resist such temptations was Emma Robinson, self-educated daughter of a London bookseller. To her, history was a stimulus to high-spirited invention, resulting in the kind of work which Thackeray burlesqued in *A Legend of the Rhine* in 1845. (As the *Legend* is given a female *nom de plume* and an acknowledgement to Dumas, it may well be Miss Robinson whom he has in mind.) In *Whitefriars* (1844), for example, her reading in Burnet and L'Estrange has excited her imagination with the possibilities of life for a young man in the late seventeenth century. His disputed parentage (as so often with Scott) opens to him a range of rapid adventures among the main events of some twenty years, necessitating frequent, if hardly profound, acquaintance with Titus Oates, Charles II, and the daughter of Algernon Sidney. The author's naïve lack of self-consciousness is both captivating and absurd, while the absence of the moral protestations which characterize the male novelists almost to a man enables her to touch with farce episodes like the prevented seduction of Sidney's daughter or the possibility of the application of 'my Lord Rochester's experiment' to the hero in female disguise. Of the many who attempted historical fiction in the 1840s, Miss Robinson perhaps comes the closest to Dumas in thoughtless facility.

It was commoner for English historical novelists to preach a contemporary moral as in Bulwer's two best examples, *Rienzi* and *The Last of the Barons* (Section VIII below). J. Palgrave Simpson[6] may plead guilty in *The Lily of Paris* (1849) to 'borrowings from Dumas', but his prolixity and pretentiousness show how little he is capable of imitation. His tendentious aim, to represent 'the

[6] John Palgrave Simpson, 1807–87, son of the town clerk of Norwich, was educated for the church at Corpus Christi College, Cambridge, but did not take orders. He became a popular London dramatist, basing some of his work (*Lady Dedlock's Secret*, for instance) on that of Dickens.

dangers of letting loose the torrent of popular convulsion', is topical enough, in Bulwer's manner, but the vague, awestruck picturesqueness (faintly reflecting Carlyle's *French Revolution*) of the background of Paris in 1418 and the melodramatic action in the foreground, alike, fail to say anything about 'popular convulsion' except that the author fears it. But he does at least give attention to his story, however shrill and derivative. In *Gisella* (1847), a similar moral translated into a contemporary setting (for which a more remote place, Hungary, offers compensation) is treated with the same archaic language, rather in the manner of *Agathonia*, or of Disraeli's *Sybil*, and with the same simple melodrama. Novelists tended, naturally, to give the first French revolution a symbolical significance. This is the effect John Sterling[7] produces in *Arthur Coningsby* (1833): the Terror all seems like a contemporary bad dream. Similarly, when Mrs Marsh in *Mount Sorel* (1845) sends the narrator to Paris for the taking of the Bastille and again for the execution of the king, these are shadow-shows to dramatize a state of mind which is still of importance in 1845.

The risks of melodramatic inflation to which all historical novelists were prone could have been reduced by choosing less exalted and momentous history; but most of them, again like the historical painters, tended to fall victim to the large subject. An exception was Charles Whitehead.[8] If it cannot be said that his *Richard Savage* (1841–2) shows the 'very high original power as a writer of Fiction' which Dickens claimed when appealing on Whitehead's behalf to the Secretary of the Literary Fund, it nevertheless shows the superiority, as models, of Dickens and Ben Jonson over Bulwer and Scott. Filling out the eighteenth-century lives of Savage with characters from the comedy of humours and with the pathos, caught from *Oliver Twist*, of the unacknowledged child, Whitehead could not avoid melodrama in inventing a private life for Savage; and yet he conveyed something of the feckless, hopeless jumble of his career when the intermittent

[7] John Sterling, 1806–44, born on the Isle of Bute and educated at Glasgow University and at Trinity College and Trinity Hall, Cambridge, became a friend of F. D. Maurice and of Carlyle, who wrote his *Life*. (See Vol. XIV, Chap. 10.) After taking orders in 1834, he resigned his curacy in 1835.

[8] Charles Whitehead, 1804–62, lived precariously as editor, publisher's reader, and dramatist. He emigrated in 1857 to Australia, where he died.

possibility of aristocratic patronage inhibited both hope and effort. The book is drab, however, less because of this subject than because of the author's inability to show any appetite for life, though even honest depression seems a virtue by contrast with the spurious exaltations of routine historical fiction.

III

A second dubious inheritance from the previous decade was the novel of fashionable life (Vol. XII, Chap. 8). As a type it seems to have little of the distinctiveness of the historical novel, partly because the comedy of manners which it essayed at its best might be found equally in other kinds of fiction, partly because it had no such dominating writer as Scott. Those who might claim such a position, like Bulwer and Disraeli, had been merely using the genre to secure their first successes (Section VIII below) and had not been able to resist the temptations of parody. It is as parody that the genre lives on, if at all, in Thackeray's skit in *Punch* in 1847 and in Mrs Gore's two best novels about Cecil, where it almost amounts to historical fiction, as she revives in 1841 the days of the Regent.

The reason Bulwer gave in 1833 for the success of the sub-genre (in *England and the English*, Book IV, chapter 2) was the 'revolution' which allowed the newly rich to hope to 'be quasi-aristocrats themselves', so that they sought 'representations of the manners which they aspired to imitate, and the circles to which it was not impossible to belong'. Here the upper classes were presented as a source of standards (which were, paradoxically, often seen as false) in behaviour, dress, and physical surroundings. Repeatedly the subject is the breaking in of aristocratic ways of life upon the sobriety of the classes immediately beneath, so that the term 'fashionable' seems to be used quite loosely. Mrs Lowndes's *The Child of Mystery* (1837) is subtitled *A Tale of Fashionable Life*, although most of its action takes place in circles with which Jane Austen was familiar. The aristocracy intrudes in only two or three chapters out of thirty, but then in the persons of 'the profligate Lord Arranmore', who presses his attentions on the wife of the story, and of the Sicilian Count who disastrously marries her aunt. James Justinian Morier appears to be satirizing such topics in *Abel Allnut* (1836) when his tale of a lordly seducer foiled includes a publisher dealing 'entirely with the nobility and with persons whose names are known in the world', advertises for a 'Manufactory' of 'skeleton Novels, upon all subjects, from the gossiping fashionable to the coarsely vulgar', and centres the action in the third volume upon Gower Street, 'a sort of frontier position on the confines of gentility'. Mrs Marsh in the first of *Tales of the Woods and Fields* (1836) has the archetypal situation of a country clergyman's daughter gradually seduced by fashionable ways (including perfume from an establishment in the Haymarket which the author carefully names) and 'flinging herself beneath the feet of the idol which shall destroy her' by marrying Lord Melville. It is characteristic of this author for fashionable life to be merely one ingredient among many, though an important one, as with her best-selling *Emilia Wyndham* (1845), set on the frontiers of gentility, where one of its main relationships may be endangered by a Duke.

Nevertheless, such a frontier position appears to be essential for the fashionable novel. The work of Emily Eden,[9] for instance, hardly belongs to the sub-genre at all because she is unselfconsciously a member of the upper classes she presents, and simply pays no attention to any such frontier. In *The Semi-Attached Couple* (revised for publication in 1860, though mostly written 'nearly thirty years before', as is clear from internal evidence), her people are seen for what they are rather than prejudged according to their social position. If there is a snob in the book it is Lady Portmore rather than Mrs Douglas, whose sarcasms are not class-conscious. Where the author presents social customs of which she disapproves, her tone is cool and detached, not excited and denunciatory: her heroine is too 'innocent to understand' that 'the first moment in which a woman lets it appear that she and her husband are at variance, is the last in which she is safe from the impertinent admiration of others'. The generalization is not, as so often, illustrated by a lord, in order to indict his class, and the heroine escapes by simply refusing to understand a would-be seducer's 'air of deep compassion'. At the same time it must be said that the book lacks direction and point compared with its successor *The Semi-Detached House* (1859), where a frontier

[9] Emily Eden, 1797–1869, friend of Lord Melbourne, was more closely attached to her brother George, second baron Auckland, whom she accompanied to India when he was Governor-General, 1835–42.

between fashionable and non-fashionable life is recognized, though only to mock it, and indeed the fashionable world itself, as the false creation of snobs. The villains are a Baroness 'with a mania for fashion and fashionable people' and her husband whom she praises for 'his wit and vivacity and air of fashion', while to the author he is no more than a 'poor dear' with his 'vulgar jokes and flashy appearance'—from which it will be plain that Miss Eden is herself fairly innocent as a novelist at a time when techniques for the management of the distance between author and character are multiplying. The main characters, who start out with comical anxieties when the barrier between the upper and the middle class is reduced to the main partition of a semi-detached house, are presented so as to give the reader rather more of the experience of seeing for himself. With the Mrs Hopkinson who is surprised to discover that 'the aristocracy are not so bad as we are told' and that Lady Chester and her sister are '*real* ladies to my mind, looking so simple and so quiet, . . . while that great storm of a woman swept over them', Miss Eden comes close to embodying her subject in persuasively imagined life, and not merely telling us about it. She became capable of this anti-fashionable novel, one may think, only after her considerable powers of observation had been turned on Indian and Anglo-Indian society as an outsider.

The separate and baneful existence of the fashionable world is essential to the polemics of Caroline Norton.[10] Her loyal long-suffering women are regularly set at a disadvantage when brought into this 'strange' world, never having had, as she says in *Stuart of Dunleath* (1851), their 'senses gradually blunted to the frivolity, pomposity, soul-deep (not manner-deep) vulgarity and the wonderful profligacy of opinion which is combined with outward formalities' there. It is a pity she has quite so firm an axe to grind, since she is capable of sporadic satirical characterization worthy of her ancestry. But the praise of female devotion, here as in *The Wife* and *Woman's Reward*, tends to drown the attack. Throwing the

[10] Caroline Norton, 1808–77, grand-daughter of Richard Brinsley Sheridan, supported her husband, George Norton, by her writing and editing. Her quarrels with him, her friendship with Lord Melbourne, and Melbourne's position as Prime Minister, caused Norton to bring an action against him for adultery in 1836. The trial, in which Melbourne was acquitted, suggested to Dickens that of Mr Pickwick in chap. 34 of *Pickwick Papers*. Mrs Norton wrote influential pamphlets on *English Laws for Women in the Nineteenth Century*, on Lord Cranworth's divorce bill of 1855, and on the right of mothers to the custody of children.

action of *The Wife* (1835), her briefest and most pointed work, back into the reign of George IV allows a greater freedom of comment. The heroine makes an aristocratic marriage against her own inclinations and preserves it when her phenomenal social success leads to the unwelcome attentions of members of the court, even of a royal Duke. The husband falls under the influence of an experienced widow while the wife steadily refuses any kind of relationship with her old suitor. Her long vigil of meditation in chapter 12 makes us think of *The Portrait of a Lady* and recognize that there is an important subject latent within the opposition of personal morality to the behaviour accepted in a selfconsciously superior class.

The potboilers of Lady Blessington[11] seem more a part of fashionable life itself than a fictional comment upon it. The reports of one who has broken into the charmed circle, they contain portraits of actual participants, and consider how such people 'get on': Lord Albany gives 'a receipt to make a woman of fashion' in volume III of *The Repealers* (1833)—'strong nerves are essential'—and the villainess offers, in volume II of *The Victims of Society* (1837), her views on London fashionable life, with lengthy quotation from the *Quarterly Review* and expressed endorsement of the opinions of Mme de Stael. The author supplements all this with brief guides to current literature, from the 'giants', Byron and Scott, in volume III of *The Two Friends* (1835), to the very latest, in volume II of *The Repealers*, where Caroline Norton appears as Mrs Grantly, 'the very beau ideal of what a Corinne should be'. The result tends to be a rag-bag, as the reviewer of *The Repealers* in Jerdan's *Literary Gazette* saw when contrasting its 'republican admixture of all classes of composition, . . . abounding in dialogue, personalities, and reflections', with 'the hereditary sway of hero and plot'. In trying to develop more 'plot', the authoress invented recklessly, contrasting two types of hero in *The Two Friends*, one conscientious and one dissipated, and six possible types of heroine in *The Confessions of an Elderly Gentleman* (1836), and then moving into blackmail and criminal violence in *The Victims of Society* and rough comedy of humours in *Memoirs of a Femme de*

[11] Marguerite Gardiner, née Power, Countess of Blessington, 1789–1849, daughter of a small landowner in Ireland, became a famous London hostess to writers and politicians—though not to their wives, because of her friendship with her daughter's estranged husband, Count D'Orsay.

Chambre (1846)—all without being able to make the reader feel that the events moved her or were of the slightest importance, despite what was often shrilly claimed for them. It is conceivable, all the same, that her friend Thackeray remembered chapter 6 of the last-named work, in which Lady Blessington's Lord Willamere, after allowing his secretary to be imprisoned as a result of endorsing his lordship's bills, makes advances to the secretary's wife. Lady Blessington's work is a good example of 'the genteel rose-water style' parodied in the first version of chapter 6 of *Vanity Fair*; indeed the repetitions of the phrase '*femme de chambre*' in this chapter look like a direct reference to the *Memoirs*; and certainly the sketched-out 'fiery chapters' of 'romance' are suspiciously like Lady Blessington at her most melodramatic. She was a conspicuous manufacturer of this kind of effect, writing, as she admitted to Landor, for money, and 'obliged to select subjects that would command it'.

Mrs Gore[12] could, no doubt, say the same. But her pursuit of what would sell led her beyond the range of ordinary silver-fork fiction. Already in volume II of *The Repealers* Mrs Gore could be pointed out as 'the authoress of half the popular novels of the day' and praised for 'her fecundity of imagination' and 'facility of language'. Although her fictions by this time numbered a dozen or so, she was only just getting into her stride. Her best work was to be produced in the next ten years, from *The Hamiltons* (1834) to *The Banker's Wife* (1843). 'Fecundity' and 'facility' are the words one still wishes to use about them, though with some doubt whether the fecundity is really imaginative or the facility limited merely to the language. Both words would more properly refer to the range of her acquaintance with the habits of contemporary society, which amazed Thackeray when reviewing her *Sketches of English Character* in 1846. 'She knows some things,' he wrote, 'which were supposed hitherto to be as much out of the reach of female experience as shaving, duelling or the bass viol.'

Although the commonest setting for all these novels is that part of society which can afford to be fashionable, Mrs Gore's aim is far from merely snobbish. It is true she believes her audience will wish to know about life in these regions, so that she sometimes

[12] Catherine Grace Frances Gore, née Moody, 1799–61, after marrying, 1823, a soldier who promptly retired, produced over fifty works of fiction, and became a successful dramatist and a lively social acquaintance of Bulwer, Disraeli, and Dickens.

gives more attention than is strictly necessary to the food or the furniture; it is true too that she plays the double game of titillating the appetite for such details while warning that upper-class life is heartless; but what she has to say is not limited to this. With equal ease and assurance she can satirize the Tory placemen of 1830 in *The Hamiltons, or the New Era*, or oppose 'the extension of the rights and privileges of the female sex' (as she explained when *Mrs Armytage, or Female Domination*, of 1836, was reprinted in 1848) and then recount, in the first person, the *Adventures of a Coxcomb* (the subtitle of *Cecil*, 1841) in the time of the Regent, or describe, in a series of letters, aristocratic life in Germany, Russia, and France (*The Ambassador's Wife*, 1842). She had as little trouble with the calamities of a speculating banker in *The Banker's Wife* (1843), or the extravagance of an undergraduate at Cambridge in *Castles in the Air* (1847). It is not to be expected that all this should be done with the acumen or the detailed accuracy of *Middlemarch*. But it has a persistent air of unquestionable command; as with most popular fiction, the author *knows* and the reader has only to acquiesce. It also has unflagging spirit: as late as 1853, in *The Dean's Daughter*, Mrs Gore can say, 'Of all excitements, that of building is, next to gambling (or novel-writing), the most absorbing.'

Moreover, she does try to make what she has to say emerge from the way the story unfolds. This is done, of course, very simply; the novel is to her primarily a contraption for getting and holding a reader: if dialogue among the personages or comments by the author will serve the purpose for the time being, she sees no need to inform the dialogue with a strong sense of character or to keep her plot in view while she comments. Nevertheless her improvising is not reckless like Lady Blessington's. Indeed the point of the whole novel may be pressed too hard—at least until the necessity to write the third volume leads to padding with unrelated material, like the mystery of Rosamond and the O'Morans in *Mrs Armytage* or the adventures of and with Byron in volume III of *Cecil*. For most of *Mrs Armytage*, the author is trying, as relentlessly as Dickens is in *Dombey and Son*, to illustrate one simple abstraction, and with a similar difficulty in turning it into a character. Where the abstraction takes a simple metaphorical form, as with *The Diamond and the Pearl* (1849), virtually the nicknames of a pair of contrasted sisters whose Christian names, Helen and Blanche, echo them, the contrast is equally obvious and unrelenting throughout the book, the plot serving to illustrate what is given rather than illuminate what is unknown. But when, in *The Hamiltons*, she enters into the successive fears of an innocent thrown uncomprehending into the fashionable world, the course of the novel appears much less premeditated. The infidelity of the Tory placeman whom the innocent heroine marries and his death in a duel are commonplaces of fiction, no doubt, and their presentation is frayed with cliché and rhetorical exaggeration ('child of shame', 'a tear [from the wife] scorched her [the mistress] like a drop of liquid fire'), but such things form only a subsidiary part of the new world the heroine is discovering, as is made clear when she receives corroboration of the infidelity on a day of threatened insurrection in 1830 and when one of the reasons why she is unwilling to leave her husband is that the Whigs have just come to power. Her purity and docility may be too much insisted on and insufficiently demonstrated, but we are not expected merely to praise them. When she is shown learning by experience, we are asked to note what she has failed to learn, about 'the influence of a political administration upon the happiness and security of a people', and she is dismissed at the end to a second marriage where she may learn to participate in her husband's success as a landlord and an employer of factory and mine labour. It is the crudest sketch beside *North and South*, but the function of the 'fashionable' part of the novel is comparable to that of the quiet southern existence which Mrs Gaskell's heroine must leave in order to discover, with pain, an unsuspected larger life.

When *The Hamiltons* was reprinted in 1850 the 'new era' had been confirmed by sixteen years of transformation. Thackeray read the novel then and confessed himself, in a letter to the author, 'not only pleased by it, I mean for the cleverness and so forth, but amused by it; and that is a great thing for a professional man to own . . .'. He was happy to disown the criticisms implied by his parody in *Punch* three years before. With *Vanity Fair* behind him, he no doubt approved of the nudge and snap of the satire, when, for instance, the villain of *The Hamiltons*, invited to the Windsor of George IV, instantly 'professed himself an enthusiastic advocate of Moorish architecture . . . Say, dear reader! *could* a man be better qualified to become Secretary of State?'

This kind of amusement is the main object in Mrs Gore's most famous novel, *Cecil* (1841). At the end of the third volume, the hero's name appears in a long list of 'master-spirits . . . who might revolutionize the country', including Henry Pelham and Vivian Grey, and that is where he belongs, among the equivocal Narcissi of the fashionable world of the twenties. Except for his friendship with Byron, everything about Cecil and his fashionable environment is subject to delighted mockery by implication. By the end of the novel, the nostalgia with which he looks back is beginning to seem slightly ridiculous and in the sequel published in the same year he takes his place on a 'superannuated Olympus', 'exclusivism, fashionable novelism, Nashism and fifty other fribbleisms of the West-end' being 'utterly extinguished by the Reform Bill' and the advent of a virtuous Queen. Both novels would be better if the author could think of more significant things for her hero to do than repeatedly to escape matrimony, though it is plain that, for her, if he is condemned by the misapplication of his intelligence, he is justified by his wit. Sometimes this is enough: his difference from the stiffness of a courtier of Louis XVIII is plain when, bidden to be polite, he can say, 'I began to beat my brains for something square and solid enough to utter . . .'.

It must be said that, for Mrs Gore, to evade matrimony is a significant action. This is clear not only when she again makes a novel out of this evasion (when the contrast is between building *Castles in the Air* and founding a solid house) but also in the very frequency with which, elsewhere, she finds her main subject not in the approach to matrimony but in its subsequent course. Titles like *The Ambassador's Wife* or *The Banker's Wife* show this, while others, like *The Debutante* (1846) or *The Dean's Daughter* (1853), are concerned with the kind of marriage the titular heroines make. They are not placed in predicaments of any great originality—in the latter two, a marriage of convenience and the affliction of second thoughts after the knot is tied—but the presentation is repeatedly detached and sharp-eyed. This stands out when the polemics of Mrs Norton are compared, for instance, with the quarrel, prompted by trouble of conscience on both sides, between the former debutante and her husband, or the coldness of the Dean's daughter toward hers, which the author understands but avoids condoning.

Not that Mrs Gore makes such a situation yield all it might. She is the prisoner of her own facility—of what she calls, writing to

Bulwer in 1832, her 'ormulu railroad'. The facility extends to language as well: ready cliché, rhetorical inversion, smooth epithet and noun all blur the elucidation and subtract from the immediacy. In *Cecil* such language can always pass for part of the thing satirized. Moreover, the main point is in the limitations of the intelligence through which the action is refracted and these can often be simply revealed by subsequent events. So *Cecil*, a *tour de force* which makes a virtue of superficiality, manages to give an impression of strong discernment by repeating Disraeli's trick of quizzing the autobiographer. It still secures for the author an attention which is denied to her less obviously amusing works. Yet the latter are more representative of this active but limited mind which, rapidly improvizing, with little of the artist's concern to realize all the potentialities of her subject, entertained readers for thirty-five years. Her work was, finally, to be appreciated only by those who could supply from their own experience the facts which most of the novels by themselves failed to make real. Later readers find she too often tells them what to make of it all without possessing them of the evidence, so that she fails to invigorate as the novelists do who supplanted her for good in the forties and fifties.

IV

If the fashionable novel seems to be concerned largely with the effect of the aristocracy upon the middle classes, the novels of R. S. Surtees[13] give the opposite impression; at any rate, a great deal of their comedy depends upon exposure of the upper classes to steady pressure from beneath. With no illusions himself about the gentry, so that the splendid roughness of his Lord Scamperdale matches the shrewd vulgarity of Thackeray's Sir Pitt, Surtees observes this exposure without being dismayed by it. His sequence of novels from *Jorrocks's Jaunts and Jollities* (1831–8) to *Hillingdon Hall, or the Cockney Squire* (1843–5) makes a London grocer, Jorrocks, acquainted with 'hupper crusts'; he becomes a Master of Hounds and eventually a Conservative MP—'"Tory men with Vig

[13] Robert Smith Surtees, 1803–64, after training as a solicitor in Newcastle and London, edited, 1831–6, *The New Sporting Magazine*. He inherited Hamsterley Hall, County Durham, in 1838, and became Deputy-Lieutenant of the county in 1842 and High Sheriff in 1856.

measures," as Conin'sby says'. The exaggeration of his accent and behaviour results most frequently in farce, yet his lack of snobbery and his persistent good sense where the facts of hunting, if not of farming, are in question give to the character and the social criticism a substantiality which the public of 1838–45 was unprepared to welcome. Reprints since then, most frequent between 1888 and 1903, mark the change in attitude of readers for whom hunting was the centre less of their present society than of their nostalgia for one which was disappearing.

Whether Jorrocks is destined for steady revival with the general reader may be doubted. He is too often a heavily artificial literary creation—when exhibiting, for instance, his prime vulgarity at the dinners of the great. His foil, the huntsman Pigg, shows more of actuality but on a slighter scale: Pigg's authentic Tyneside speech, seldom overworked, is the index of thought processes which the author understands intimately, but for which he cannot find sufficient scope in the fiction.

It may be claimed with some justice that the excellent hunting episodes of these early novels are scope sufficient. In these, it is true, the need to be accurate for the knowledgeable reader keeps the outline and much of the detail specific in a way which also secures the attention of the less informed. Like Marryat, Surtees can enliven on the page not only what is done but why. When the reasons include not only a countryside which Surtees knows thoroughly but the opposed temperaments of two leading characters, round whom representative figures from country society are grouped, the result promotes both relish and judgement of part of the human spectacle, and this despite the tenuousness of the total action into which the episodes are fitted.

It is only when he develops his rogues and gives them centre stage that Surtees achieves a novel which will stand up as a whole. His first popular success, *Mr Sponge's Sporting Tour* (1849–53), has the Jonsonian subject of competitive petty egotism, and his final triumph, *Mr Facey Romford's Hounds* (1864–5), shapes a whole Jonsonian plot out of the simple effrontery of impostors, unquestioning in their belief that society owes them a living. The high comedy, in which their bubble bursts but they carry on undaunted in Australia, has an undertone of mordant pessimism a little like Jonson's own. Indeed Surtees goes beyond Jonson in showing the ease with which a quite acceptable lady may be made out of a fortunate actress while the less fortunate one must turn impostor to gain a position, or in showing the excluded, resentful servants ready to break into the alchemists' halls and the ease with which Australian gold realizes their dreams. It makes an imaginative commentary on the social changes Surtees observed with amusement but with limited hopes. Thackeray's admiration for him rested on the secure foundation of a shared outlook.

The social changes which Surtees had begun to observe in the late thirties, and on which Bulwer gave his informed commentary in *England and the English*, were recognized to be important for the development of the novel as early as the death of Scott. Writing in 1832, Harriet Martineau[14] criticized Scott (not quite justly) for neglecting the middle classes. 'What', she asked, 'can afford finer moral scenery than the transition state in which society now is . . . Heroism may now be found not cased in helm and cuirass, but strengthening itself in the cabinet of the statesman, guiding the movements of the unarmed multitude . . .'. Seven years later, her own first novel, *Deerbrook*, offered a humbler theme than this, but certainly gave prominence to the middle classes and to their guidance of the classes beneath them. The refusal of the novel by Murray, prompted perhaps by 'Mr Lockhart's clique', made her believe there was a more general 'prejudice against the use of middle-class life in fiction'.

But is there any evidence of such a prejudice? Even Mrs Gore herself has plenty of good-humoured presentation of the middle classes, into which Mrs Armytage's son marries, for instance. Dickens was an exception, Miss Martineau believed, because readers would accept 'life of any rank presented by Dickens, in his peculiar artistic light, which is very unlike the broad daylight of actual existence'. In that case, Miss Mitford must have been able to throw the same illumination, in view of the way in which she continued to delight the public with *Belford Regis: Sketches of a Country Town* (1835) and *Country Stories* (1837). Anne Manning's *Village Belles* (1838) catered for the same kind of taste, though with nothing like the same skill. Dawdling and prosy, with very little sense of place, though with an occasional mild maliciousness,

[14] Harriet Martineau, 1802–76, daughter of a Norwich manufacturer, made her name as the popularizer of political economy. (See Vol. XIV, Chap. 14.) Carlyle spoke of her in 1837 as 'swathed like a mummy in Socinian and Political-Economy formulas; and yet verily alive inside of that'.

it might have justified a good deal of prejudice; but Jerdan's *Literary Gazette* found it 'delightful' and Jerdan, while hardly unbiased, was not a man to countenance swimming against the tide. The Howitts were, in the same way, well aware of public taste, as is shown by their miscellany of popular natural history, *The Book of the Seasons*, in 1830 and later in numerous reprints; when Bentley contracted with Mary Howitt[15] for a novel, the result, *Wood Leighton* (1836), set a melodramatic piece of middle-class family history in a frame of people and places she had known in Staffordshire early in the century, as the daughter of a surveyor. This frame forms the strongest part of an otherwise mediocre book. Much of the mediocrity itself arises from adherence to the detail of ordinary life, financial detail in particular, while human motives and reactions receive, by contrast, a merely routine attention. Mrs Howitt's calm Quaker impartiality toward social classes and religious denominations is attractive but hardly exciting. Yet Bentley reprinted the novel in 1850. A similar phenomenon can be seen in the stories, mainly of middle- and lower-class life, which Mrs C. I. Johnstone wrote for the Edinburgh periodicals her husband acquired in 1832–4. One of the two longer ones was concerned with social climbing and the other made comedy (with an appropriate glance at Goldsmith) out of 'Exclusivism as it exists among the minor orders of the middle class', but the greater number of her stories had not even this sort of connection with the fashionable world, while her 'artistic light' was very dim indeed; yet they were popular enough to be collected, along with others by, for instance, Mrs Gore, in the three volumes of *Edinburgh Tales* in 1845–6.

Nor can such fiction be said to have gained its audience by complacency about the middle-class life it presented. The criticism of its intellectual and religious narrowness which is to become a major mid-Victorian theme is already present, if in a somewhat unsubtle form, in Sarah Stickney's[16] *Home, or the Iron Rule* in 1836. It is criticism from a pious standpoint, by means of an action

[15] Mary Howitt, née Botham, 1799–1888, collaborated with her husband William in a great variety of popular literature and herself translated the work of the Swedish novelist Frederika Bremer.

[16] Sarah Stickney, d. 1872, was primarily interested in women's education on the principles of her series of books on *The Women* . . ., *Daughters* . . . , *Wives* . . ., *Mothers of England*. She married William Ellis in 1837 and shared his concern with temperance and overseas missions.

involving a daughter whose joyless home leads her to make a marriage of convenience of which her father, like Dickens's Gradgrind, approves. She takes 'stimulants' in order to face the matrimonial rigours but (in this also, like George Eliot's Janet) repents and redeems herself by nursing her husband and by further good works after his death.

When Harriet Martineau's own novel appeared, it cannot, then, have had to encounter much resistance from the predispositions of moderately thoughtful readers. She was, in any case, under-estimating in the *Autobiography* (1877) the effect she had already had in 1832–4 with her stories illustrating economics, taxation, and the poor law, which Empson in the *Edinburgh Review* (1833) believed were read as fiction, quite apart from the principles they illustrated. With a few of them, like 'Ella of Garveloch', this claim may still be credited: the careful economics of survival in the western isles are an important part of such a story and give it a flavour not unlike Defoe, while there seems to be a greater generosity than the principles of political economy require in the centring of so much of the interest upon a Wordsworthian idiot boy whose innocent eye and heart are to be kept from contact with money. By July 1839 the *Edinburgh Review* could speak, as of a long-acknowledged fact, of the 'nice discrimination of feelings and motives which so peculiarly distinguish the writings of Miss Martineau'. Not even the most determined didactic intention had been able to quell this ability; the modern reader of *Deerbrook* is repeatedly amazed at a penetration which no lurking sententiousness can mar. Yet the understanding of the detail of human behaviour which makes this an outstanding novel of the 1830s, comparable with the last work of an old master, Maria Edgeworth's *Helen* in 1834, did not result in the vivid awareness of people which *Helen* communicates and which the *Edinburgh* reviewer's comparison of *Deerbrook* with Jane Austen implied. Partly this is because the characters are shown reacting rather than acting. With the self-preoccupation and jealousy which small occasions reveal in her major character, Hester, Miss Martineau is masterly, but her jejune plot fails to give Hester and the other two heroines enough of the initiative, while the male character, Hope, who has the initiative, does not believe he ought to use it to propose to Hester's sister and spends the rest of the novel coping with the consequences of having married Hester herself. It is life all right, but somehow the spring is broken. The only character who really makes things happen in the novel is the petty traducer whom the whole village apparently believes; the melodramatic results which she produces mark the difference from *Middlemarch*, where causes and effects are more carefully linked and proportionate to each other. Yet small scenes—those in which, for instance, dilatory tradespeople show their disapproval of Hope and his wife—rival George Eliot herself, and there is a persistent power of giving crisp actuality to everyday things which takes the book far beyond most of its immediate contemporaries. She retains this power in her stories for children, particularly 'The Crofton Boys' (1841), although she is still tempted in them to take a short cut to her catastrophe ('The surgeon took off his foot . . . 'Never mind uncle! I can bear it."'), satisfied as long as she can fulfil her didactic intention.

The trouble was that the author could not, finally, bring herself to believe in fiction, despite her admiration for Jane Austen as 'the Queen of novelists'. The connection of cause and effect in human affairs, perception of which underlies the success of nineteenth-century fiction, was already beginning to have a debilitating as well as an energizing effect upon it. Harriet Martineau, when contemplating the novelist's task, seems to have been overawed by a philosophical determinism which is distinct in the *Autobiography* and only with difficulty countered by her theology in the full-scale discussion at the beginning of the third volume of *Deerbrook*. The problem of 'thoroughly incorporating the doctrine in the tale' in the *Illustrations of Political Economy* had, she says in the *Autobiography*, confirmed her in the belief that 'the creating a plot is a task above human faculties. It is indeed evidently the same power as that of prophecy: that is, if all human action is (as we know it to be) the inevitable result of antecedents, all the antecedents must be thoroughly comprehended in order to discover the inevitable catastrophe.' This task being beyond the human mind, 'the only thing to be done' was 'to derive the plot from actual life, where the work is achieved for us'. Yet after rejecting, as subjects for her novel, a story from the police-court news, and the tract of history she returned to in 1840 in *The Hour and the Man* (where she had no idea, apparently, how far the epic aggrandizement and the moral abstraction of her treatment of Toussaint L'Ouverture took her away from history), she pitched on a story which afterwards turned out not to be true. But she remains unconscious of the irony in the *Autobiography*, and has no idea that the inevitability of a 'catastrophe' (that is, a denouement) depends upon the author's power to convince us that it would follow from characters of a certain kind, in circumstances the force of which we are made to feel. Her earlier writing of fiction to illustrate certain given principles seems to have inhibited her power to contemplate the particularity of the consequences following from particular actions.

In the novel, her success in imagining instances of moral victory or defeat tempted her to move too quickly from these to the principles they exemplified, rather than to their detailed outcome; she tended to rely, for the forward movement of the plot, upon communal reactions, the simplified nature of which is all the more apparent from the sharpness of the detail in the more personal foreground. The reader cannot help recalling that mob violence was topical in the winter of 1838–9, when she was writing the second and third volumes; torchlight meetings of the followers of O'Brien and Feargus O'Connor were taking place in the north, while a Convention in London of the Industrious Classes, in support of the People's Charter, was being planned for the very month in which she completed the book. The attack on the house in the second volume and the comedy when a piece of charred stick resulting from children's play is found near the church door and exaggerated into a plot to burn the church down seem to originate much less in the plot of the novel than in the intention to make it yield comment on a pressing issue. In the third volume, the hardness of the times, evidenced in an impressive, quiet robbery by hungry men, is of the same kind: the reader wishes that the novel were great enough for the author either to forgo such things or to make them an integral part of its fabric.

V

In itself, however, Harriet Martineau's inclusion of such material is symptomatic of the cognizance fiction is beginning to take of the condition of the people. Two early examples which gave it central importance were appearing serially when *Deerbrook* was published, Mrs Trollope's *Michael Armstrong: The Factory Boy* (1839–40) and Mrs Tonna's *Helen Fleetwood* (1841). Mrs

Trollope[17] writes from the sheer shock of having discovered what factory labour was like, particularly for children, but she is too little of a novelist to make anything out of her feelings. She is not even a lesser Harriet Martineau, writing to illustrate a simple set of predetermined ideas; she has only her sudden feelings, aroused, as Dickens's were, by Lord Ashley's speech in the Commons on 20 July 1838, and fed, like Dickens's, by visiting the satanic mills in person. Mrs Trollope's more famous son said in his *Autobiography* that, 'with her, politics were always an affair of the heart,—as, indeed, were all her convictions. Of reasoning from causes, I think that she knew nothing.' Her fiction too is an affair of the heart, knowing nothing of reasoning from causes. Even vehemence of feeling is made to seem ridiculous by the sheer ineptitude of every constituent of the fiction in *Michael Armstrong*: the iniquitous wealth of the manufacturers appears in pastiche of the fashionable novel, but with an ill-directed violence of caricature which deprives the parody of all satiric force; the plot can do nothing but remove two or three juvenile characters from factory labour or from apprenticeship by the intervention of the wealthy.

Mrs Tonna's[18] Evangelicalism makes her simplify character and action, but keeps her from falling into Mrs Trollope's irresponsible inventing. Her subject in *Helen Fleetwood*, the spiritual dangers to which children are exposed in factory labour, is no doubt imperfectly dramatized: unable really to get an action going as a result of the conditions she contemplates, she allows her heroine merely to 'sink under' them and die. But there has been sufficient of commonplace actuality in the listlessness of the children during breaks from work, the 'moody vacant looks' of their elders, or the malicious harassment of Helen, the newcomer, merely because she is quiet, for the author's convictions to be plain by other means than her steady preaching. The breaking up of a family unit by the Factory System, a frequent charge against it by reformers, is presented in archetypal contrast with a rural felicity which owes more to literature than to the understanding of social change: the return of the remainder of the family to rural employment (like that which Michael Armstrong obtains after

[17] Frances Trollope, née Milton, 1780–1863, supported her husband and family with an extensive list of novels and travel books like that of her third son Anthony.

[18] Charlotte Elizabeth Tonna, née Browne, 1790–1846, was born in Norwich and wrote fiction as a sideline in a hyperactive career as a Protestant controversialist.

escaping from his apprenticeship) shows how little, finally, such fiction has to do with curbing the inhumanities of the 'System'.

A rather more informed response to what Carlyle called 'this grand Problem of the Working Classes of England' is found in novels by Disraeli, Charles Kingsley, and Mrs Gaskell which are discussed in their place amongst the rest of the work of each of these. A fourth novelist, Geraldine Jewsbury,[19] is equally informed (as herself the daughter of a mill-owner). After writing in Jerrold's *Shilling Magazine* in 1846–7 on such topics as 'The Civilisation of the Lower Orders', Miss Jewsbury attempted to run them into fiction as the opinions of the benevolent employer whom her heroine marries in *Marian Withers* (1850–1). Such matters may be only a little less incidental here than in *The Half Sisters* (1848), but at least her work gains from large-scale manufacturing, its processes and its human relations, a concern with something beyond her own over-heated fancy. For the lack of some such concern, her novels set in the eighteenth century, *Zoe* (1845) and *Right or Wrong* (1859), lose themselves in pointless invention and hectic feeling which is not made any more precise by its connection with religion. Elsewhere it is only in *The Sorrows of Gentility* (1856), the heroine of which eventually marries the tanner she had scorned as a girl, that mundane detail and the mixed emotions of ordinary life are allowed their proper place. There and in the parts of *The Half Sisters* which are concerned with a Carlylean discipline of work, Geraldine Jewsbury is an open partisan of the feminine point of view, and this too gains weight from the presence (even if only intermittently) of an everyday world in which survival and status have to be earned.

That part of the condition of the people which concerned Ireland was taken for his province by William Carleton,[20] who was at home in both of the languages of the country, so that he could, as he said in his *Autobiography*, 'transfer the genius, the idiomatic peculiarity and conversational spirit of the one language into the

[19] Geraldine Endsor Jewsbury, 1812–80, was born into a Derbyshire mill-owning family who moved to Manchester in 1818. She herself lived near her friends the Carlyles, in Chelsea, from 1858. Her work as reviewer and publisher's reader was of greater effect on contemporary fiction than her novels.

[20] William Carleton, 1794–1869, was the son of a tenant farmer in Co. Tyrone. After gaining such education as he could and aiming to enter the priesthood, he became a Protestant, made a precarious living in Dublin as a novelist, and was granted a civil list pension in 1848.

other'. The sketches in which he first did so, collected into the two series of *Traits and Stories of the Irish Peasantry* in 1830 and 1833, attracted attention on both sides of the Irish Sea. By becoming a Protestant Carleton had added to his ability to bestride the fences between the lower and middle classes and between the two languages. His talent as a writer of fiction did not quite match this unique equipment. A labourer in one of the longest of the *Traits and Stories*, 'The Poor Scholar', when questioned by the bishop, says, 'It wint agin the grain wid me, to tell him the lie, so I had to invent a bit o' thruth . . .'. It is this that Carleton finds hard. He has more ability to transcribe what he has seen and heard than to invent in order to clarify and condense the truth. The *Traits and Stories* repeatedly take the form of lively reportage in alternation with pamphleteering on the landlords' neglect, the ignorance of the gentry, and the evils of pauperism. The fiction which should be invented so as to make these points itself, rather than merely to illustrate them, tends to be trite, as in examples like the landlord who is moved, by hearing how the family of a 'poor scholar' were evicted, to exclaim, 'By heaven, we landlords are a guilty race'; or the Colonel who investigates their case, secures their restitution, and helps the scholar to an education; or the traducing of McCarthy, helper of the poor of Tubber Derg, while he is absent trying to confront the landlord in Dublin. W. B. Yeats was right to praise Carleton as a historian: 'The history of a nation is not in parliaments and battlefields, but in what people say to each other on fair-days and high days, and in how they farm, and quarrel, and go on pilgrimage. These things Carleton has recorded.' He succeeds best when his story is made out of people's talk—the servant's complaint of a stingy mistress, 'Why the stirabout she makes would run nine miles along a dale board an scald a man at the far end of it', or the tippling midwife whose story is told entirely in dialogue, culminating with the oath she had given not to tell it. In the famous 'Lough Dearg Pilgrim' (1829) he makes exact physical detail cut deep and gives shape to the whole by his emphasis on the first pilgrims he overtakes on his way to the Lough, one of whom robs him at the end. He is proud to report in the *Autobiography* that the identity of this character 'was at once recognized by the whole northern public'; to him the distinction of the story is more a matter of transcription than invention.

In his longer works he had the problem of making the terrors of Irish life into the groundwork of fiction that would hold the attention of those who might be successful in pressing for improvement. The material was itself a temptation to melodrama and the expectations of his readers removed all incentive to understatement. His first long serial, *Fardorougha the Miser* (1837–9), had the makings of a powerful story in the use of the Ribbon movement for purposes of private vengeance by a man who became to some degree caught in the movement's own rules; but he was the mere villain, of little interest in himself, while the miser against whom his attacks were directed had nothing to do but react to them. That Fardorougha did so with a terrifying shrill vehemence, in the idiom of the country, was memorable, but the actions of his which had raised enemies against him all belonged outside the book, while what he did within it, especially his niggardliness towards his son, had no bearing on the calamity that overtook them both. Even his paroxysm at hearing what it would cost to defend his son against a charge of arson had no effect, for his banker had absconded. The son, the pattern hero, falsely accused, firm in adversity, and refusing to retaliate, made too didactic a contrast with the villain. There were moving moments, for instance, when the father refused, in court, to give what looked like crucial evidence and was only saved from prison by the boy's own confession of it, but these were too few to relieve what was after all a tale of mere exciting incident. Moreover, for the last third of it, Fardorougha, the only character of importance because the only one of memorable speech, was absent.

Even Yeats acknowledged that, in *Valentine McClutchy: The Irish Agitator* (1845), 'continually the intensity of purpose lowers the art into caricature'. At least the intensity was directed against both Ribbonmen and Orangemen, Carleton having declared himself in the preface 'a liberal Conservative'. This did not prevent the fiction from being stiff and clumsy. A more domestic story, of superior abilities destroyed by alcohol, *Art Maguire* (1845), contained vivid dialogue which was less melodramatic, despite the frequent flavour of a temperance pamphlet and the appearance of the anti-alcohol campaigner Father Mathew. *Rody the Rover, or the Ribbonman* (1845) gave a picture of multiple duplicity denying to a village the prosperity which industry, in this case mining, might bring: it was a matter not only of recruitment to the secret society but of a recruiter in the pay of an Orangeman and perhaps of the government. The main action, centring upon the hero's doubts both about Ribbonism and about striking for better wages, was lucid enough, but there was altogether too much subsidiary intrigue (Rody was also an attractive bigamist) for it all to be well worked up. Carleton was writing too fast in this short period in the middle forties to which the best of his longer fiction belongs.

The Black Prophet (1846–7) contained some of his most mordant glimpses of suffering and petty profiteering. He spoke, in the preface to the volume edition, of a narrative 'founded upon' the famine of 1817 'or, at all events, exhibiting' it. There was the rub. To found it upon what he called 'so gloomy a topic as famine' was to risk losing readers, while merely to take the opportunity of 'exhibiting' the topic was in this case to risk making it incidental to a murder mystery. He apparently could not allow his story to arise from what he had seen and, even more important, from the kind of talk he had heard. Lively material appeared, for instance, in the relation between the magistrate Dick o' the Grange and Jemmy his butler, whose service and advice were essential to his decisions; but all this was of no importance to the crucial disclosures or the charge of murder arising from them. All the same, although both characters and events were kept incidental to the plotting of a fake prophet, there was something appropriate in the vehemence to which the heavy Gothic mode lent itself, when Carleton was reporting a whole country 'in a state of dull but frantic tumult', its inhabitants 'like creatures changed from their humanity by some judicial plague, that had been sent down from heaven to punish and desolate the land'.

The success of *The Black Prophet*, which the author seemed to measure by the subscriptions it brought to the fund for famine relief, set him more confidently on the propagandist road, so that by the time of *The Squanders of Castle Squander* (1852), a story which was not without promise, of corrupt horse-racing and riotous living even in jail, became a pamphlet, partly historical, about the evils of the 40-shilling freeholder franchise, the consolidation of farms, and the sufferings of the virtuous poor. Carleton could boast that the subterfuges needed to serve a writ upon the squire came from actual cases and that the whole was 'written in a spirit of strictest impartiality and truth'. But his conscience as a reporter was no substitute for talent as an artist. It was, however, only a matter of degree. In all his work the talent is so marked as to make the reader long that, instead of being content with passable fiction for the contemporary market, he would invent more of the truth, to concentrate, focus, and render into art what he knows so well.

Charles Lever has the same problem as Carleton in adapting what he knows of Irish life to the current conventions of fiction. He is freest in the improvisatory *Confessions of Harry Lorrequer* (1837–9). His understanding of Irish character is not inferior to Carleton's, but he is a little too good-humoured about it; although he can be shrewd, he is seldom sharp; and his view tends to take too much account of stereotypes. He shares the habit of Mrs Samuel Carter Hall and Samuel Lover of plundering the place for amiable oddities, without any of the proud anger of Carleton, at his best, at the way Irishmen treat each other. The violence of Irish life breaks through but there is little effort to clarify or comprehend it in relation to the comedy: Lever passes quickly on, just as he does when terminating with public riot the delightful farce in which O'Malley's friend Webber engages in conversation with the fictitious occupant of a Dublin drain. Farce is his strongest form of humour, but the sort of farce which is engineered by a dominant character, like Frank Webber, Major Monsoon, or Micky Free, in *Charles O'Malley*, and which it requires some wit to engineer.

His episodic manner having come under criticism, he is claiming by the time of *Roland Cashel* (1848–50) that his 'story, like a stream, has one main current'; but this is supplied merely by a villain of simple ambitions and endless plots: the simplicities of crime fiction have been added to those of the novel of fashionable life (he is eloquent on the 'soulless dictates of fashionable existence'). *Sir Brook Fossbrooke* (1865–6) develops a more sustained mystery, although here and in *The Bramleighs of Bishop's Folly* (1867–8) a free hand with coincidences in the foreground action and with important discoveries offstage deprives the reader of the sense that events are inter-related, and makes even the characters' reactions to them seem mere contrivance. In his last and arguably his best novel, *Lord Kilgobbin: A Tale of Ireland in our own Time* (1870–2), he comes nearest to integrating current problems into the restless invention of his plot. Its principal situation is resolved through the power over the populace of Donogan, a condemned nationalist on the run, who makes secure the inheritance of the fiancé of one heroine and marries the other himself. The presentation of all this is fairly limp compared with the farcical energy of an impudent improviser and chameleon journalist, straight out of early Lever, Joe Atlee. 'I don't know,' says Donogan, 'that anything worse has befallen us than the fact that there are such men as Joe Atlee amongst us, and that we need them . . .'. The liveliest things in the book—discussions of the confused state of the country, mimicry of Gladstone's impenetrable rhetoric, fraternization between Irish Whig and Irish Tory—tend to be those of least relevance to the central situation, the relation of the Kearneys, the family of 'Lord Kilgobbin', to the O'Sheas, which moreover takes too long to declare itself. But the book is the strongest example of Lever's peculiarities, the episodic verve and the architectural flaccidity, and the tendency to rely for narrative interest upon the changing feelings of women while showing insufficient interest in them as characters, so that they stir into individuality and fade into lay figures with disturbing alternation.

By the 1840s other popular romance writers were making current social problems part of the background to their stories. G. P. R. James, for instance, in *The Convict* (1847) brought in not only 'physical force' Chartism but also the 'great national sin' of not trying to reform transported criminals before allowing them to become 'the fathers of a population about ere long to overspread the wide uncultivated tracts around them'—even though the intrigue had little use for the first of these issues and none at all for the second. In *Beauchamp, or the Error* (1845–8) his chief character took a far from superficial interest in the social changes caused by large-scale manufacturing in the time of 'one of the four Georges':

> One of the great problems of the day is this: what proportion of the profits accruing from the joint operation of capital and labour is to be assigned to each of those two elements?

He feared the power of the employer 'when I see even a temporary superfluity of labour', for

> wealth will always take advantage of poverty, and the competition for mere food will induce necessity to submit to avarice, till the burden becomes intolerable!—and then—

and this even though the story was about quite other things.

Even novels and tales which announced quite general didactic intentions often showed a clear relation to the condition of the people, like F. E. Paget's[21] *Caleb Kniveton* (1833) which preached 'the fearful depravity of a revengeful spirit' by means of a topical example of rick-burning. The very lack of 'thrilling incidents' in the same author's *Tales of the Village* (1840), justified in the introduction, was a kind of protest against the troubled times. William Gresley's novel *Charles Lever: The Man of the Nineteenth Century* (1841) was more directly polemical, against what Paget called, in *St Antholin's* (1841), 'Chartism and socialism, and their kindred abominations'. Gresley's plea for more churches, in *Colton Green: A Tale of the Black Country* (1846), and Paget's, in *Milford Malvoisin* (1842) and *St Antholin's*, 'lest the masses of our manufacturing population ... become entirely heathen', brought piety into relation with the changing conditions of life in the Midlands. These conditions made Gresley's fiction, but not Paget's, descend to particulars. Deploring the way in which coal-miners formed a race apart, Gresley sent his hero down the pit to see their conditions of work for himself, and the description was specific enough for the reader to see them too.

VI

In the 1840s, lower-class life appeared in fiction much more often in the form of what was called 'low life', that is, crime. In part this was no doubt because the lower classes had earlier gained attention mainly as the victims of a savage penal code. The fashion was set by Ainsworth and Bulwer in the thirties. By 1840 *Dearden's Miscellany* could take *Jack Shepherd* and *Oliver Twist*, which both appeared in *Bentley's Miscellany*, as representative of a 'gallows-school of novelists'. An imitator of Dickens like Albert Smith,[22] when attempting what he called in *The Fortunes of the Scattergood Family* (1844–5) the 'novel of everyday life', took crime as an unavoidable ingredient: in this novel, smuggling, housebreaking, and forcible detention; in *The Pottleton Legacy: A Story of Town and Country Life* (1849), the plotting of a railway ganger to

[21] Francis Edward Paget, 1806–82, and William Gresley, 1801–76, were Tractarian clergymen.

[22] Albert Richard Smith, 1816–60, was born at Chertsey and qualified as apothecary and surgeon but gave up his practice for a career as a dramatist and solo entertainer.

discredit the hero and deprive him and his sister of a legacy. *The Struggles and Adventures of Christopher Tadpole* (1848) was obviously modelled on *Oliver Twist* but took the hero into a greater variety of low life, in the salt mines as well as the Liverpool slums, and made the reader acquainted with a more picturesque range of crime—highway robbery and Italian brigandage as well as the intrigues of an unscrupulous lawyer. Smith was somewhat given to using the narration of entertainments as a short cut to being entertaining himself, in 'trips', picnics, private theatricals, and showman's patter; but he was careful to draw attention to the harsh conditions of life in the lower reaches of society, from the sufferings of a governess to those of workers in the salt mines which he claimed, in the Envoi to *Christopher Tadpole*, to have visited himself. As much as in his attempt to rise to the demands of high life and historical romance in *The Marchioness of Brinvilliers: the Poisoner of the Seventeenth Century* (1845–6), crime held these novels together. His less sedulous manner in *The Adventures of Mr Ledbury and his friend Jack Johnson* (1842–4), which he calls at the end a mere 'performance', is more distinctive, from its very indulgence in miscellaneous novelties. But even they include a cousin who is a coiner, a murder followed through right to the guillotine, and the 'heartless' treatment received by a 'fallen girl'. The pleasant habits that made Smith successful as a solo entertainer outside of the novel can just be glimpsed despite the materials he believed obligatory for success in fiction.

Catherine Crowe's [23]*Susan Hopley* (1841) and Richard Cobbold's[24] *Margaret Catchpole* (1845) are both concerned with working-class girls drawn into contact with crime, the first as detective, the second as an accomplice while under unusual emotional strain. But the first is the merely titular heroine of a cheap and cloudy melodrama, while everything about Margaret Catchpole has the stamp of precise practical detail. The simple moral scheme, in which the author is at one with his chief characters and his readers, is of such obvious practical application

[23] Catherine Crowe, née Stevens, 1800–76, was best known for her writing on spiritualism. Her *The Night Side of Nature, or Ghosts and Ghost-Seers* (1848) interested Baudelaire.

[24] Richard Cobbold, 1797–1877, was born at Ipswich and educated at Caius College, Cambridge. A sporting parson who was also chaplain to the workhouse, he produced many didactic works in prose and verse.

that it becomes one of the means by which a specific way of life is communicated and the responsibility of the chief characters is asserted, all the more strongly because the circumstances which expose them to temptation are so clear and definite. The morality, far from inhibiting, actually enhances the presentation of the heroine's two lovers (respectively, colourless but exemplary, and spirited but shady, rather in Trollope's manner). Margaret herself, exposed to almost invincible temptation by her attachment, is not viewed any less sympathetically (though far from indulgently) for being incapable of a superhuman firmness of mind. Many readers, then as now, would no doubt go further and regard her as a victim. Against this implication the author obviously takes himself to be clad in complete steel by the Christian terms of his statements in person, in the body of the book no less than in the preface, even though in at least two crucial incidents his heroine is as much at the mercy of accident as Hardy's Tess. Yet, unlike Hardy, Cobbold makes a concluding accident unite her with the exemplary suitor long after the death of the other one. The scope left, in this way, for the reader's attitude to vary strengthens the hold of the book upon him. It is a modest work, much of it wooden and obvious, but repeatedly the author's very unpretentiousness protects him: unwilling as well as unable to probe motives and feelings beyond his pious purview, he allows them to be inferred from actions; so the involuntary force of Margaret's obsession becomes plain and the specific adventures lead to something beyond the excitement of the moment, strong though that often is. It is a small triumph, but one which Cobbold was never able to repeat.

No touch of unpretentiousness is admitted to Samuel Warren's[25] dealings with the lower classes. It is on the very inadequacies of his imagination that his fame rests. *Ten Thousand a Year* (1839–41), on the strength of which he regarded himself as 'a high-minded rival' of Dickens, was little more than a pamphlet nearly twice the length of a twenty-part Dickens serial. Only readers politically predisposed, like those of *Blackwood's Magazine*, to endorse its heavy satire of social climbing could have demanded so many editions of this interminable compendium of snobbery, with its abuse of 'the monkey millionaire' and its flattery of 'the acute

[25] Samuel Warren, 1807–77, studied medicine in Edinburgh and law at the Middle Temple. Called to the bar in 1837, he became QC in 1851 and was Tory MP for Midhurst, 1856–9.

reader', its lavish exclamatory sentiment, and its legal footnotes and appendices, lengthening in successive editions. Occasionally hatred stirs the author to a sort of lumbering, grotesque fancy, though for the most part he is content with tendentious label and rough illustration. It was natural that a sententious journalist with Warren's interests should attempt a detective novel: *Now and Then* (1848), aptly subtitled *Through a Glass Darkly*, shows by its nullity how much this form of fiction relies on crisp detail to make the reader wish to find out exactly what happens; as the subject gives little scope for snobbish approval and depreciation, Warren has nothing with which to attempt to make good the deficiency.

Samuel Phillips's[26] *Caleb Stukely* (1842–4), also written for *Blackwood's*, is less of a pamphlet than *Ten Thousand a Year*. It attempts to enter into a wider range of lower-class life, even though the treatment accords readily enough with the known prejudices of the magazine's readers. The political apocalyptic and dissenting emotionalism to which the hero successively succumbs are not merely denounced, as in Warren's Revd Dismal Horror; they are actively shown, in the Chadband-like harangues of the absconding Mr Rational and in the friendship, implausibly connected with nefarious dealings on the Stock Exchange, of a less unorthodox preacher. Although Phillips relies on a conventionally lachrymose love story to hold the book together, its central part communicates a strong sense not of the physical surroundings, but of the confusion of mind in which a youth may fall to the bottom of the social pile and fight his way back into the middle classes. Dickens was sufficiently impressed by the first half of the book to write to the author praising it.

VII

It was hardly to be expected that the novel would remain untouched by the religious controversies of the decade or more after 1833, when J. H. Newman, John Keble, and Hurrell Froude formed in Oxford their Association of Friends of the Church and began to publish their *Tracts for the Times*. During the same years,

[26] Samuel Phillips, 1814–54, son of a London merchant, studied briefly at the universities of Göttingen and Cambridge. From 1845 he wrote influential literary articles and reviews for *The Times*, including 'The Literature of the Rail' (1851) and 'Literature for the People' (1854).

the brothers of two of the Tractarians, F. W. Newman and J. A. Froude, began to question aspects of orthodoxy, and George Eliot ceased to subscribe to it, while attention was drawn to the tolerant Broad Church emphasis in the preaching of Thomas Arnold by his early death in 1842 and the publication of his biography by Dean Stanley in 1845, a book which even so untheological a novelist as Dickens could declare to be 'the textbook of my faith'.

Among the bizarre results in fiction was the *Hawkstone* (1845) of William Sewell[27], a High Anglican who could apparently justify its fustian adventures because the villain was a Roman Catholic who believed the *Tracts for the Times* to be 'doing us effectual service'. Geraldine Jewsbury could make the central male character in *Zoe* (1845) a priest who loses his faith (though, for a Professor at the English College in Rome, his reasons are not of the clearest). Curiosity aroused by a title like *From Oxford to Rome* (1847) could gain readers for an anonymous work of the most minimal interest as fiction, though full of the melodrama of sentiment; the newspaper self-justification of the authoress, Elizabeth Furlong Shipton Harris, attempting to distinguish her own position from that advocated in the novel, added to its notoriety.

For many readers and writers there was no firm boundary between fiction and life, so that novels which appeared to present personal experience had an added excitement. In the case of J. A. Froude[28] this was a justifiable reaction. The only impressive part of his *The Nemesis of Faith* (1849) consisted of the 'Confessions of a Sceptic' which were transparently his own. But he went on, though quite without talent as a novelist, to conduct by means of fiction an 'experiment in life', trying out what might happen in the case where a faith which was completely dependent upon confidence in the Bible became undermined when that confidence was lost. The story brought the central character to the verge of suicide, to be rescued by a priest and a temporary confidence in the Church which 'doubt soon sapped'. The fiction was not substantial enough to constitute a comment on the kind of faith which,

[27] William Sewell, 1804–74, an elder brother of Elizabeth M. Sewell, was educated at Merton College, Oxford, and became a fellow and tutor of Exeter College, Whyte's Professor of Moral Philosophy, 1836–41, and an early Tractarian. He withdrew from the movement after Tract 90 in 1841.

[28] James Anthony Froude, 1818–94, was ordained deacon but refused to take priest's orders and resigned his fellowship at Exeter College, Oxford. He became a major historian. (See Vol. XIV, Chap. 16.)

needing such breakable support, became retribution rather than salvation to him, but the mere presenting of the difficulties caused scandal to all parties and a second edition was required at once.

In all Froude's fictions he can be seen experimenting with events. *The Spirit's Trials* (1847) attributed to its hero a life much like the author's up to 1840 and from there tried out a possible sequence leading to his final testimony and death. *The Lieutenant's Daughter*, published with it under the title, *Shadows of the Clouds*, which suggested afflictions like those of Wordsworth's *Margaret* ('My apprehensions come in crowds; . . . The very shadows of the clouds . . .'), followed one sequence through and then offered a happier one, produced by altering the date of a single event. In neither case is the fiction worth much, though the businesslike brothel-madam in the latter story is unexpected among so much that is sentimental and shrill. What is interesting is to see the future historian trying out, in these early works, the varying possibilities of given situations, actual or invented. To go beyond the mere devising of plot, however, he needed to see much further into the nature of his main personages than his interest in the opinions they held or illustrated could take him.

Lavengro (1851) by George Borrow[29] is of a similar form to *The Spirit's Trials* in so far as autobiography fades into fiction. Alternative titles like *Lavengro: An Autobiography* and *Life: A Drama* and the references in the Preface to 'this dream' are symptomatic of the strange texture which emerges, superior, in the vigour and directness of its parts, to everything else in the present chapter, though indistinct in its tenor as a whole. The reader seems invited to seek for analogies between the sufferings of the inner life, remarked but not investigated, and some of the matters which are selected for specific observation—especially when these concern, in the central chapters, another man liable, like the author, to fits of horror—before the book moves first into a brief, original love-story and then, along with its sequel, *The Romany Rye* (1857), into fantastic sectarian debate. The result is unclassifiable, and uniquely memorable.

[29] George Borrow, 1803–81, educated at Norwich Grammar School, son of a peripatetic recruiting officer, went to London in 1824. Already acquainted with many languages, he became for a time a literary hack before the roving narrated in his two semi-novels. Subsequently he served the British and Foreign Bible Society and married a well-to-do widow. (See Vol. XIV, Chap. 18.)

J. H. Newman,[30] with his slighter, more predictable plot in *Loss and Gain* (1848), imagined more subtly than Froude. This novel may be one of the least of the works of the outstanding Tractarian, but it seems to catch the insulated urbanity of Oxford talk in the 1840s and, if one can accept the limitation of subject, has a little of the quality of Peacock in its pleasant exhibition of crotchetry. Moreover, although the central character is incompletely individualized, the more important of his states of mind are movingly present to the reader, being again the author's own.

Satire which shows the professional finesse of Newman's at its best is, of course, uncommon. In *The Bachelor of the Albany* (1848), by Marmion Savage,[31] farce and a rather cruder conversational comedy than Peacock's make play with the archaizing tendencies of the Tractarians and the Young England party, in a romance the burden of which is that 'there is no living in society without taking one's fair share of its cares and duties'. The story of the conversion to marriage and civic responsibility of Barker, the bachelor, with his 'almost fanatical repugnance to duty', takes as tendentious analogues for his way of life an idle pluralist of the Church of Ireland and 'the lives of deans and even of bishops', now that they may be 'discussed in the newspapers, and the vulgar doctrine of the wages of labour . . . applied to the relations existing between the priesthood and the people'. The life of a dean is said, in an argument Trollope will dramatize more tellingly in the next decade, to be 'the closest approximation to doing nothing attainable unless the Puseyites and Lord John Manners shall succeed in reviving monastic institutions'. When Barker enters parliament, his opinion about the reform of abuses also anticipates Trollope's: unless an 'objectionable' institution can be 'utterly abolished', its 'deformities' constitute 'the beauties of the system'. Peacock's manner is carried into names like the Revd Bartholomew Owlet and Lord John Yore, and into readily identifiable characters like Mrs Martin, author of *The Matrons of England*, *The British Wife*,

[30] John Henry Newman, 1801–90, son of a banker, became a fellow of Oriel College, Oxford, in 1822, vicar of St Mary's, Oxford, in 1828, and a Roman Catholic in 1845. (See Vol. XIV, Chap. 11.)

[31] Marmion Savage, 1803–72, was educated at Trinity College, Dublin. After some thirty years' government employment in Dublin, he moved to England and succeeded John Forster as editor of the *Examiner*, 1856–9.

The British Stepmother, on the model of Sarah Stickney's well-known series.

The work of Newman's sister, Harriett Mozley,[32] indicates the direction which the best novels prompted by religious concerns will take later in the century. As she explained in the preface to *The Fairy Bower* (1839), for her the 'scenes and situations of life' are the important thing, and 'characters as they really are', rather than 'moral portraitures for unreserved imitation or avoidance'. In *The Fairy Bower* and its sequel *The Lost Brooch* (1841), self-conscious opinions are as amusingly presented as any in *Loss and Gain*, though verging on farce since they are the opinions of children. But the point of the fiction is that they should be tested in the kind of 'scenes and situations' which constitute the 'actual sphere and trial' of 'young persons', that is, in domestic life. Grace, the heroine, only 10 years old in the first book, and over-ready to admire the intelligence of her friends, is uneasy when she cannot admire their conduct. Her own intelligence appears from her unhappiness when drawn in to acquiesce in a small deception because she herself would be the gainer if it were revealed. A less protracted deception over a broken cup causes her equal misery. Although it is all viewed as 'a sort of *rehearsal*' for the 'real stage of life', it is real life all right, to be appreciated by the old as well as the young for whom it was written. Maturity, lying in the future, and made present to the heroine when she more than once receives the shock of the possibility that her mother might remarry, seems an analogy for the infinite moral background to the action, remote though with 'footstep instant now'.

The sequel, as 'curiously clever', is less '*covertly* satirical' (the terms, though not the emphasis, are Charlotte M. Yonge's). Its comedy is rather more coarse-grained, and even its moral point less subtle: exposure leaves the culprit's self-righteousness intact to a degree which is comical but also frightening; the physical danger to the servant falsely accused is on another plane and not funny at all. Emphasizing the satire of strict evangelical education, the author has less time for the inner dialogue of conscience where she is at her best. *Louisa* (1842) takes her further away from such exploration. The best of the religious novelists who follow her in

[32] Harriett Elizabeth Mozley, née Newman, 1803–52, married a Tractarian clergyman. She 'set the fashion', according to Charlotte Yonge, for 'books on child life'.

the second half of the century show and acknowledge how well they understand her revealing of this inward dimension in the apparently small dramas of domestic life.

VIII

Two minor authors stand apart, Bulwer and Disraeli. Neither is an innovator, though they gave fresh impetus to some already established kinds of fiction, to the fashionable novel and the social and political thesis novel in particular, and Bulwer to the crime novel and the historical novel as well. Both authors are important for their effect upon Thackeray. Bulwer stirred his indignation, Disraeli a kind of half-contemptuous amusement, and in each case Thackeray's reasons had something to do with the nature of his own first fictions.

Although Bulwer's[33] work repelled Thackeray for its 'sentiments ... big words ... premeditated fine writing', it forms a document which no historian of the novel can overlook. At first it shows clearly the kind of credentials a novelist thought he had to offer to readers brought up on the Romantic poets, especially on the early Byron. And once Dickens is established, the linking of his name with Bulwer's, even by Forster in his *Life of Dickens*, tells something about them both. For Bulwer makes his main appeal to the simplest sympathies and antipathies and, provided he can stir them, is not very particular about the means. He likes to use wide social contrasts, the rich with the poor or the respectable with the criminal, and to simplify character and event so that no reader can mistake the effect intended. On the other hand he has an unremitting literary selfconsciousness which makes genuine simplicity impossible. Only occasionally does he allow himself to look directly at some aspect of the life of his own time and forget that he is a man of letters.

Bulwer studied what he called 'the essentials of practical popularity' all his life. So he is, for the historian, a kind of barometer of taste. Beginning with the Byronic egocentricity of *Falkland* (1827), he quickly saw how Disraeli in *Vivian Grey*

[33] Edward George Earle Lytton Bulwer, later Bulwer-Lytton, 1803–73, was born in London and educated at Trinity Hall, Cambridge. Dandy and friend of Disraeli, he entered parliament as a Whig in 1831 but joined Disraeli's party in 1852. He was created a baronet in 1838 and baron in 1866.

(1826) had developed the possibilities of the novel of fashionable life and went one better in the baroque affectations of *Pelham* (1828). Public interest in the punishment of criminals led him to attempt a thesis novel on the model of Godwin and Holcroft in *Paul Clifford* (1830) and to present a famous crime from the point of view of the criminal in *Eugene Aram* (1832). This interest in crime, already apparent in *Pelham*, became a steady ingredient in his work and by 1838, writing 'On Art in Fiction' in the *Morning Chronicle*, he gave his reasons. He not only found 'the true secret of Creative Genius' in 'the intenseness of the sympathies', and the merit of the greatest fiction in the power to depict 'the metaphysical operations of stormy and conflicting feelings', but he claimed that the author who was in this way 'versed in the philosophy of the human heart' would find his 'widest scope' in 'the portraiture of evil and criminal characters'. But when the public found his criminals merely repulsive, he was ready to forgo this formula and give them the 'agreeable emotions' of middle-class life, in the Caxton novels of 1849–59, with the same glib and tireless fluency. They still ran to melodrama as parsley runs to seed.

Pelham is the only one of these books in which the writing has any life. Bulwer follows the strategy of *Vivian Grey*, with a hero whose impudent success is exposed to satire. But whereas Disraeli makes the eventual failure of Vivian Grey's plots into the culminating criticism of his impudence, Bulwer, after allowing that impudence to reveal itself with equal high spirits and implausibility, sets going a serious action—the political ambition of Pelham and his love for the sister of a man who appears to be a murderer—from which his hero emerges with reputation if not reward in the first case, and complete success in the second. No amount of self-depreciation ('I was a consummate puppy . . .', 'pampered and spoiled as I was . . .') will conceal that the author admires him. He was meant, as the 1840 preface said, to 'grow wise by the very foibles of his youth'. In the first edition he has an extravagant, sensuous effeminacy *pour épater*, but even this does not hamper his unusual capability in self-defence or prevent him from enjoying, more than the reader does, the humours of the far from effeminate rogues of the last chapters. The obvious attempt to satirize the excesses of the dandy has to work all the time against the run of a plot which justifies him. Moreover, the plot becomes steadily cruder. The promising beginning, in a terse and spirited account of the interrupted elopement of Pelham's mother, is not sustained; the hard-boiled hyperbole which Pelham transparently affects—'... we all have our foibles, as the Frenchman said when he boiled his grand-mother's head in a pipkin'—gives way to simple melodrama, when the murder in the sixty-fourth of the eighty-six chapters allows Glanville, with his apparent villainy and his Byronic despair, to take centre stage. In novel after novel, whatever the literary formula Bulwer at first attempts to follow, sensational violence is his main method of bringing characters into developing relationships with each other. The superficial literary sophistication does not conceal the essential poverty, the substitution of crude stimuli for an interest in what particular imagined people are and do. This is the secret of Bulwer's popular success despite much denigration from the critics.

In *Paul Clifford* (1830), he advances his bid for popularity by the claim to a thesis. He wrote it, he said in the 1840 preface, as 'a satire on the short cut established between the House of Correction and the Condemned Cell', to draw attention to 'a vicious Prison discipline, and a sanguinary Criminal Code'. He may imitate the irony of Fielding, as Paul is accompanied to the House of Correction by 'a very old "file"' and 'a little boy, who has been found guilty of sleeping under a colonnade'; but he forces upon us at once, and without any further corroboration from the narrative, his main point that 'the peculiar method of protecting the honest' which English law adopts is 'to make as many rogues as possible in as short a space of time'. He avoids treating 'in long detail of the moral deterioration of our hero' and subjects him instead to swift temptation by the dull life-story of a journalist whom he has already met outside the prison and who could just as easily have tempted him outside as in. It makes a very slight hinge on which the moral must turn.

Acrimonious literary satire and good-humoured lampooning of public figures, from George IV down, in the guise of highwaymen, make the novel into something of a topical miscellany. The political satire, in particular, asks for a different kind of reading from the realistic fiction. It is characteristic of the way Bulwer writes as a selfconscious man of letters that he can imitate *The Beggar's Opera*, writing new words for popular songs for Gentleman George, fighting Attie (Wellington) or Old Bags (Lord Eldon). The ingenuity and high spirits naturally disturb the tone of his handling of crime. For if the avowed moral is to mean anything, Clifford must seem worse off as a result of his fall. Instead, he is—and has—better company than before. His adventures as gentleman and highwayman, in fact, have all the attraction which Dickens forswears in the 1841 preface to *Oliver Twist* (despite his attempt to keep *Paul Clifford* out of the indictment): 'merrymakings in the snuggest of all possible caverns, ... embroidery, ... lace, ... jack-boots, ... the dash and freedom with which "the road" has been, time out of mind, invested'. The thesis at the end, '"Circumstances make guilt," he [Clifford] was wont to say: "let us endeavour to *correct the circumstances*, before we rail against the guilt!"' leaves the reader comfortably to believe, if he will, that there is no such thing as guilt—for what is it, if circumstances 'make' it and the individual has no responsibility?—or, on the other hand, that such a thing will remain to be railed against after the circumstances have been corrected. This invitation to 'believe what you list' is another source of Bulwer's power.

In *Eugene Aram* (1832), the moral ambiguity in the method of presenting this famous murderer eventually entrapped even the author; having convinced himself that Aram was not guilty, he had to rewrite the conclusion for the edition of 1851. After writing a comedy of crime in *Paul Clifford*, he was now aiming, as he pointed out in the original preface, at 'something of the nature of Tragedy'. But it was tragedy as Byron had attempted it, discussing, rather than dramatizing, the largest issues. The greatness of the hero is a matter of the merest rhetorical advertisement, suggesting, as early as the middle of Book I, that he could well stand with Socrates or Napoleon, since greatness may 'be shown in wisdom, in enterprise, in virtue, or even, till the world learns better, in the more daring and lofty order of crime'. The caveat 'till the world learns better' hedges characteristically, as the author sets about elevating the crime. In the first edition the 'lofty order' of the crime is found in the motive, to gain financial independence and leisure for study (so that the deed becomes 'a great and solemn sacrifice to knowledge') and in the alleged turpitude of the victim (so that the killing may be considered as in war—according to 'whether mankind would not gain more by the deed than lose'). Of Aram's 'burning desires ... resplendent visions ... sublime aspirings', the reader knows nothing because the author knows nothing. The *frisson* is sufficient; recollections of Faust will apparently supply the rest. The author can formally separate himself from the hero's utilitarianism, while evoking sympathy for his lavish claims to a Byronic self-sufficiency at war with the world. Thackeray in 1832 found 'The sentiments ... very eloquent clap-trap,' and trusted that 'when my novel is written it will be something better'.

Godolphin (1833) brings forward into contemporary London the same difficulty of reconciling the hero's inner and outer worlds. It is a subject caught from Goethe's *Wilhelm Meister* (as the opening of chapter 20 acknowledges, in the first edition). The hero, as much 'enamoured of the Shadowy and the Unknown' as he is dissatisfied with 'the petty ambitions of the world', becomes a selfish solitary. But Bulwer's own political interests pull against this subject, and his hero's marriage to a widow of strong political views eventually stirs him to do something with his life—just before his death. He allows her to procure him a seat in parliament. The centre of the novel, in the effect upon an uneasy marriage of the passing of the Reform Bill, gives it a certain distinction despite the even more uneasy surroundings of occultism and melodramatic death. Godolphin's wife is a bold sketch, for 1833, of a woman thinking for herself, and suspecting the Whigs in 1830 of 'only playing with democratic counters for aristocratic rewards'. This part of the novel is in touch with real things in a way which is rare with Bulwer. But then this is also the year of one of his more solid works, the social commentary *England and the English*.

In *Ernest Maltravers* and its sequel *Alice* (1837–8), politics are a subsidiary field for the scheming of the villain or for the diversion of a hero exhausted by the labour of imitating Byron. With an eye also to *Wilhelm Meister*, as the preface points out, the centre of the novel is in the hero's literary apprenticeship, to which the villain is made to contribute by steadily impeding his attempts to marry for love before the last page. The impudent egotism of a villain 'not having much vanity nor any very acute self-conceit' and the occasional pointed but unobtrusive irony in Bulwer's treatment of him (meditating a break with the hero, for instance, he keeps 'an impromptu sarcasm in reserve') are worth volumes of romanticized crime, but the high stilts of the 'towering and haughty' hero, 'always asking for something too refined and exalted for human life' (at least until the last page), take the greater part of the novel away into the romantic stratosphere, despite its satirizing of an Italian poet who *will* 'talk fine' and has 'no commonsense'.

Bulwer had already found in 1834 a better formula for success, in the historical romance. Here he laboured for accuracy of detail but did not need otherwise to alter his methods from those of *Eugene Aram*. All his four historical novels are conceived of as tragedies, with the words 'the last' prominent on every title-page—and all proceed by the confecting of character from abstractions spiced with rhetoric. The first one, *The Last Days of Pompeii*, was successful at once and, unlike everything else of Bulwer's, still has readers. Its melodramatic plot, involving a Christian convert and a necessitarian (as victim and murderer), the suppression of the evidence and the discovery of it as the hero, falsely accused, faces the lions in the arena and Vesuvius prepares to erupt, is cheap but has a clarity which Bulwer seldom reached. No activity is expected of the reader: the ideas discussed are to be prejudged as Christian or not, and seen, like the characters, in their simple relation to the crime and the love-story.

Rienzi (1835), on the other hand, which was begun during the same visit to Italy, has more resemblance to a novel for adults, in so far as it requires them to see topical reference in 'the Situation of a Popular Patrician in Times of Popular Discontent' (the title to Book II, chap. 3). But the situation of the patrician is exaggerated into that of a tragic hero sustained by 'his OWN SOUL . . . and contented, therefore, to be *alone*!' and his politics are so high-minded as to make allegory predominate over any more subtle concern with his life. In *The Last of the Barons* (1843) the interest for the author of the hero, the Earl of Warwick, lies in his inability to resist the policy of Edward IV which 'based a despotism on the middle class'; Warwick's fall shows the fate of an aristocracy 'associated with all such liberty as had yet been achieved since the Norman Conquest' and prepares the way, we are told, for the tyranny of the Stuarts. The introduction of an unhistorical inventor with his steam-engine (the possibilities of which Warwick alone sees) even allows a glance, in this mid-fifteenth century context, at the importance of manufacturing industry. The inventor's daughter is called Sybill and, like the heroine of Disraeli's novel of two years later, is faced with a symbolic choice between suitors of opposing classes and political views. In these two novels

Bulwer at last sustains a connection between his literary and his political concerns. He takes the plot from history, and its emphasis and significance from the politics of his own time. Although he works up the whole by means of the familiar sentiment and declamation, the result is a pair of popular novels which are capable of prompting thought. The thought is rudimentary, no doubt, and rests upon fear of disorder, but the nullity of *Harold* (1848) shows how little a sedulous acquaintance with the historical evidence can do in the absence of some such current of contemporary interest. *Zanoni* (1842), too, belongs with the historical novels in so far as its focus is upon the French Revolution. Dickens perhaps recalled the hero's sacrifice of himself on the guillotine. Everything, however, is subordinated to a kind of moralized spiritualism which, however seriously Bulwer himself took it, derives from Gothic romance. All that remains of the serious concerns of the two principal historical novels is the derivation of the Reign of Terror from French free thought, which is, of course, repudiated as the very antipodes of Zanoni's vacuous pseudo-philosophy.

Bulwer makes these excursions into the fictionalized past in the thirties and forties mainly in order to apply the simplifying methods of melodrama to subjects which come from contemporary life. For the remainder of his career, except for the unfinished *Pausanias the Spartan* (1876), he is content with melodrama which is set predominantly in contemporary society and is not organized so as to make reference to public questions. *Night and Morning* (1841) and *Lucretia* (1846) are crime novels, perhaps written because of the success of Ainsworth's treatment of Dick Turpin and Jack Sheppard, but already showing Bulwer's imitation of himself—in the opposition, in *Night and Morning*, of a good-hearted criminal and a vicious aristocrat, recalling Paul Clifford and Lord Mauleverer, and in Lucretia's gradual corruption through guiding her conduct by 'the hard calculations of the understanding' as Eugene Aram had done. In *Lucretia* the difficulty Bulwer has in the orderly narration of his complicated intrigue, the violence of the declamation and of the action, and the bathos which results from the attempt to show a villainess who is said to make 'intellect' her 'sole god' but seems to the reader to have only low cunning, mark the novel as the climax of his mistaken attempt to give evidence of genius by 'the portraiture of gigantic crime' (his phrase in the preface). The preface also claimed that the crimes related had actually taken place; in some of them the activities of the forger and poisoner Wainewright were recognizable. This hardly made any less repellent Bulwer's claims for his treatment of them as tragedy. Most of the reviewers were revolted and they frightened the publishers; *The Times* attacked in a leading article; even Macaulay, praising the book in a letter, begged the author to write something 'more cheerful'. Bulwer attempted to lighten the gloom by mitigating the catastrophe for the 1853 edition, but his principal answer to his critics is *The Caxtons* (1848–9), which, he alleged in the 1853 preface to *Lucretia*, he had begun at the same time as that novel.

'The art employed in *The Caxtons*', he writes disarmingly to a friend, 'is just that of creating agreeable emotions . . . That is one branch of art and rarely fails to be popular.' For this purpose he chooses a middle-class 'family picture' as his subject and second-hand Shandyism for the selfconscious attempt to treat it with humour, at least for the first third. But the change from his earlier novels is considerably mitigated once Uncle Roland's disowned son enters the action, with his romantic origins and the melodrama of his attempted abduction of an heiress. Nevertheless, he and the rest of the younger representatives of the family eventually make good in Australia and India and there is a rousing talk of the 'marvellous empire . . . annexed to the Throne of the Isles'. Once again, when changing direction, Bulwer published anonymously. He carried on the pretence by putting out the next two novels as 'by Pisistratus Caxton'. In the first of these, *My Novel*, the disguise wore thinner the longer it appeared in *Blackwood's Magazine* (1850–3). For the first half, village life is the subject, with a refugee from Italy to stimulate the local humours. The second half produces, from the hat of the experienced conjuror, the past lives of characters who are seen in the first as mere 'Varieties in English Life', turns one of them into the now familiar type of villain, 'the spirit of Intellectual Evil', and supplies him with accomplices in the refugee's enemies, fresh from Italy and in touch with the secret societies. The busy fabrication of intrigue reaches an unusual climax when Bulwer manages to involve, as contestants in an election in the same constituency, not only the villain and the man who has treated him as a son, but that man's actual (and of course long-lost) son and the aristocrat who had long before been the father's rival in love. The appearance of topics which were preoccupying Dickens at the same time (*Bleak House* was appearing in monthly instalments from March 1852 to September 1853) shows Bulwer's eagerness to produce the complete novel of middle-class life. So there is a pushing but attractive business tycoon to rival Dickens's Ironmaster and some sentences of the best purple are devoted to the state of the drains. Bulwer returns to politics while this novel is appearing: he justifies his support of Disraeli in the *Letters to John Bull* in 1851 and enters the house, as member for Herefordshire, in 1852. In the political part of the novel (which stands in sharp contrast to *Bleak House*, with its ridicule of parliament), he refashions the problem of *Rienzi* into that of the Conservative who at one point sympathizes with a popular cause. Setting the action back about twenty years, however, keeps him clear of public questions which are of any importance to the fifties and the discussion seldom descends from generalities. Nevertheless, a touch of the excitement of politics is a relief from the strained intrigue.

What Will He Do with It? (1857–9) has only a touch of this kind of relief. It begins quietly with an old man and his granddaughter, who recall *The Old Curiosity Shop*, and a crusty but warm-hearted hero in late middle age who owes a little to the author's self-idealization and a great deal to Dickensian sentiment. But after only the first quarter, with the introduction of the child's supposed father, there begins a burrowing into the past for the key which will explain the present and retain the magazine readers, until, months later, the father dies, 'a mountain . . . burned out by its own inward fires', and the child is proved to be 'no more the Outlaw-Child of Ignominy and Fraud but the Starry Daughter of POETRY AND ART!' Influence from Dickens is superficial (and often clumsy in detail, as with the old nurse, accused of passing letters, who 'immediately wanted to take her Bible-oath and smelt of gin'). It only highlights Bulwer's reliance on the formulas of his first successes, even though their content is now modified, with chameleon accuracy, to suit middle-class demands: so the worthy mentor of the hero is not content to advise, 'Mark this! never treat money affairs with levity', but must go on, 'MONEY IS CHARACTER'.

For the remainder of his life Bulwer devises further variations on his own themes. *A Strange Story* (1861), written at Dickens's request for *All the Year Round*, refurbishes the elixir of life and the pseudo-philosophy of *Zanoni* by means of an observer devoted to scientific materialism and a lurid plot which may well have stimulated H. Rider Haggard. *The Coming Race* (1871) is a new departure, looking with temperate hostility at the possible consequences of that materialism, in a Utopia from which even an American visitor finds he must escape; but though it is less turgid than anything else he wrote, it is also, as fiction, extremely slight. *The Parisians* (1872–3), which was unfinished at his death, reverts to type; it returns, shortly after the Franco-Prussian War and the Paris commune, to the horrified contemplation, as in *Zanoni*, of the excesses of French free thought and politics, superimposing this upon an unusually ill-managed investigation of disputed ancestry. *Kenelm Chillingly* (1873), which was written at the same time, attempts again to mix Shandyan whimsicality with Victorian pathos, but also contains some crisp and humorous identification of the characters in a manner which has not been found since the opening of *Pelham*.

The pretentiousness of much of this immense and popular output may seem to place Bulwer at the head of those Victorians who debased the literary coinage, but its range of subject and its literary aims undoubtedly gave the novel standing and repute at a time when it lacked them in the eyes of many readers. To some at the time and to most since, his strutting assumption that apparent literary sophistication justifies the result has seemed insupportable. Since he has, after *Pelham*, hardly any humour, his glib rhetoric and stereotyped sentiment lead repeatedly to unintentional self-parody. He has concerns of considerable interest, from Godwinian polemic to the elevation of criminals into tragic heroes, from the pleasures of the middle classes to the magic and mummery of *A Strange Story*, but they all come to us in a language which seems unable to escape from the approximate and the second-hand, so that each novel gives not a sequence of events, thoughts, or speech of which we are expected to judge for ourselves, but a series of crude simulacra, our attitude to which is prejudged by the author.

At first sight Disraeli[34] appears to have many of the same faults

[34] Benjamin Disraeli, 1804–81, was born in London, son of Isaac D'Israeli (Vol. XII, Chap. 14), and baptized in 1817 following the death of the Jewish grandfather after whom he was named. Notorious as a dandy, he entered parliament in 1837 and in 1846 made his name by attacking Peel's abolition of protection for home-grown grain. He became leader of the Conservatives in the House of Commons, Prime Minister, 1868 and 1874–80, and Earl of Beaconsfield, 1876.

as Bulwer. He does not mind dealing out extreme and commonplace praise or blame to his characters even though he may fail to activate them sufficiently to afford corroboration; he does not mind melodrama if that is the only way these simplified characters can be made to move. He is, in addition, capable of a kind of helpless absurdity which Bulwer, with his greater literary aplomb, avoids. Disraeli's naïveté as a novelist may lead him to direct wish-fulfilment; but at the same time there are unmistakable indications that, though he may hope to deceive the reader about his heroes, he is not quite successful in deceiving himself. And in this way he is capable of surprising the reader into thought.

His first hero, in *Vivian Grey* (1826), is an unknown boy who gulls a stupid Marquis into believing himself able to lead a new political party. The outline and much of the detail come from Disraeli's own unsuccessful attempt to get J. G. Lockhart to edit a new daily paper for Murray, the publisher. Yet Vivian's impudence is relentlessly exposed. The first readers of volume i were also invited to share in an amused scrutiny of the polite society which he was deceiving and which he lovingly and inaccurately described. The tone, at once ingenuous and self-aware, even gave the impression that the author belonged to this society and was divulging a current political intrigue (likely enough in a year in which Lord Liverpool's government was on the way out and his most obvious successor, Canning, was distrusted by leading Tories). So the novel was read as if it came from that Society for the Diffusion of Fashionable Knowledge which Disraeli invented in his second piece of fiction, *The Voyage of Captain Popanilla* (1828), that is, as if it had been written 'to complete the education of the Millionaires'. *Popanilla* was dedicated to Robert Plumer Ward, whose *Tremaine, or the Man of Refinement* (1825) had given a cumbersome example of the kind of verisimilitude apparently required for such purposes (see Vol. xii, Chap. 8). Unfortunately, in *Vivian Grey*, Disraeli could not keep up beyond the first volume his impudent pretence of contributing to the process. In the second volume, melodrama reduced the hero's responsibility for his defeat without making the polite world responsible for it either, and at the same time mitigated the criticism of his impertinence. A further three volumes, published in the next year, were mere picaresque travelogue, lacking that play with the attitudes and opinions of author, hero, and reader which had distinguished the first. In revising the whole for the 1853 collected edition, Disraeli corrected some of the solecisms in the account of society but also cut out some of the satirical high spirits.

His second attempt at a fashionable novel, *The Young Duke* (1831), combined mockery of high society ('. . . a lady or a lobster salad. Ah! why is not a little brief communion with the last as innocent as with the first!') and of the literature which described it ('Take a pair of pistols and a pack of cards, a cookery-book and a set of new quadrilles; mix them up with half an intrigue and a whole marriage, and divide them into three equal portions'). But he also gave to the Duke's political opportunities an attention which had to be taken as serious, though it did not inhibit humour. The improvised mixture can still be read, partly because of the reader's uncertainty as to what is coming next—Disraeli had read *Don Juan* to some purpose—partly because of the author's light-hearted attitude towards himself and what he is up to, thus avoiding Bulwer's tendency to strut and peacock.

The not wholly serious first-person confidentiality was a far stronger ingredient in the editions before 1853 than after. Some of it came from the actual author with a 'holy ancestry', a 'lettered father', and 'a pretty style, though spoilt by that confounded puppyism'; some of it was spoken in the persona of an illustrious Roman recluse. Thackeray, recalling with amusement this 'prodigious novel' when reviewing *Coningsby* in 1844, found it 'impossible to help admiring the intenseness of the Disraelite-ego'. The affectation and egotism ('all very heinous,' Disraeli called it in the first edition, 'and painfully contrasting with the imperturbable propriety of my fellow-scribblers,—"All gentlemen in stays, as stiff as stones"') took a form which may have had its effect upon the mode of novel-writing Thackeray eventually adopted, with its pretended arch confidences and its pointed contrasts with other authors and modes. Byron and Sterne may have given the precedent but Disraeli had shown what might be done with it when writing for a public which so steadily abandoned its mind to light reading.

But again Disraeli was seduced, by the picturesque pleasures of *Childe Harold* and the eastern tales, into deserting the model of *Don Juan* and submerging his characteristic self-scrutiny in pseudo-heroics. An epic poem, of which, fortunately, only three books were published (1834), was allegedly conceived 'on the plains of Troy'; a 'Psychological Romance', *Contarini Fleming* (1832), was written as the 'development of my Poetic character'; and two long oriental fantasies, *The Wondrous Tale of Alroy* (1833) and *The Rise of Iskander* (1833), exhibited that character in deplorable action. Only in Parts 1 and 2 of *Contarini Fleming* does Disraeli preserve the ability to mock himself—as the poet who, the day after writing rapturous lines, finds that they are 'crude, silly stuff', and accepts advice that his 'poetic feeling ... is not a creative faculty ... but simply the consequence of a nervous susceptibility that is common to all'. Much in even these earlier sections of *Contarini Fleming*, and more still in the remainder, is inflated in the manner of Bulwer and suffers as he does from the disastrous example of half-read *Wilhelm Meister* (to which Disraeli acknowledged his debt in the 1846 preface). Even simple amorous romance, in *Henrietta Temple* (1837), was better than this, particularly with the entry of a comic character like Mirabel (drawn from Count D'Orsay). *Venetia* (1837) reverted to literary romanticism, turning into sentimental fiction what Disraeli had learnt of the lives of Byron and Shelley from Byron's ex-gondolier. Yet even here, among much absurdity (including blank verse pastiche) he could not keep a straight face in defence of his hero. When Carducis-Byron protests that, married to Venetia, he would be faithful, 'for my imagination could not conceive anything more exquisite than she is', he is told sharply, 'Then it would conceive something less exquisite.'

The novels of the forties, *Coningsby*, *Sybil*, and *Tancred*, are apparently more serious, the work of a Tory MP supported by a group of romantic High Churchmen in opposing the majority of his party and Peel its leader. The polemic here is only too confident but almost everything that is positive in it is inept as fiction, from the impeccable heroes and heroines and the vatic excesses of Sidonia, the financial grey eminence of Europe, to Sybil singing in 'the ruins of her desecrated fane' when the two nations are identified as 'THE RICH AND THE POOR' or the angel who waves a palm-tree sceptre at Tancred and announces 'the sublime and solacing doctrine of theocratic equality'. On the other hand the satire repeatedly succeeds. Lord Monmouth in *Coningsby* (1844), the representative of the upper classes who control the political machinery of England in their own interests and with whom the 'New Generation' of the subtitle must break, is the most impressive of the characters. Disraeli's attitude toward him resembles Dryden's toward some of the great figures of *Absalom and Achitophel* in that censure takes the form of amazement, even tinged with admiration. Lord Monmouth's discriminating concern for his own comfort and advancement is clear in the heartless comedy of his marriage to the stepdaughter of his mistress; an urbane and pointed commentary on it runs the satire into the verbal forms appropriate to praise. Disraeli allows Lord Monmouth to be 'good-natured, and for a selfish man even good-humoured'; so his 'fine manners' are seldom clouded by 'caprice or ill-temper'. His contempt for mankind is 'absolute'—he 'never loved anyone, and never hated anyone except his own children'—but he has a kind of respect for a very rich man whom 'you perhaps could not buy'. Even the graceful and ingenuous hero gains stature from the reaction to him of such a personage:

> It would be an exaggeration to say that Lord Monmouth's heart was touched; but his good nature effervesced, and his fine taste was deeply gratified. He perceived in an instant such a relation might be a valuable adherent; an irresistible candidate for future elections: a brilliant tool to work out the Dukedom.

In his review of *Coningsby* in the *Morning Chronicle*, Thackeray praised Disraeli's 'admirable humour' but found that he showed 'scorn for *many* things which are base, not for all; and, in the midst of his satire, coxcombry intervenes, and one is irresistibly led to satirize the satirizer'. Although Thackeray in 1844 did not need to be taught the trick of leading the reader in such a direction, it is hard to believe that *Vanity Fair* did not gain something from his having observed in *Coningsby* the advantages of an urbanity which verged upon impudence, for an author who presented his action by means of comment in varying tones of voice. Thackeray's parody of Disraeli in *Punch* in mid-1847 concentrates on simpler matters than this, but that there is a connection between the 'outrageously fashionable' *Coningsby* and the fourteenth and fifteenth numbers of *Vanity Fair* (chapters 47–53), written in mid-1848, is suggested by the presence in them not only of another character, Lord Steyne, modelled like Monmouth on the third Marquis of Hertford, but also by the matching of Steyne's attendants Wagg, Wenham, and Fiche and his mistress the Countess of Belladonna with Monmouth's Gay, Rigby, Villebecque, and the Princess of

Colonna, and by the parallel episodes in which Monmouth's friend Mrs Flouncey and Becky Sharp both find themselves ostracized by the ladies in the great lord's drawing-room and take refuge by a table full of prints. No doubt Mrs Flouncey has as little to do with the original of Becky Sharp as Mrs Gore's Colonel Hamilton in *The Banker's Wife* (1843) has with Colonel Newcome, but it is tempting to believe that in the first case Thackeray may have been stirred to go one better, using his own recollections.

Thackeray characteristically enjoyed the political impartiality of the satire. 'The hits at both parties', he said, 'are both severe and just.' What is even more surprising is that the satire does not spare the hero or his friends. In the crucial scene, in Book VIII, chapter 3, in which Coningsby refuses Lord Monmouth's arrangements for him to become Conservative candidate for Darlford, on the grounds that that party is 'unequal to the exigencies of the epoch, and indeed unconscious of its real character', my lord's misunderstanding of the words is still high comedy, because he can give them exact meanings with reference to the past, whereas Coningsby can only explain his notion of 'political faith, instead of political infidelity' by reference to 'principles' some of which are remarkably vacuous: 'Let me see authority once more honoured; a solemn reverence again the habit of our lives . . .'. These are hardly the instruments with which to face the power his lordship has to do as he will, which is an important part of the power of the Conservative party. The mockery continues when Coningsby, disinherited, takes chambers in the Temple and one of his Young England visitors is enchanted by the name. 'The tombs in the church convinced him that the Crusades were the only career. He would have himself become a law student if he could have prosecuted his studies in chain armour.' Even Disraeli's own notion of the 'Venetian constitution' of England is the subject of humour when the same friend denounces it 'to the amazement of several thousand persons, apparently not a little terrified by this unknown danger, now first introduced to their notice'. The novel does not merely follow the author's avowed purposes. This gives it life despite much cliché in admiration of the titular hero and his romantic sentiments.

Sybil (1845) has less of this pull between avowed purpose and actual presentation. All the same, the aristocrats whom the book attracts are, as Thackeray said when he reviewed it in turn, 'most brilliantly hit off, more so than the plebeian likenesses, the men and women of the mines and factories,' whose cause Disraeli was taking up. He had used, we now know, the 1842 report of the Children's Employment Commission in order to get the details of working and living conditions in Wodgate right, but documents could not conceal his 'want of experience and familiarity with the subject ... It is', said Thackeray, 'a magnificent and untrodden field (for Mrs Trollope's Factory story was wretched caricaturing, and Mr Disraeli appears on the ground rather as an amateur): to describe it well, a man should be born to it. We want a Boz from among the miners or the manufactories ...' They had in fact to wait for Mrs Gaskell's *Mary Barton* and in the meantime to put up with the implausibilities of Sybil and her father, a physical force Chartist, who owed more to Bulwer's Sybill and her father the inventor, in *The Last of the Barons*, than to actual experience. Both of Disraeli's 'social' novels owe something, too, to Scott's use of the picturesque and symbolic contrast between the feudal Jacobite highlands and the Whig world to the south of them. In Disraeli the contrast was between the old centres of culture and power and the unknown manufacturing regions to the north (beginning at Birmingham), which were generating new wealth and new forms of life. But only too often it seemed that a tawdry rhetoric was clouding the interest it was meant to kindle—'The Age of Ruins is past. Have you seen Manchester?' The personages symbolic of his main contrasts only too often failed to emerge into the clarity required of a novel, however sufficient they might be for romance. Even the epigrams of his intermittent political pamphleteering, in both books, failed to dissipate a certain mistiness about what he really wanted. No doubt to be more specific about his positive programme would have been, as Leslie Stephen said of his theory of the 'financial' revolution of 1688, 'to destroy its charm, and to cast pearls before political economists', but it does seem as if, in his two most famous works, Disraeli is a novelist only by the accident of the form he chooses to get a hearing for his cloudy political case. What cannot be denied, however, is that he does, by the coruscations of his advocacy and the pointed felicities of his satire, prompt the reader to think, and so make the novel, however entertaining, more than an entertainment.

In *Tancred* (1847) he settled for comedy—some of his very best—in the first two books, and some of his very worst of wild romance in the remaining four. At least he knew what he was doing sufficiently to keep the two apart. When he returned to fiction with *Lothair* (1870), he seemed to have forgotten, and to be mixing them incongruously. The mixture was more than a little troublesome to a polemical purpose concerned with the secret societies, with which, in their Irish forms, he had gained close acquaintance while in power in 1867–8. They were, said a character in the novel, 'hurrying the civil governments of the world to a precipice'. Yet revolutionary conspiracy was treated not unfavourably and its leader, Theodora, with positive adulation. She enslaves the hero, rallies her followers against the papal troops in the battle which kills her, and before dying makes Lothair promise never to become a Roman Catholic. The dubious and entertaining manoeuvres of people in Rome to convert him seem to justify her warning; yet it is a disingenuous English cardinal who warns him, in turn, against secret societies. His view of the revolutionaries as irreligious has been belied by the presentation of their leader, Theodora, herself. Theodora's religion turns out to be not unlike that of the undogmatic Syrian Christian with whom Lothair talks at length, after escaping from Rome and making—where else?—for the Holy Land. The crucial advice that he should settle down in England to do what good he can as a member of the landed gentry comes from the male leader of the revolutionaries; yet it is from satire of the landed gentry that the best of the comedy of the book arises. The fiction, far from persuading the reader to adopt certain opinions, sets them in paradoxical circumstances that strip them of all solemnity. Although the effect is odd and original, it is also uncertain: the increasing complexity of the author's concerns is not matched by any development in the technique of presenting them and he has had, on his own confession, no time to read any of the major contemporary novelists whose methods might have helped.

In his final complete novel, *Endymion* (1880), the varieties of political opinion are less important than the possession of power. In order that his mind may play freely over the delights of 'the great game', as he calls it, he makes his hero a Whig, in a period extending from the end of Lord Liverpool's government in 1827 to the beginning of Disraeli's own ascendancy, when he himself, 'a gentleman without any official experience whatever', was not only 'placed in the cabinet but was absolutely required to become the leader of the House of Commons'. Smythe, who was the original of Coningsby, appears as Waldeshare, to whom the House of Commons is 'a mere vestry', Jacobitism the cause of the rightful dynasty, and 'all diplomacy since the Treaty of Utrecht . . . mere fiddle-faddle'. The fragment of another novel, hidden in Monypenny and Buckle's *Life*, displays with sweet-tempered mockery a hero whose religious dedication, torrential rhetoric, and hawk-like appearance are those of Gladstone. It seems the final argument for Disraeli's possession of the talents of an excellent minor novelist and not merely those of a major politician.

Most of Disraeli's fiction has reference to politics and some of it is concerned to persuade, but, apart from his romantic devotion to the landed interest, Disraeli's political views have so little of the fixed and definite that even persuasion does not greatly hamper him as a novelist. He has a freedom which he uses to make humorous play with possible actions and opinions, in a fashion which was appreciated by Thackeray. The reasons which Thackeray gave, in a small, almost casual review of *Coningsby*, in the *Pictorial Times* in May 1844, apply to most of Disraeli's fiction. 'The volumes', wrote Thackeray, 'cannot fail . . . to make [the reader] think and laugh, not only with the author, but at him.' It is a back-handed tribute, but its terms suggest the place of this minor talent in the prehistory of *Vanity Fair*.

3. Thackeray

In the first of the *Roundabout Papers* he wrote for the *Cornhill* during his last years, Thackeray[1] praised 'the novels I like best myself—novels without love or talking, or any of that sort of nonsense, but containing plenty of fighting, escaping, robbing or rescuing'. Many earlier remarks show, in the same half-serious way, that the novels which gave him what he called 'the most fearful contentment of mind' were quite unlike his own, in which love and talking are of all things the most important. Strong as he is in sheer narrative drive, he is stronger still in his power to present the action so as to suggest ways in which the reader may arrive at a judgement of it, or often at more than one. This means, for him, talking incessantly. The talk has a period flavour which distracts many later readers and much of it seems to be small talk, hard to overlook even when turned to purposes which are far from small; it seems to draw attention to a persona which is simply the author's and, even when inconsistencies appear and suggest that in each case the persona is a mask adopted for a purpose, this purpose may easily be missed if the reader has not already been prompted to start judging the action for himself. Everything in the novel is meant to take the reader in the same direction, by means learnt partly from Fielding, who had to educate a public for a new form of pleasure, and Sterne, who developed the art of making the character of the narrator himself part of the game.

At first Thackeray's methods look relatively crude. The per-

[1] William Makepeace Thackeray, 1811–63, was born in Calcutta, son of a senior servant of the East India Company. Educated at Charterhouse and Trinity College, Cambridge, in the expectation of living as a gentleman of independent means, he turned to art and to writing when his fortune disappeared in the collapse of certain Indian agency companies, 1833. He married in 1836. Separated from his wife, from 1842, by her insanity, he settled at 13 Young Street, Kensington, in 1846, continuing as a free-lance journalist until the success of *Vanity Fair* brought financial security. For a time from 1848, a romantic friendship with Mrs Jane Brookfield mitigated his solitude. In 1853 Carlyle saw him as 'a *big*, fierce, weeping, hungry man; not a strong one'. He suffered from the mutual animosities of his own admirers and those of Dickens and was not on speaking terms with Dickens from 1859 till shortly before his own death. Public lecturing in England ('The English Humorists of the Eighteenth Century', 1851, 'The Four Georges', 1856–7) and America, 1852–3, 1855–6, marked his celebrity. He edited the newly established *Cornhill Magazine*, 1859–62.

sonae he adopts, Yellowplush, Ikey Solomons, Fitz-Boodle, even Titmarsh, seem, as one looks back, to be rather beneath him. They are used to cultivate a hard clear-sightedness about roguery, understandable as a reaction against the magniloquent moral ambiguity of Bulwer in *Eugene Aram* and *Ernest Maltravers*, but nevertheless a wanton curbing of the author's sensibility, to those who understand its later forms. The personae, however, often turn out to be less simple than they appeared to be at first, touching the hard outlines of plot and character with the kind of uncertainty which marks the mature novels, prompting the reader to pause and consider.

Ikey Solomons, a famous London fence, is made the narrator of *Catherine* (1839–40) so that he may denounce 'thieves as they are', from inside knowledge: there is to be no confusion of vice with virtue, no hint of their coexistence in one character such as the mature Thackeray was to excel in portraying. But even Solomons is soon allowed to see that 'our novel-writers make a great mistake in divesting their rascals of all gentle human qualities; they have such—and the only sad point to think of is, in all private concerns of life, abstract feelings, and dealings with friends, and so on, how dreadfully like a rascal is to an honest man'. The author had set out to show the folly and worse of doing what Ainsworth and Bulwer had done, making criminals into heroes or heroines. 'The triumph of it,' he reassured his mother, 'would have been to make readers so horribly horrified as to cause them to give up or rather to throw up the book and all of its kind.' But he could not help, he confessed, 'a sneaking kindness for his heroine, and did not like to make her utterly worthless'. Although the reader's freedom of judgement about her is limited, he can see how her upbringing and associates turn her to crime and what relation her crimes have to ordinary human feelings. By supplementing his source (the *Newgate Calendar*) with material of this sort, Thackeray took one step in the direction of the double vision he was later to exercise on good characters as well as bad, and one step away from the unmitigated harsh outline which had made characters in the *Yellowplush Correspondence* (in *Fraser's Magazine*, 1837–8) like Deucace or his father, Lord Crabs, so jejune. Ikey Solomons remained little more than a ghost in *Catherine*: the author took some pains with him, but could not refrain from lecturing in his own person or treating the narrative as an excuse for distracting literary parody, particularly at the finish. He could not define his own role in the performance: was he to be a critic of the action or a character within it? The pseudonyms were his method of experimenting, allowing him to be both by turns and to see if the one role could be made to interact with the other.

He was trying, in fact, in this early fiction, to use all Fielding's methods at once—parody, as in *Joseph Andrews*, direct critical discussion, as in *Tom Jones*, as well as the creation of characters to make the whole into a fiction where the reader might mix participation with detachment. Thackeray had himself been a headstrong reader of all kinds of contemporary fiction, but so little of it was of the first rank that his parody lacked a single target of stature such as Fielding had had in Richardson; Bulwer was a poor substitute. The criticism which supplemented parody lacked Fielding's clear affirmative aims; the characters were still relatively insubstantial; but Thackeray was experimenting to see how all these things might be combined so that the reader might discover the pleasure of being made to think.

The *Yellowplush Correspondence*, his first fiction of any length, had hardly succeeded in this. It was entertaining, if you could tolerate the facetious spelling, to see a footman learn so quickly from the gentlemen-rogues; but the movement of the actual story had little need for him and none at all for his viewpoint: how could these neat, Surtees-like stories of roguery have been written by one who aspired to be a fashionable novelist? The gorgeous hilarity when Bulwig entered, speaking in resplendent pastiche, faded as soon as his comical adorer became endowed with good sense, reasoning with the great man or even going on to detect the 'moral man of fashn' in *The Diary of the Times of George IV*. In *Catherine*, the gaol-bird narrator was better managed than this, and the literary parody, though hardly his, was more important to the whole, mocking whatever might obscure the simple truth—from the Gothic diabolism of Maturin to the name-dropping of Bulwer's historical romances. Of course, the best way to make the truth appear would have been to tell it. It was characteristic of Thackeray rather to approach the matter obliquely, turning it in the light, telling it indeed, in places, and with a simple verve, but also parodying the varied ways in which it might be told. The self-mocking, allusive, apparently indecisive author of *Vanity Fair* was already at work.

He tried out various disguises, of which two are of importance for the fiction. Michaelangelo Titmarsh, first assumed when writing on art in *Fraser's* in 1838, became his middle-class persona, a 'potato' compared with the gentry-tulips, in *The Second Funeral of Napoleon* (1841); Fitz-Boodle, who first appeared in *Fraser's* some four years later (and found 'Tidmarsh' vulgar), was the clubman and pseudo-gentleman.

Titmarsh was only on the periphery of the early fiction, 'editing' and illustrating his cousin Samuel's *History of the Great Hoggarty Diamond* in 1841, but he did make a difference to it. Without him, in *A Shabby Genteel Story*, the previous year, sentiment and farce, humorous condescension to the oddities of the lower classes, and satire at the expense of the gentry and their hangers-on had all sat rather uneasily together. The middle-class position common to both 'editor' and narrator in *The Great Hoggarty Diamond* helped Thackeray to frame a self-consistent story, as if in the light of his later claim (in the lecture on 'Charity and Humour' in 1853) 'that humour is wit and love; . . . that the best humour is that which contains most humanity'. It was the first sustained example of a kind of fiction he was to write perhaps to excess. The 'humanity', shown in the treatment of the villain as well as of the ingenuous narrator and of the upper as well of the middle classes, was not incompatible with mockery, but the mockery left the reader with something to do because it was suggested to him rather than forced upon him. Where he was meant to see further than the narrator (in his first meeting with Lady Jane and Lady Fanny, for instance) the small revelations were made with a deftness and surety which no Thackeray narrator who was at the same time a character was able to regain until *Esmond*.

Titmarsh was perhaps even more important as the author of journalism, for it was here that Thackeray developed his crucial powers of cordial familiar address to a reader who was assumed, above everything, to hate humbug. And Titmarsh could do this most disarmingly because of his slight self-depreciation, so that his most strutting prejudices could be taken as weaknesses. By the time of *The Irish Sketch Book* (1843) and *Notes of a Journey from Cornhill to Grand Cairo* (1846), in which the pseudonym was kept on the title-page while the preface was signed in person, the pseudonym was less a disguise than the sign of something more important—of the author's concern to see himself as well as the scene before him and to recognize the difference between the limited self so created, Titmarsh, and the actual ego stretching beyond him into the unknowable. So each aspect of the limited self seemed exposed to criticism from some other, slightly less limited, along the line of an infinite regression. Even today this small drama of amiable egotism keeps the reader of *The Irish Sketch Book* engaged.

Fitz-Boodle was a simpler figure, used for simpler purposes, and mainly in fiction. Bachelor, club-man, *flâneur*, he could satirize himself, but in a way which was much more conventional, much less individual than Titmarsh's—suggesting certain genteel *Professions* 'to the unemployed younger sons of the nobility', or treating as farce his own amorous *Confessions*, in 1842–3. In *Men's Wives* (1843) he seemed merely the bachelor smugly interested in discomforts he had avoided. Parody was beyond him (except for 'the style of the very best writers of the sporting papers'), since 'a man, to be amusing and well informed, has no need of books at all' and, without them, could be 'a match for any of you literary fellows'. So the way was open back to the earlier barren pretence of telling the truth which the 'literary fellows' would sugar over. The cardinal example was the truth about Denis Haggerty's wife, whose folly, though literally blind, was displayed without compassion. Here the pseudonym encouraged an over-simple attitude towards a marriage in important respects like the author's own. The venom still shocks.

It was hardly a good omen, then, that Fitz-Boodle should be made the author of the first work of which Thackeray said (in a letter to his mother in October 1843), 'I think I am writing a novel.' The pseudonym was meant at first to keep this work, *The Luck of Barry Lyndon*, separate in the public eye from *The Irish Sketch Book*, in both a simple sense and a more sophisticated sense. 'The dashing, daring, duelling . . . whiskey-drinking people' who had appalled and amused Mr Titmarsh were to be satirized out of the mouth of a representative from the previous century, a representative with all the energy of Jack Sheppard, a great deal more intelligence, and even, at first, some charm. At first, his lies were piquantly exposed, in footnotes, by the experienced English club-man. But the sense that one type of wastrel was commenting on another was lost as the momentum of the Irishman's adventures increased and the interest shifted to the contrast between what he claimed for them and what the reader could see them to be for himself. Footnotes underlining this contrast were ascribed to the editor of *Fraser's*, or, where they were not, showed more literary sensitivity than was compatible with the persona of Fitz-Boodle. In reminding 'some delicate readers of the present day' that 'this is an authentic description of a bygone state of society, not a dandy apology, or encomium, such as some of our rose-water novelists invent ...', Thackeray himself was reacting against Bulwer's romantic treatment of the past. When he forgot Fitz-Boodle in the pleasure of allowing Barry Lyndon as first-person narrator to show himself up, this narrator (in fact another persona) in turn gave trouble. It was not only that he was allowed opinions about war more appropriate to the 1840s than the 1760s, but that, once he had deteriorated into the hopeless villain, his mendacity made him monotonous. More varied lines of interest were suggested in the last quarter of the novel, but a lying narrator could not develop them. The reader needed to know more of the heiress he terrorized and fascinated, or of the stepson who played Hamlet to his Claudius. It is arguable that Thackeray's difficulty in grinding out the concluding chapters came at least in part from the awareness that the novel could not be satisfactorily completed under the terms with which he had begun it. Lying narrator and pseudonymous author, even when supplemented by footnotes both sharp and stupid, were insufficiently flexible as a method of dividing himself and going to buffets.

After *Barry Lyndon*, particular pseudonyms were laid aside until the fiction itself had generated one for *The Newcomes*. In *The Snobs of England* (*Punch*, 1846–7) the acknowledged author, being 'one of themselves', that is, a snob, challenged readers to recognize snobbery by catching the narrator at it. No longer limited by a name or an occupation, the narrator was thus set free for his more protean tasks in *Vanity Fair*. *The Snobs of England* also developed his powers of affording glimpses of a whole society seen in sharp, representative figures. They formed Thackeray's most inclusive work to date, reinterpreting the whole range of middle-class society to itself. To this end he reinterpreted the word 'snob', turning the mere swaggering vulgarian (the sense hitherto) into the careful imitator of his superiors, the sham gentleman. The thing without the name had appeared already in Barry Lyndon's spurious pedigree, false gentility, and admiration for rank, but there it had been, on the whole, comfortably remote from the 1840s: 'I speak of the good old days in Europe, before the cowardice of the French aristocracy (in the shameful Revolution, which served them right) brought discredit and ruin upon our order.' Now, Thackeray moved uncomfortably close to his reader, though with a disarming claim to be no better than one of the snobs himself and with an almost uninterrupted good humour, restraining satiric excess.

So *Vanity Fair* (1847–8), which suddenly took a first place in Thackeray's work, and, to many readers then and since, in the whole of English fiction from 1832 to 1848, was the product of a decade of experiment with fictional method and tone. He found a subject capable of harmonizing all his gifts in the contrast of the most intelligent of his rogues, Becky Sharp, with a group of worthier but less intelligent characters, no less fully realized than she and no less exposed to criticism, even though their ultimate moral distinction from her remained clear. From the beginning, the contrast between them was dramatized with exceptional brilliance and point while the specific events, described as occurring before the reader's eyes, merged into a hovering, tentative, anonymous consideration of their nature and meaning. Thackeray kept the consideration from becoming heavy by his practised management of conversational prose rhythm, as well as by turning the writer into a number of unnamed characters with a common lack of concern for their own dignity, and by bringing into the discussion the ways of viewing life which were implied by current kinds of novel. Above all, commentary and narrative were repeatedly left incomplete, tempting the reader to the pleasure of completing them himself. To this end all his previous roles as parodist, literary critic, pseudonymous author, and first-person participant were subjugated.

His talent for parody was given fullest rein outside the novel in *Mr Punch's Prize Novelists* (1847), and was kept under firm control where it appeared within the novel. The page-long examples of 'the genteel rose-water style' and 'the romantic manner', in the second instalment, were eventually excised from the 1853 edition; the remaining parody was either more economical or more integral—a touch of Bulwer in the conversation of the superannuated sentimentalist Miss Crawley ('Lord Nelson went to the deuce for a woman') or of Carlyle in mockery of the author's own moralizing ('. . . yet, look you, one is bound to tell the truth as far as one knows it . . .'). For its climax the novel 'soared' into 'the genteel heights' where it came closest to parody of Disraeli's 'outrageously fashionable' *Coningsby* (with which it had an episode, and an original for the villain, in common). Parody merged seamlessly into literary criticism, to suggest ways of judging the action: a disclaimer to 'rank among the military novelists' appeared in the midst of economical campaign impressions (dependent on Gleig's *The Story of the Battle of Waterloo*, to which a footnote made reference) which contrasted admirably with Lever's rhodomontade and carried the criticism of attitudes to war a good deal further than the previous novel's incongruous preaching; phrases for his own work, like 'a homely story' or 'only a comedy', looked like courteous self-depreciation when in fact they made flattering hints, as from one wide-awake reader to another, that the story was much less simple but that it was for the reader to judge; persistently the fashionable novel was invoked or suggested, for while this kind of novel had described the Fair and its vanities, presided over here by the First Gentleman of *Cecil*, it was inadequate to 'exemplify . . . the Vanity of this life'; beyond this, the reader of this 'Novel without a Hero' was to take amused account of the tendency of characters in fiction (Bulwer's especially) to strike heroic attitudes.

The narrator was enabled to become 'Equilibrist and Tightrope dancer' (a title Thackeray himself claimed, as a lecturer, in a letter to the Carlyles in 1851) through his very freedom from a fixed name and character. By turns the bachelor and the family man, he could reflect this multiplicity of masks in the terms he used to address the reader, 'every well-regulated person', 'every reader of a sentimental turn (and we desire no other)', 'the observant reader', bringing the clumsy footnotes of *Barry Lyndon* into the text in a deceptive, confidential murmur, making the interaction of reader and supposed writer into part of the comedy. As a result, Thackeray could dramatize the business of judging the characters by prompting the reader to share it—with admiration of Becky's astuteness and spirit even when the narrator did not express it, or reservations about Amelia's simplicity even where he seemed to praise. And whenever praise and blame *were* expressed, the reader who was entering into the game could never take them as final: to call Dobbin, hurrying on the marriage to someone else of the girl he loved himself, 'this absurd and utterly imprudent young fellow' was at first oblique praise though it was to turn out to be true blame (and Dobbin to have known it all along), just as Amelia's urge, when widowed, 'to sacrifice herself' to George's memory could be compassionated and then questioned, as she comfortably accepted Dobbin's sacrifice too, until eventually it was dismissed altogether, as she recognized 'a new, a real affection'. Praise for Becky's astuteness was similarly equivocal: 'our readers will recollect, that, though young in years, our heroine was old in life and experience, and we have written to no purpose if they have not discovered that she was a very clever woman'; but her cleverness was to be 'proved by her after-history' to have distinct limitations—in her expectation that a rich aunt, however romantic, would forgive a nephew for marrying a penniless adventuress, however entertaining, or in her open scorn for the virtue which was to outwit her—after she had excused herself from any effort to release her husband from a spunging-house, in a letter which even he, with all his lack of brains, could see through at once. The result was not so much to arouse the moral energies of the reader as to refine and redirect them, making impossible the extremes of both approval and repudiation. It was the entire opposite of the result achieved by popular novelists like Bulwer, with their habit of telling the reader what to think; it was strikingly different from the practice of Dickens, bearing the reader unresisting on the strong current of his own hatred and affections.

The voice of the preacher was heard, but it was one voice among many, which was right in a novel with such a title as *Vanity Fair*. For vanity meant not only the foreground comedy of idle amusement and insubstantial display, but also Bunyan's town where Beelzebub and Apollyon set up their fair to traffic in wives and husbands as well as in whores, in souls as well as bodies, where Faithful was killed and Christian barely escaped. Behind this was the vanity of *Ecclesiastes*, the futility of all things; and behind this again the preacher had other, ironically inflected words for the destinies of Christian and Faithful, 'the sphere whither we are bound', 'the hidden and awful Wisdom which apportions the destinies of mankind'. The Manager of the Theatre, as he called himself in his preface, was clearly expecting the various senses of his title, like his own various voices, to interact, sharpening the reader's awareness of the way good and bad coexist, in varying degrees, in each one of the characters and in the developments of the action which contrasted the marriages made by Becky and Amelia. The title was so good because it formed yet another means of turning the action to the light—which was the light of 'the Author's own candles' (in the preface), though there were hints in the novel that a steadier day, if it were attainable, would extinguish them. Unless the reader could see this turning and these varying lights, a great deal of the novel was passing him by.

Even so, what remained was very large. Even a reader blind to the glancing lights of interpretation could see that the claim that the novelist knows everything was placed beside the tale told by the narrator as a character, encountering some of the others in Pumpernickel or hearing stories of them at dinner: so his account of them became a matter of suggestion and speculation, giving them a paradoxical extra reality, as if they were objectively there to be discussed. This formed the groundwork of the reader's acceptance that the multitude of walk-on characters, swiftly identified, were there too, that the whole of a society from First Gentleman to scullery maid stood solid around him. However richly comic the close-ups, the conspectus was melancholy, and moving. Yet it questioned its own apparently comprehensive conclusion, '*Vanitas Vanitatum*!' The final comment rested not with the author but with the audience—unless of course they were children, and the characters puppets, and the novel mere play.

In the serial fiction after *Vanity Fair* Thackeray was tempted, by the high rates he could command, to dilute the material in order to keep the novel going, now that it had virtually become a periodical. 'It is a sort of confidential talk between writer and reader,' he said in the preface, after looking over *Pendennis*. The trouble was that no amount of talk could delimit a subject when each novel merely took a colourless young man and followed his career. (*The Virginians* even split him into two halves in two narratives told separately and only tenuously linked.) None of these books is without some interest; all have extraordinary passages of narrative strength or contemplative charm; after reading them consecutively one is conscious at the end of the cessation of a highly individual voice; but they all fall short of the sustained quizzical and compassionate vision of *Vanity Fair*, the irony within irony, catching the reader as well as the writer and his characters. *Esmond*, the exception, was not a serial but a three-decker.

Of the serial novels, *Pendennis* (1848–50) came nearest to the method of Thackeray at his best, because it was a comedy. The hero's development from helpless immaturity was seen with charity and delight from the viewpoint of a narrator who generalized the experience of maturity, sometimes from within the story, where the characters were his 'friends', sometimes from outside it in a position beside the reader's own, with genial and gentle derision as he fudged up a conventional position for himself or claimed to know the reader's. What he regarded as a false kind of maturity, that of 'the world', he dramatized with comic detachment in Major Pendennis, who walked in from specific associates and associations in *Vanity Fair*. Balzac's habit of cross-reference between novels was thus applied to the evaluation of the action. The double view of the Major, both admiring and depreciative, found its parallel in the extremes of jealousy and devotion presented, though without benefit of comedy, in Mrs Pendennis. Unfortunately both characters were there for their effect upon a young man whose interest was not quite sufficient for his central position. And the selfconscious literary concerns, which had been assimilated into the liveliest part of the fiction in *Vanity Fair*, tended, when centred upon so slight a hero, to become mere discussion, just as they had in the very early work. When, in chapter 41, Pendennis paraded his immaturity by writing a novel, the content of which the reader had already 'become acquainted with ... in the early part of his biography', he was shown discussing the propriety of doing so with George Warrington. Now Warrington's marital situation mirrored, if somewhat inadequately, that of the mature Thackeray and the concern about 'sell[ing] his feelings for money' appeared much more the author's than the character's. Comedy seemed to be in abeyance and much of the argument to be continuous with the advice long ago given to Bulwig by Yellowplush: 'Why must we always ... be talking about ourselves?'

Contemporary readers were prepared to grant Thackeray Yellowplush's justification, 'You've a kind and loyal heart in you ...' Later readers have tended to add, as he did, 'a trifle deboshed, perhaps'. Yet it carried him, untainted by the brittle farce of Fitz-Boodle's *Confessions*, through the first instalments in which the young and infatuated Pendennis faced the statuesque self-possession of his idol, the Fotheringay, witless but far from senseless. Whether his history is, as has been claimed, 'the first true *Bildungsroman* in English fiction' is debatable. Immeasurably better than a host of attempts from Disraeli's and Sterling's in the early 1830s to Geraldine Jewsbury's *Zoe* and Samuel Phillips's *Caleb Stukeley* in the mid-1840s, *Pendennis* is arguably not a novel of the hero's 'formation', since Thackeray claimed, as late as the end of chapter 59, that 'it was the same Pendennis . . . We alter very little.' The illusion of change could be said to have been 'all done by mirrors'—the mirrors of memory, in a character who 'having a most lively imagination, mistook himself for a person of importance very easily'. It was a question of his sense of himself and his place in the world as built up by his assimilation of his experience in memory. Yet his slightness as a character made his introspection, for instance at the beginning of chapter 73, rather insubstantial as fiction. What life it had was caught from the author's own self-admonition in the letters. Pen seemed a less adequate remembering persona than Dr Solomon Pacifico, the 'Proser', Thackeray's mouthpiece at this time in *Punch*, whose happiness 'On being a Fogey' consisted, like Pen's, in a 'proper management of his recollections'.

Such habits of avuncular reminiscence, as of a man whose life was over, had something to do with Thackeray's return to historical fiction; for the vistas of memory could be lengthened and made even more poignant through the reader's knowledge of facts the characters could not know. Such vistas glimpsed in *Vanity Fair* ('the Corsican monster locked up at Elba . . .') had fallen short of making it into a historical novel; chapter 51 had stressed, rather, that the 'manners' at the centre of the *Fair* 'were not . . . essentially different from those of the present day' (and Thackeray's own illustrations showed the fashions of his own time). In *The History of Henry Esmond, Esq.* (1852) he took a longer view and concerned himself with a difference not only of dress and manners but of hopes. It was only after the reader's interest had been aroused by specific differences that the underlying melancholy resemblance was to become plain. Thackeray chose a period carefully, in order to dim the lustre of both losers and gainers in the Glorious Revolution (which Macaulay's *History* had begun to celebrate only a few months after the last number of *Vanity Fair*) and to bring into view the first of the Four Georges whom he was to satirize himself, in lectures delivered in America in 1855–6. By that time, moved by the Crimean disasters, he was joining Dickens in the

Administrative Reform Association to protest, as the latter put it, that 'our political aristocracy and our tuft-hunting are the death of England'. *Esmond* went back to a crucial stage in the development of these 'corruptions which weaken and degrade the country' (Thackeray's phrase in a speech he prepared for that Association) and showed the hero, a viscount, calling himself only 'Esquire' on the title-page of his 'history', written after emigrating to America. There 'the signs of the times' made him 'think that ere long we shall care as little about . . . peers temporal and peers spiritual, as we do for . . . the Druids'. From the far side of the failure of the Jacobite cause, Esmond could look back on his hopes of gaining Beatrix, and on the military and political ambitions she had incited, as symptomatic of a whole nation's misguidance. In this way, and not merely by the easier claim that such behaviour was typical of men everywhere, the over-orchestration of his melancholy was to be justified. 'We alter very little' was now rephrased (with a verbal tact characteristic of the book, using 'develop' in the now archaic but still intelligible sense of 'disclose') as the belief that 'fortune, good or ill, does not change men and women. It but develops their character.' Esmond, looking back, by means of the third-person names for himself, at the boy, the lad, Harry, Mr, Captain or Colonel or Knight in arms, took pride in being the same person. Before it was finished his narrative suggested that it was the same nation in 1852 as in 1714.

There is no evidence that this reinforcement of the subject of memory, after its striking development in *Pendennis* and *David Copperfield*, was affected by the recent publication of Wordsworth's *Prelude*. Thackeray has left no record of an interest in *The Prelude* to match that of Elizabeth Gaskell and Charlotte Brontë and his treatment of the subject is in any case continuous with that in his first full-scale historical novel, with, for instance, the moving passage in which Barry Lyndon returned to Ireland in chapter 14 and, as he thought of the past, had one last chance, before his final hideous triumphs, of redeeming the time. His discovery, 'I believe a man forgets nothing,' was matched now by Esmond's: 'We forget nothing. The memory sleeps but awakens again . . .' Thackeray's practice in the art of revealing the action by commenting on it enabled him to give the subject of retrospection a fine resonance and an unusual symmetry. The action moved forward, as Esmond concealed his legitimacy in order not to imperil the inheritance of those he loved, but the comment which clarified it looked insistently back, to the orphan, found, when seeking his parents among the family portraits, by one who came to him as *Dea certé*. By the end of the novel, and still more on successive readings, it seemed as if its Augustan narrator meant this allusion to Venus, appearing before Aeneas, to be capped by the conclusion to the fourth Eclogue, where the boy smiled on by parents might be deemed worthy to sleep even with a goddess:

> Incipe, parve puer, risu cognoscere matrem . . .
> Incipe, parve puer, cui non risere parentes
> Nec Deus hunc mensa, Dea nec dignata cubili est.

The whole novel showed how such divine favours might be attained and how painful the process, when Beatrix, the woman, was in many ways preferable to Rachel, the goddess.

George Eliot, who found it 'the most uncomfortable book', saw that it contained 'the same characters' as *Vanity Fair*—'Lady C. is Amelia, Esmond is Dobbin, and 'Trix is Becky'. In both, one might add, the destructive powers of the little domestic goddess appeared as plainly as the virtues, however tarnished, of the mortal woman. Nothing was quite what it seemed. Harry, as Prince Hamlet, early found the goddess's possible remarriage to the rector 'as monstrous as King Hamlet's widow taking off her weeds for Claudius', but in the *Spectator* pastiche, in the second volume, it was the mortal woman whom he addressed as Jocasta, signing himself 'Oedipus'. Discomfort such as George Eliot's was intended, and intended to be felt in relation to the narrator's commentary as well as his actions. It bore in fact upon the centre of the novel's meaning. Esmond's commentary showed him determined to be grateful for the goddess's paradoxical worship of her devotee (embarrassingly dramatized in the third volume), although he remained unconvinced that he would be able, even in the 'last day' and 'last hour' of which he spoke in chapters 1 and 9, to put off all his devotion to the mere mortal woman. In the most anachronistic of his reflections on memory, 'such a passion once felt forms a part of his whole being, and cannot be separated from it; it becomes a portion of the man of today . . . Our great thoughts, our great affections, the Truths of our life, never leave us.' But his god-like warnings to his grandson, 'What can the sons of Adam and Eve expect, but to continue in that course of love and trouble their father and mother set out on?', which were given when he was talking about Beatrix, applied equally to Rachel, whose jealousy, it was clear from the preface, was as lively and dangerous in her second marriage, to Esmond, as in her first. The final word about Esmond was to be found outside of those 'Written by Himself'. Thackeray approached it in chapter 3 of *The Virginians* when he wrote of 'some bankruptcy of his [Esmond's] heart, which his spirit never recovered' and again in the letter of November 1851 which called Esmond 'as stately as Sir Charles Grandison'. It was for 'the observant reader' in 1852 to write the novel's conclusion; only he could see what lay beyond the reticence and the rectitude of its grave and fallible narrator.

Readers in 1852 were not ready for this. The extreme of the misunderstanding appeared in reviews. Samuel Phillips in *The Times* was outraged by the marriage: it was out of character for 'Esmond the importunate and high-souled, the sensitive and delicate-minded'. Thackeray could shrug this off ('He does not understand what I write any more than poor Jack Forster . . .'), but he believed damage had been done to his sales. Only a rich man could afford to 'take pains and write careful books' when people could so easily misunderstand them. For a man who was still trying to make '10000 for the young ladies', his daughters, there was some justification in returning to instalment publication with *The Newcomes*. He had now established a complete dominion in the realms of 'love and talking'. (Carlyle could even blame him for 'spreading [love] over our whole existence'.) In the huge canvas of *The Newcomes* (1853–5), love and talking appeared in the simpler forms associated with less extraordinary novelists. He adopted in the second instalment a named narrator, Pendennis, only in order 'to talk more at ease than in my own person'; he was trying to avoid the charge of usingautobiography, but it was a transparent device and, Pen being what he was, quite without interest. Once Pen was obliged to supply from conjecture a few points of the history', a half-serious apologia for the novel as history, in Fielding's manner, made Pen even more like the actual author in devising for the characters speeches 'as authentic as the orations in Sallust or Livy'. No doubt 'in this manner Mr [G. P. R.]

James, . . . Robinson Crusoe, and all historians proceeded', but at his best Thackeray had offered something more, exposing what the characters said to the criticism of his own 'orations', and the remarks of both to the arbitrament of the event—as, for instance, with the talk of 'heart' and heartlessness in *Vanity Fair*. Now, although phrase and rhythm still gave steady delight, the overtones tended to be simpler, relevant only to the immediate occasion; and this occasion tended to work simply with, and not against, the force of other occasions in the book. Those who stood in the position of parents, whether Colonel Newcome or Lady Kew, were seen urging on the same calamity, the arranged marriage. Clive was seen to fall into the trap and Ethel to evade it, not only because she had more spirit and sense but also because her resistance had not been sapped, as his had been when she passed him by in favour of others for whom her regard was less. The trap and Vanity Fair which baited it were insistently denounced, never mocked; and the terrors of Clive's arranged marriage became part of the denunciation. It seemed a reversion to the venom of the earlier work, despite the enormously increased powers of artistic realization. In *Barry Lyndon* the notion that 'nothing comes of first love' was made amusing; now the narrator's question ('Suppose we had married our first loves were we the happier now?') recalled even Pendennis with a poker face. Mockery, however, would have needed another narrator than Pen, now very much married to Laura. The characteristic play of mind had given way to a less glancing mode of conversation with the reader; the benign and mocking small talk became serious, the few jokes sad: 'poor little Clara . . . poor little fish (as if she had any call but to do her duty, and to ask *à quelle sauce elle serait mangée*)' was the destined victim; when Ethel, after escaping the same fate, was cut by the Colonel, she was 'the poor thing', spoken with pity only (though admittedly a complicated pity, since neither character was more than half right, and forgiveness over-rides all rights).

It is true that Ethel was presented with a characteristic uncertainty which, like Heisenberg's, allowed only one variable at a time to become clear. It is true that the reader was kept guessing and invited to guess. But if this process was to carry as much of the book as the debate with the reader about Esmond or Becky Sharp had done, there was needed at the centre someone less null than Clive. And the confident slanging of Ethel, as panther, zebra, *lusus naturae*, should have come from personae less superannuated and predictable than Pendennis and Warrington, particularly when such terms were used also of other characters—Barnes as a reptile, or Rosey as a bird before Mrs McKenzie the boa-constrictor—characters about whom there was no uncertainty at all. The beast fable in fact was simple denunciatory satire, not parody calling in question its own simplicities in the way that parody of other kinds of fiction had done in *Vanity Fair*. To be sure, poetic justice, such as some fables might show, was specifically refused, but this only repeated the selfconscious conclusion to *Barry Lyndon*. There was life in *The Newcomes* (as well as monstrosity of size and scheme), as Henry James acknowledged in the preface to *The Tragic Muse*, but an important Thackeray dimension was missing.

A reader might conceivably supply it from the contrast between the astringent view of marriage which seemed to be latent in the book, and the smug Pendennis domesticity. But this contrast was displaced from the centre by a simpler one between the needs of the young in love and the evils of matchmaking, of which Clive's disasters were a particular instance. All the same, the direction of Thackeray's interest was plain from the wish his daughter recorded, to go on in his next novel to show a subsidiary character from *The Newcomes*, J. J. Ridley, with the 'trials of a wife and children'. Thackeray, however, did not follow up this plan; he only allowed himself to add to the novel he did write, *The Virginians* (1857–9), (in the manuscript of the completed chapter 85) a passage on the topic of marriage which did little more than prose about the tiresome prospect from 'the summit of felicity', much as Dobbin might have done. The tendency to repeat himself was by now plain. He did, however, succeed in developing out of *Henry Esmond*, with some of the old energy, the characters of the ageing Beatrix and her half-sister, Esmond's daughter. There was genius in these parts of *The Virginians*, but not in the whole. Warrington as the lugubrious narrator was no help, and even when the number of the author's digressions was reduced for the volume edition, there were still far more left than served to illuminate the sluggish action.

As if to show that his hand had not lost all its cunning, he gave a virtuoso performance in *Lovel the Widower* (1860), his first piece of fiction in the *Cornhill Magazine* after he became editor. Its plot was given, in a play which he had written some five years before without finding a manager to take it. By developing the bachelor onlooker he turned farce into comedy. It remained the story of a Becky Sharp-like character, called Bessy, whose background in poverty and the stage, however respectable in fact, was not respectable in the eyes of the other characters, but this story was so managed as to bring forward Mr Batchelor's warring wishes to marry her and not to marry her, his ability openly to defend her for her goodness and patience in 'many trials', even though at the crucial moment the revelation that she had previously known the family black sheep made him draw back and wonder, 'Do I know all about her, or anything; or only just as much as she chooses?' The arch self-mockery of Batchelor was perhaps too much of a mere game, but there was no doubt about the skill with which it was played.

After this, *The Adventures of Philip* (1861–2) was a regression. Another colourless young man impeded this promising development of the early and unfinished *A Shabby Genteel Story*. Pendennis's patronizing small talk about the characters he approved of remained small talk, with little satire implied. Even he, however, could not quell the greater liveliness of the villains and humbugs. The thread of the whole was slender except where such characters gave it strength by making the comedy raise questions of interestedness and disinterestedness. This gave piquancy to the penitent elocution of Philip's father, now that the woman whom he had seduced had the power to impugn the legitimacy of his son from a later, regular marriage. A remarkable climactic scene allowed her not only to defeat, without thinking of it, the clergyman who had celebrated her counterfeit marriage, but to do so while acting not in her own interest but in Philip's. The raffish company of the early fiction was there but the comedy was less simple, even if less complex than in the major novels. The same mind was at work as when Becky Sharp had acted apparently against her own interests in showing to Amelia, George's widow, his old note of assignation with herself: although, by dethroning the idol, it had urged Amelia back to Dobbin and so out of Becky's own clutches, it had also expressed her contempt and through it her envy of Amelia's virtues. At the same time it gave her her revenge on George, who had impeded her pursuit of Amelia's brother, and left her free to batten on this brother now.

It was simplicity, however, rather than complexity which Thackeray was seeking in this final phase. *Denis Duval* (1864) was to have, he claimed, 'no moral reflections and plenty of action'. The eight chapters published take the familiar form of reminiscences by the chief character 'now the curtain is down and the play long over'. They keep the freedom to bring in reports of the action from more than one narrator, but it is not an action which admits of a variety of interpretations. Thackeray was now attempting what he had always regarded as the chief virtue of a novelist, to keep the reader engaged with 'surprises, disguises, mysteries, escapes and dangers', on the principle of the first of the *Roundabout Papers* that 'All people with healthy appetites' love novels as 'sweets'. He was forgetting the work for which he is now best remembered. There, in *Vanity Fair* and *Esmond* particularly, even Johnson's 'reader uncorrupted by literary prejudices' finds his appreciation of life increased not only by the illusion that it is going on before his eyes but also by participating himself in the attempt to understand it, while the reader who has literary prejudices finds that these are made, by means of parody and allusion, to serve the same end. The principal drama is staged between author and reader in their shifting attitudes towards each other and towards the material of the story, while the story is diversified by the disguises the author may assume, and by the varying attitudes and expectations he may attribute to the reader. It all subtilizes the simple relationship in which the author has the power to devise the whole show and the reader the power to refuse to watch. The author moves towards the position of a fictional character with limited freedom and his characters towards that of free agents, while the reader becomes, as he is drawn into the drama, both more responsive to guidance and more aware of his freedom to judge. So the action gains its fullest life not when it is being apprehended as if without intermediary and not, on the other hand, from the simple force of the author's over-mastering intention for it, but from the activity of the reader, excited by the pleasures of apparent uncertainty and the difficulties of decision.

4. Dickens

I

Although Dickens,[1] like Thackeray, began with short pieces for newspapers and periodicals, there was much more of his characteristic quality in these from the beginning. Of course what is, to many, still Dickens's most typical novel, *Pickwick Papers* (1836–7), was begun little more than two years after the first of the *Sketches by Boz* (1833–6), while the *Sketches*, moreover, were revised before appearing in book form in 1836, and even the first version of some of them, like 'The Drunkard's Death', being specially written for book publication, overlapped with work on *Pickwick*. Further revising of these early short pieces for subsequent editions up to 1850 reduced the amount of vulgarism and semi-topical detail and so brought them closer to the novel, though revision might sometimes work the other way, as with the omission of expressions of the form which Sam Weller had made his own, like '"raything warm" as the child said when it fell into the fire'.

Similarities between the *Sketches* and the first novel go beyond matters of detail. The earliest *Sketches*, now to be found in the section 'Tales', use the conventions of the popular stage farce—misunderstandings and discoveries, such as later mark Mr

[1] Charles John Huffam Dickens, 1812–70, was born in Portsmouth, whence his family moved to London, 1814, to Chatham, 1817, and again to London, 1822. His father, employed in the Navy Pay Office, was in the Marshalsea prison for debt, February-May 1824, accompanied by all of his family except Charles, who was in lodgings in Camden Town and working (till June) at Warren's Blacking Warehouse, Hungerford Stairs, Strand. After brief education in Chatham and at Wellington House Academy, Hampstead Road, 1824–6, Dickens learned shorthand, while employed in a solicitor's office, and became a reporter, first of legal proceedings and then of parliament, 1832–6. He wrote *Sketches*, collected in 1836, and was invited to write others to accompany a series of comic plates; these became *The Pickwick Papers* and made his name. Dickens visited the United States 1842, 1867–8, Italy 1844–5, Switzerland 1846–7, and France repeatedly. He edited the *Daily News* for a short time in 1846, and then two weeklies of his own, *Household Words*, 1850–8, and *All the Year Round* from 1859. In 1858 he separated from his wife, Catherine, whom he had married in 1836. In public readings from his novels, 1858, 1861–3, 1866–70, he became virtually a professional actor—he had long been a keen amateur and had early been attracted to a career on the stage. 'Charlie,' Carlyle is reported to have said, after hearing him read in 1863, 'you carry a whole company of actors under your own hat.' Never robust, though capable of unusual physical exertion, he died of a cerebral haemorrhage while still in full productivity.

Pickwick's crucial conversation with Mrs Bardell before his friends find her fainting in his arms, or the cutting across each other of two lines of conversation such as will later make it seem as if Bob Sawyer is eating the leg of a child. Revision may supplement these conventions with those of narrative, as in the present conclusion to 'Mr Minns and his Cousin', which takes the place of the theatrical 'tag' at the end of its earlier form, 'A Dinner at Poplar Walk', the first published *Sketch*. Nevertheless the impress of stage farce can still be seen. Quite late 'Tales' like 'The Tuggses at Ramsgate', published after the writing of *Pickwick* was begun, develop by means of blocks of self-echoing stage dialogue such as will make individual episodes spruce and sharp long after Dickens has begun to acknowledge himself as a novelist. Such passages, too, gain from the theatre an air of festive pretence, sometimes accentuated by the introduction of actors as characters, much as he will later introduce into *Nicholas Nickleby* the Crummleses, whose stage language is only a little more melodramatic than that of Nicholas himself. There seems to be a kind of genial theatrical compact between author and reader which in turn affects the pretences of the characters, transforming the mere hypocrisies which a satirist like the early Thackeray would scornfully unmask, into the struggles, only half successful, of the self to gain expression. The whole performance is an entertainment, as in similar work of Albert Smith's, but goes beyond him in hinting at the self behind the pretence, as with the would-be 'funny gentleman' in 'The Steamer Excursion' who strikes the attitude of a harlequin but of an 'awkward harlequin'. Repeatedly it is the characters who, like Nicholas Nickleby later on, most resemble the author himself, the chivalrous or sentimental young gentlemen, who have most obvious need of the theatrical conventions: they are learning what forms of behaviour may be possible for them, like the 'mysterious' and 'romantic' Horatio Sparkins in the *Sketches*, with his 'theatrical air' and his lavish line of talk to gain the ear of an heiress. Touched as this talk is with what we now read as the idiom of both Jingle and Skimpole, it is simpler in its theatrical derivation: Horatio Sparkins, whose real name is Samuel Smith, belongs to the line of romantic clerks in plays by John Poole or J. R. Planché, characters who mix the pleasure of given names like Narcissus or Lothario with the business of surnames like Stubble.

Poole had also written, for the periodical press, sketches which

Dickens had read, as is shown by his taking over, either now or later, of the occasional name (Snodgrass is the most famous) or episode (in 'The Steamer Excursion', for instance); but Dickens from the start had ambitions to go beyond such work and write a novel. In December 1833 he spoke of a plan to 'cut my proposed Novel up into little Magazine Sketches', perhaps the first hint of *Oliver Twist*; just over a year later the same end, the making of a novel, was to be reached, he said in the conclusion (later excised) to 'Mr Watkins Tottle', by means of further chapters of 'his wanderings among different classes of society'. The looks of this hero of 50, 'plump, clean and rosy', were not unlike Pickwick's.

By 1834, Dickens was developing his interest in a kind of descriptive and reflective writing which would supplement what he owed to the stage. His aim, expressed in the preface to the first series of the *Sketches*, was 'to present little pictures of life and manners as they really are'. Yet his brisk and decisive mode of seeing attracted quite as much attention as what was seen. Already in 'Mr Watkins Tottle' the lock and the nails which gave the door of a spunging-house the 'appearance of being subject to warts' showed with what idiosyncrasy, despite his admiration for Washington Irving, Dickens was avoiding the pleasant pictorial charm of the famous *Sketch Book* of 1820—though his reviewers could not avoid comparing them, and even claiming Dickens as an imitator. Topics and visual details were selected so as to yield the strongest contrasts of light, shade, and moods to match them—the brilliance of the morning into which a boy emerged from prison, or the contrast of dark streets with the blazing gin-shop. Sometimes the contrasts were complicated by a curious amalgam of satirical observation and mock-sentimental nostalgia, as in 'The First of May', much less distinct in tenor, for Dickens aimed not so much to sketch as a visual artist might, succeeding by the deftness of the fewest strokes in Thackeray's manner, but rather to use what could be seen in order to suggest what could not. Indeed, what the author 'only' imagined sometimes mattered most of all; he could construct the essentials of a man's life, down to the kind of dinner he would eat, from a single glimpse, or imagine wearers for a whole shop-full of second-hand clothes. Yet he was presenting a far from imaginary middle- and lower-class London, that is, a more limited social range, as well as a sharper focus, than Pierce Egan in *Life in London*, with which the *Sketches* were also compared at the time;

and this particular London was shown 'as it is' in the sense that its present state in the 1830s was at the centre, unlike *The Town* in Leigh Hunt's essays of 1834–5, in the monthly supplements to his *London Journal*, where the emphasis fell upon 'the past . . . traceable' in the present. Dickens had little concern with the past, except when, as in 'Scotland-Yard', it could be made humorously to sug-gest how persistent was change. Phrases like 'the ancient simplicity of its inhabitants' were affectionately mocking in their context, and, like the 'degradation and discontent' which, in the original version of 'The First of May', had succeeded 'the merry sports of the past', were intended to make the reader face the present.

The phrase 'as they really are', however, could also carry in some cases an implied contrast with things as they might be. Writing, in both series, in the aftermath of the first Reform Bill, he voiced defeated hopes for wider reform than that of parliament. They can still be seen in the ironic reference to the chimney-sweep 'interdicted by a merciful legislature from endangering his lungs by calling out'; they were once to be found also in the specific parliamentary references made by the ludicrous 'Parlour Orator', as well as in the conclusion to 'Gin-shops' which, until the 1850 edition, despaired of their decrease unless 'an antidote against hunger and distress' themselves could be found.

Dickens's manner of proceeding from these sketches to a full-scale novel was probably much less haphazard than has been supposed. When Chapman and Hall made their proposal to pay him to write a regular series of sketches to go with Seymour's plates of a Cockney sporting club's adventures, Seymour was proposing a tall, thin man as its hero. But once Dickens had been approached, the publishers urged Seymour to go back to the short, stout figure he had already introduced in his illustrations to Richard Penn's *Maxims for an Angler* three years before: this man clearly resembled Boz's Mr Watkins Tottle, whose papers, 'the materials collected in his wanderings among different classes of society,' Boz had proposed, in his original conclusion to the sketch, to 'publish, . . . carefully arranged, . . . from time to time'. So emerged the 'book illustrative of manners and life in the Country' which the publishers agreed upon for *Pickwick*, rather than the adventures of a Nimrod club which Seymour wished to produce. Moreover, Dickens secured that Seymour should illustrate his work, not he Seymour's. The continuity with Dickens's own early plans for a novel is clear.

Soon the countenance of the 'immortal' hero changed from the irascibility of Seymour's plate for the first chapter to one which 'glowed', in the second chapter, 'with an expression of universal philanthropy'. His nature owed something to the three great ingenuous benevolents of the novel of the previous century, Fielding's Parson Adams, Sterne's Uncle Toby, and Goldsmith's Dr Primrose. It was this which made the difference from Watkins Tottle, who, for all his resemblance in shape to Mr Pickwick, 'had a clean-cravatish formality of manner, and a kitchen-pokerness of carriage, which Sir Charles Grandison himself might have envied'. In both cases the eighteenth-century precedents were important; but Pickwick owed a great deal also to the innocence about 'manners and life in the Country' of the author himself, who always tended to see them through the clichés of nostalgic pastoral. The nostalgia was accentuated by setting the action back in the days before railways (days about which Tottle might have 'materials'). The addition of a Cockney attendant, Sam Weller, acute, resourceful, apparently disenchanted, kept a certain kind of city life before the reader's attention even when the narrative was set in the country—a trick caught perhaps from Jorrocks the hunting grocer. The contrast of Sam's nature with his master's was developed until, again with eighteenth-century precedents, the pair could be seen by T. H. Lister in the *Edinburgh Review* for 1838 as 'the modern Quixote and Sancho of Cockaigne'. Genius entered the novel when all these things became concentrated into a particular verbal form, once, that is, Mr Pickwick's grave and guileless courtesy had begun to contrast with the witty extension of well-known techniques from fiction and the stage in Sam's pungent and good-humoured kind of gruesome similitude and Jingle's impudent staccato fragments (such as had marked the speech of Goldfinch in Holcroft's perennial play from the 1790s, *The Road to Ruin*, and had already been revived by Surtees and Marryat). Such language in Dickens's hands made directly dramatic the clash between innocence and experience, and transformed affectionate farce into universal comedy. It was, as Dostoevsky later said, a 'weaker idea than Don Quixote, but nevertheless immense'.

Superficially, farce was the dominant mode, but, like the logical sense beneath the nonsense of Lewis Carroll, there was a steady human and moral substratum to the hilarity. Mr Pickwick might seem too gentle and gullible and the antiphrastic descriptions of him too obvious ('that colossal-minded man', 'a man of his genius and observation'), but he was capable of an occasional second-hand cynicism and eventually proved determined to go to gaol rather than pay a farthing of costs or damages—until induced to give way by the sufferings of others, including those who had wronged him. Sometimes the substratum appeared to show a tendency, as T. H. Lister put it, 'to make us practically benevolent—to excite our sympathy on behalf of the aggrieved and suffering'. It was not that particular reforms were being recommended but that narrative and dialogue took account of prevalent opinion, as incidental topicality might do on the stage, and with as little direct didacticism. Mr Pickwick might express alarm when, on a shooting expedition, boys were set to climb trees, as if 'the distress of the agricultural interest, about which he had often heard a great deal, might have compelled the small boys attached to the soil to earn a precarious and hazardous subsistence by making marks of themselves for inexperienced sportsmen', but the attention which was drawn to the troubles of the times obviously took second place to Mr Pickwick's mistake and even to light-hearted parody of the heavy style of radical denunciation.

So the materials which might be collected by a mere Watkins Tottle 'in his wanderings among different classes of society' became a circus where society seemed often to be interpreted as convivial enjoyment and its different classes as the varieties of men judged by their comparative fitness for such cordial purposes. Sam's salty humour of burglars, Bluebeards, and executions exaggerated the brutal facts (which the inset stories exaggerated too, but without comedy) as much as Mr Pickwick's innocence exaggerated the possibility of misinterpreting them, so that even the business of innocence cured by experience became itself a carnival. This gave Mr Pickwick his triumph over the other famous business man to whom some of his adventures, including his trial, were indebted, Jorrocks. Except on the hunting field, which Dickens was right to avoid from the start, Surtees's hero was lumbering in speech and action compared with the sustained verbal circus of Sam and his master.

In *Oliver Twist* (1837–8) the carnival quality persisted, for all that the book was to lead many of its cast to prison or the gallows. It was begun in January 1837, just after the ninth number of *Pickwick* had set its two heroes on a course which would finish eventually in the Fleet. In the greater part of *Oliver Twist* those bound for prison were presented with an incongruous geniality, apparently misunderstood at the time even by Thackeray. Although Oliver was the hero and had been intended, according to the author's preface to the 1841 edition, 'to shew . . . the principle of Good surviving through every adverse circumstance', Dickens, as a downy bird himself, could not resist the opportunity to extract as much humour from his innocence in the underworld as from the ignorance of Tommy Chitling and Noah Claypole. The assumption, for such purposes, of a genially ironical courtesy would have been funnier in the thirties when the fashionable novel was a recognized sub-genre and language like 'Mr Sikes's residence', 'Nancy's wholesome perfume of Geneva', and Toby Crackit's shame 'at being found relaxing himself with a gentleman so much his inferior in station and mental endowments' would read like parody. The comical climax was secured by the tendency of the thieves themselves to use the same courteous language: for instance, in Bates's lament that the Dodger could not 'go out *as* a gentleman' (which meant that with his small indictment, for a 'twopenny half-penny sneeze-box', he would perhaps not figure in the *Newgate Calendar*). In court, the Artful triumphed by assuming this gentility, boasting of his 'appintment with a genelman in the city' or his 'numerous and 'spectable circle of acquaintance', making the gaoler's claim to know him well into 'a case of deformation of character', scorning the witness whom he was entitled to cross-question, with 'I wouldn't abase myself by descending to hold no conversation with him', and going off with the magniloquent assertion of his own superiority to the Bench. As a turning of the tables it was mere rhetoric, but a good deal of its success derived from the fact that it dramatized Dickens's own steady ironical mode of speaking about the criminals. Anyone might see that the Dodger's claims were hollow, as he did himself by 'grinning in the officer's face, with great glee and self-approval' once he was in the yard; his guilt was never in doubt. But the vivacity of his rhetorical self-assertion set him beside the Pardoner in *The Canterbury Tales*, whom Chaucer enjoyed as much, and as little condoned.

Some critics, as in the *Spectator* in November 1838, could see how superior these low characters were to the abstractions of Hook, the romantic 'high-spirited blades' of Ainsworth, or Bulwer's disguised dandies in *Paul Clifford*; but other critics either found Dickens merely 'coarse' or, like Thackeray, charged him on the contrary with sentimental omission. Dickens should, he was told in the last chapter of *Catherine*, in *Fraser's* for February 1840, have passed by his 'most agreeable set of rascals ... in a decent silence; for, as no writer can or dare tell the *whole* truth concerning them, and faithfully explain their vices, there is no need to give *ex-parte* statements of their virtue'. Dickens in the 1841 preface could protest only that what he had put into characters such as Nancy was 'TRUE' and showed them 'as they really are'—this latter phrase involving the same suppressed assumption as in the preface to the first series of *Sketches by Boz*, that his own strongly emphatic methods were the ones for getting to the heart of things, intensifying, as Thomas Hardy would later say, 'so that the heart and inner meaning is made vividly visible'. It was after all not with the criminals by themselves that he was concerned, but with their lives as part of the social organism, showing what was a common outcome when the only alternatives open to a pauper child were those which appeared in this 'Parish Boy's Progress'. He would have been entitled to point to the contrasts powerfully made between the bullying coldness and petty pride of those who administered the country's charitable institutions and the cheerful good fellowship which, though only temporarily, obscured what was vicious in the criminals—the contrast between the scientific supper 'allotted by the dietary' in the workhouse and Fagin's sausages or the Dodger's 'gunpowder' tea, or that between the lack of concern shown by the board who had a specific responsibility towards Oliver and the self-endangering concern of Nancy, who knew only that she was responsible for his recapture and had foolishly 'never thought' that he would make a convenient assistant to a housebreaker. Nancy's oscillation between caring and (under the influence of gin) not caring was subtly observed: she became the first of the women whom Dickens made interesting by their gradual comprehension of what they had suffered.

The murder and its aftermath in the last four instalments were presented without moral ambiguity. Though Dickens was fascinated with the murderer's state of mind, Thackeray was unjust in overlooking the horror and claiming that we are merely 'led . . . to have for Bill Sikes a kind of pity and admiration'. Fagin, the self-congratulating disposer of the fates of his friends, was certainly deflated when, in court, his strained attention caught at sharp, irrelevant details but could fix upon nothing for long, in his helpless fear. The problem throughout was not so much the moral one that the reader might be led to confuse right with wrong as the aesthetic one that the author's imagination was not equally stirred by the bad and the good, so that only the bad characters seemed to be alive. Oliver and his benefactors were artistic failures, made acceptable only by the construction of the whole as a pregnant fairy-tale.

The fairy-tale element, as generations of readers have known, was a powerful part of the novel's success. Providence was said to be the unsleeping cause of Sikes's obsession with the eyes of the girl he had murdered. Providence wielded the 'stronger hand than chance' which had first 'cast' Oliver 'in the way of his father's friend'. The comical emphasis which was given in chapter 2 to the fact that 'nobody had taught him', in his first 9 years, to pray, and that he was to have his sole education 'in the one simple process of picking oakum', showed that Oliver had only Providence to thank, in the crucial scene before he was forced to be an accomplice in crime, for his ability to pray and his powers of resistance to corruption.

The 'spirit', said in chapter 2 to have been 'implanted . . . in Oliver's breast' by 'nature or inheritance', turned out, as the author's plan revealed itself, to have come from the nature of his parents who, though unmarried, had been capable of love: the contrast was furnished by the vindictiveness and self-seeking of his half-brother, born of a loveless marriage. It was in pursuance of this line of thought that a thief's mistress should show great concern for Oliver's welfare, while the tyrants of his childhood, the beadle and the matron of the workhouse where he was born, should join in a marriage of convenience. Those who claimed that vice and virtue had been confused by the manner in which the criminals were presented might equally have objected to the humour and zest with which Dickens displayed the beadle's covetous courtship. The delight in self-incriminating hypocrisy was once again Chaucerian. Mr Bumble's exhilarated solitary scamper through Mrs Corney's possessions, her acquiescence in the discussion of free coal 'and candles', and the cultivation by the pair of them of the appropriate sentimentalities, all came to the reader, like the good fellowship of the criminals, with a vivacity unparalleled in the novels of the thirties, but with a superadded satirical edge derived from the construction of the whole. The comedy led to an illusion of intimacy with characters and situations, as in the theatre; and this intimacy, concerned as it was with food and warmth and companionship, deepened the social criticism by connecting the evils threatening the parish boy with the details of the creaturely life we all share. This strong connection can still be felt, so that, though we laugh, the laugh is on us all. It made a joyous extension of Wordsworth's *Old Cumberland Beggar*, which had contrasted the 'cold abstinence from evil deeds' of the comfortably-off, who support the 'HOUSE misnamed of INDUSTRY', with the goodness of the unfortunate to the unfortunate.

Unlike *Pickwick*, with which its writing overlapped, the book had a firm plan. Already Dickens was putting the question posed in *Bleak House*, 'What connection can there be?' The question already concerned characters as remote from each other socially as the parish boy and his half-brother, the son of a woman of fashion, 'wholly given up to continental frivolities' (and the climax, incidentally, was staged in a ruined place which was once thriving 'before . . . chancery suits came upon it'). Now *Oliver Twist* was probably the first novel Dickens had in mind at all. The letter of December 1833 had planned 'a series of papers called *The Parish*, the first of which, published in 1835 and eventually included in *Sketches by Boz*, contained both a beadle and a workhouse master: the transition from one of these positions to the other was eventually to contain all the comedy of Bumble. The germ of Nancy's devotion to the man who had ill-treated her was the central thing in the *Sketch* of 1836, 'The Hospital Patient'. Above all, *Oliver Twist* showed clearly what was only implicit in the *Sketches*, namely that sense of the social organism with which Dickens would give unity to the whole and energy to the parts in the monster constructions of his maturity.

Nicholas Nickleby (1838–9) widened the social range while lessening the sense of interdependence. Such plan as it had derived directly from *Oliver Twist*, since it grew out of the escape from an oppressive institution of a minor, Smike, for whom it was the business of the plot to provide relatives and a father. When Dickens was gathering material for *Nicholas Nickleby* about Yorkshire boarding schools, early in 1838, he was just about to release Oliver from the underworld. The obsession with crime continued straight on into the introduction of Squeers in *Nickleby*, chapter 4, by means of a quite irrelevant glimpse of Newgate and the wretch about to hang, as in the as yet unwritten double climax of *Oliver Twist*, 'when curious eyes have glared from casement, and house-top, and wall and pillar; and when, in the mass of white and upturned faces, the dying wretch, in his all-comprehensive look of agony, has met not one—not one—that bore the impress of pity or compassion'.

Dotheboys Hall, bare and dirty like Fagin's house, took up another humanitarian cause almost as well known as that of the workhouse child. Writing to Anna Maria Hall when the novel was half done, Dickens claimed to have 'kept down the strong truth' of Dotheboy's Hall 'and thrown over it as much comicality as I could, rather than disgust and weary the reader with its fouler aspects', but Squeers, though he combined the functions of Bumble and Fagin, was at first a poor substitute, a mere rampant, one-eyed puppet which could be agitated to a war-dance of impotent malignity at the end of chapter 45, or sent stiff to bed, in chapter 12, 'with his boots on and an umbrella under his arm'. But he became more than a puppet in the acuteness of his pleasure in beating boys. The threat of recapture which had given a certain excitement even to the saccharine pleasures of Oliver's prosperity was repeated in the pursuit of Smike by Squeers and by the man who had first taken him from his father's hands to those of the schoolmaster.

The neglect of their responsibilities by parents, upon which Dickens was from now on repeatedly to insist, appeared here in the case of Smike's father, Ralph Nickleby, and of Nicholas's mother. They were treated in ways which were to become highly characteristic of the author. 'Serious' satire, in Ralph's case, resulted, for the most part, in routine melodrama, though with hints of something more arresting, which Dickens was subsequently to develop, in the consistency of the similes used—Ralph's uncultivated heart 'rusting in its cell', his voice like an iron door quarrelling with its hinges—and in the analogical suggestions conveyed by his surroundings, like the garden adjoining his office, 'unreclaimed land', 'fenced in by four high white-washed walls . . . in which there withers on, from year to year, a crippled tree'. The climactic device by which Ralph was made to hang himself below the very ceiling trapdoor which had inexplicably terrified his child, years earlier, gave even melodrama the power to suggest an inviolability, transcending time, in the connection of parent and offspring.

Mrs Nickleby, on the other hand, whom the plan of the book made equally subject to satire, was removed from the realm of mere moral responsibility into the freedom of the utterly absurd. She was the first of Dickens's great comic soliloquizers, transfiguring ordinary life, like Sarah Gamp or Flora Finching, with 'a torrent of favourable recollection'. She generated a kind of poetry from mundane detail made fantastic by its very miscellaneousness; yet, despite the freely ranging absurdity, this torrent could cast up submerged matters which bore directly upon the subject of the book, like her late husband's 'horror of babies', in chapter 41, which connected, in her mind, fine weather with roast pig. With a similar comic relevance, humour could transfigure even Squeers, in his cadenza, in chapter 38, on his position as Smike's 'feeder, teacher, and clother . . . classical, commercial, mathematical, philosophical, and trigonometrical friend', whose milk of human kindness had been turned to curds and whey by ingratitude, or the fake parent Snawley in chapter 45, descanting on 'the elevated feeling . . . of the ancient Romans and Grecians, and of the beasts of the field and birds of the air, with the exception of rabbits and tom-cats, which sometimes devour their offspring'.

The humour showed Dickens's detachment from what was personal in a subject, paternity, which in fact touched him nearly. Humour played equally over his idealization of himself in the hero and over his own new profession of novelist. In the first volume edition, Nicholas's resemblance to the author was openly suggested by the frontispiece, from an engraving of Maclise's portrait of Dickens which could be compared with the illustrations of the hero. Within the novel, the hero's romantic chivalry led him into clichés of language and behaviour scarcely less flagrant than those comically revealed in the lavish self-dramatization of the actors, among whom indeed he had most success in making his own living. Routine contemporary novels were as affectionately satirized as the theatre, in the extended parody of fashionable fiction, which Mrs Wititterly, Kate's patroness in chapter 28, received with rapture:

'So voluptuous, is it not? So soft?'

'Yes, I think it is,' replied Kate, gently; 'very soft.'

Even Dickens's own fairy-tale in *Oliver Twist* seemed to be mocked when Squeers was ready to agree that the recapture of Smike was the work of Providence in chapter 38, or when Ralph, in chapter 45, jeered at Nicholas's alleged expectations of a high lineage for Smike: 'Your romance, sir, is destroyed . . .'.

Any such suggestions of a self-awareness like that of the early Thackeray were quite absent from the novel which followed, *The Old Curiosity Shop* (1840–1). Its resounding success was no doubt helped by the price of 3*d*. per weekly number (though this factor failed to bring the next weekly serial, *Barnaby Rudge*, up to the same circulation), but more must be credited to its combination of comedy with a kind of non-dogmatic piety for which Wordsworth had set the example. When the heroine, Nell, determined in chapter 9 to take her gambling grandfather away from London, her aim was to 'walk through country places, and sleep in fields and under trees, and never think of money again . . .' and his to 'trust ourselves to God in the places where He dwells'. Once they were free of London, their feeling, voiced by the girl, 'as if we were both Christian, and laid down on this grass all the cares and troubles we brought with us, never to take them up again', showed how insubstantial the piety was. Bunyan would have made short work of this equation of Christian's burden with mere 'cares and troubles', and Dickens made nothing of the extent to which these cares remained with them and others were added. He allowed Nell her measureless stoicism and her self-reassurance, in chapter 31, that they had found the peace and contentment they sought, while himself making the journey afford glimpses of society which the London part of his action would not admit.

So he was able to include, with fear and wonder, as at a foreign country, the industrial midlands. Nell's allegorical status (she seemed, he had said, taking a hint from Hood's review of the novel, 'to exist in a kind of allegory') made the author's own fear of what was stirring in the midlands acceptable as that of the heroine herself, when the prose became afflicted with iambic convulsions, in four- and five-beat groups. Both she and the author were relieved to begin the journey which followed this, 'with no trouble or fatigue', towards a place to 'learn to die in', amid the detritus of an immemorial England. It was an embarrassing, uncontrolled vision of the present in relation to the past such as Dickens would attempt to subject to irony in his next novel. Here the best he could offer by way of irony was the comic schoolmistress, Miss Monflathers, who reproached the heroine for tending a waxwork 'when you might have the proud consciousness of assisting, to the extent of your infant powers, the manufactures of your country . . . and of earning a comfortable and independent subsistence of from two-and-ninepence to three shillings per week'—the nearest he got to striking the 'blow . . . for these unfortunate creatures', the factory children, which a letter had promised in late 1838. But ironic comedy of this kind was a very small part of the book's effect compared with repeated things like the arch antics of the Garlands' self-willed pony, or the elaborate, generalized sentiment of Nell's association with death from chapter 16 when she first found her place 'among the tombs'.

Fortunately, the novel also contained figures each demanding such a varied response as the zestful demon Quilp, attractive even to his ill-treated wife, or Dick Swiveller in his raffish, shabby pseudo-gentility. Swiveller, with his love of assuming 'gay and fashionable airs', including the habit of calling a diminutive slavey 'the Marchioness', had his origins in the same fashionable fiction as Mrs Wititterley read, but he quite transcended them as a soliloquizer whose theatrical fantasies and scraps of popular verse have an inherent joyful foolishness. The Marchioness was a creation the originality of which allowed pathos to arise, unforced, from a substratum of comedy.

Barnaby Rudge (1841) completes the experimental stage of Dickens's career. Evidence suggestive of long-range planning such as is visible in the origins of *Pickwick* and *Oliver* appears in an agreement as early as May 1836 to furnish 'the entire Manuscript—on or before the 30th day of November next'—to Macrone, who was to publish it in 'Three Volumes of the usual size'. It was an obvious bid for respectability and, with a story from the years leading up to and including the Gordon Riots of 1780, 'sixty years since', an attempt to follow Scott. Scott had shown towards the past a double attitude to which that in *Sketches by Boz* was considerably closer than to anything in the historical novels of his popular successors. The postscript to *Waverley* had recognized not only that sixty years of economic development had set an enormous gap between 'the present people of Scotland and their grandfathers', but also that the process had entailed both gains and losses—on the one hand, 'progress' and the disappearance of 'much absurd political prejudice' and, on the other, the loss of 'many living examples of . . . old Scottish faith, hospitality, worth, and honour'. Such attitudes were presupposed in both the mock-lamentation of some of the *Sketches* ('We marked the advance of civilisation, and beheld it with a sigh', in 'Scotland Yard', for instance), and the picturesque nostalgia of *Pickwick*. *Barnaby Rudge* began (in the only section to be written before 1841) as if in a venerable country inn from *Pickwick*, but went on to present the innkeeper as a petty tyrant, 'dogged and positive' in opposition to his own son, and giving way to him only after being disabled in mind when the inn was sacked by rioters from London. Dickens proceeded to build up round the reader a whole society of which, as in his late novels, one feature was relentlessly impressed, in this case resistance to change.

It was all, apparently, to make up an ironical antiphrastic study, in the manner of Fielding's *Jonathan Wild*, of the kind of non-revolution which led inexorably to its opposite. The social conspectus extended all the way from the savage, illegitimate son of a gentleman, an ostler who changed the cry of 'No Popery' into 'No Property', through the apprentices pledged, in chapter 8, 'to resist all change, except such changes as would restore those good old English customs, by which they would stand or fall', up to the Protestant lord, nominal leader of what began as a popular attempt to force the government to preserve all the legal restrictions upon the activities of Roman Catholics. But Dickens's comic invention was unstirred: he was simply not amused by his diminutive apprentice, with an inner voice 'whispering Greatness'; however sedulously he might model his angular Lord George Gordon on the eighteenth-century imitations of Don Quixote,

> Sitting bolt upright upon his bony steed, with his long straight hair dangling about his face and fluttering in the wind, his limbs all angular and rigid, his elbows stuck out on either side ungracefully, and his whole frame jogged and shaken at every motion of his horse's feet,

Lord George remained a sorry figure; Sir John Chester's affectations were as little amusing, whether addressed to his illegitimate ostler-son or to his legitimate Edward, whom he expected to keep his father in unchanged affluence by rejecting love to marry money.

Yet though the details were cruder than in *Bleak House*, they were selected, as in the later book, with an eye to a specific principle of unity. Even the preoccupation with crime, which had already made Dickens one of the Newgate novelists for Thackeray, afforded a further example. Rudge Senior was not merely a speechifying villain out of the deplorable inset stories in *Pickwick*, further inflated by Bulwer's example in *Eugene Aram* and his preaching 'On Art in Fiction' in 1838. Rudge underlined, in his very implausibility, an obsession with the unchanging past and the constraints of the past upon the freedom of the next generation: regularly revisiting the scene of the murder he had committed over twenty years before the opening of the novel, he learned only in chapter 17 of the son born on the night of it and condemned through the mother's terror to perpetual childhood.

What seemed lacking was Bulwer's ability to plot: the criminal part of the action was as creakingly carried out and as inadequately attached to the rest as was Chester's laborious intrigue to break his son's attachment to a younger woman who was neither Protestant nor likely to be rich. Dickens was already relying overmuch on the suggestive power of such conceptions in their mere collocation with each other and with one major event of emblematic significance, rather than on their connections as cause and effect. He was threatening the forms of the traditional novel in a way which has recommended him to twentieth-century readers and critics but has made him at the same time notoriously difficult material for criticism. The suggestive range of the collocations he offers is very wide and it is difficult for the critic to preserve the particular relationship among them that Dickens maintains in each novel by his own methods of emphasis.

The major event to which all was to be related, in the case of *Barnaby Rudge*, was the rioting in 1780, chosen and made typical because prompted by a futile desire not to produce change but to stop it. Only in 1841 was Dickens free to work at the story as the successor to *The Old Curiosity Shop* in *Master Humphrey's Clock*. By then the civil disturbances which had multiplied in the years since 1836, when Chartism replaced parliamentary reform as the focus of popular hopes, had made the main event of the story as topical as parts of *Deerbrook* were in 1839. Many later readers have found Dickens's febrile prose, wherever 'The crowd was the law', to be evidence of a fascination amounting to sympathy. But his preoccupation with crime was sufficient to account for the excitement with which he was moved by the violence depicted in his sources. His problem, as he said in a letter, was 'to select the striking points and beat them into the page with a sledgehammer'. The excitement inherent in the material was redoubled by his own methods of emphasis. Sometimes there was indulgence in simple horror, sometimes in the pleasures of a chase, as when chapter 67 offered, ultimately to little purpose, the same view as Sikes had had from a roof-top, of the upturned faces of pursuers. But, all the time, the vision Dickens strove to keep before the reader was that of a society which, even though its strongest efforts appeared to be directed to preventing change, was in fact changing. The strongest evidence of this to a reader in 1841 was the difference which sixty years had made in the criminal code, a process which Dickens, in his opposition at the time to capital punishment, believed had yet to be completed. The rioting mob of 1780 might, he said in chapter 49, be 'sprinkled, doubtless, here and there with honest zealots', yet it was composed 'for the most part of the very scum and refuse of London, whose growth was fostered by bad criminal laws, bad prison regulations, and the worst conceivable police'.

Consistently, then, the climax was an attack upon Newgate to release the prisoners (among them Rudge, whose fear of being lynched suggested a parallel with Captain Porteous, whom Scott's mob had pursued even into his cell in *The Heart of Midlothian*). Dennis the hangman, joining the rioters to protect his own 'sound, Protestant, constitutional English work' from foreign substitutes and opposing the release from Newgate of those condemned to hang, gave heavily ironical emphasis to the kind of public inertia (in this case about capital punishment) that Dickens had in mind. The ghastly face of one so reprieved was stressed at the end of chapter 65, from the infant recollections of eye-witnesses still living sixty years later. At the same time, the bewilderment of an indigent prisoner, 'whose theft had been a loaf of bread', when suddenly he gained 'liberty to starve and die', the misery of debtors unwillingly freed from the Fleet after being for long 'dead to the world, and utterly . . . uncared for', all made quite specific what kind of sympathy Dickens had with the crowd's activities.

The unsuccessful attempt to make a sympathetic character out of Barnaby, the idiot son, seemed to show, like Little Nell's rural nostalgia, the baneful effect of unassimilated Wordsworth. The 'delicious dream' of his life in the country in chapter 45 seemed better than Oliver's pleasures in similar circumstances only because actual objects, though still rather at second-hand, were substituted for the mere paraphrase of the poet's leading ideas in *Oliver Twist*, chapter 32. The crazed rhetoric of both son and father in their belief that a criminal, like Wordsworth's 'wretched Woman' in *The Thorn*, 'is known to every star', though oddly arresting, was none the less forced and derivative. Barnaby's disability, however, 'which rendered him so soon forgetful of the past' so long as he was journeying through the countryside, was the obverse of his father's obsession with the past; the echo of Nell's nostalgia which followed their reunion in, and release from, Newgate followed also the final destructive saturnalia of the riots in a lake of burning liquor, making it merely foolish of the son 'to think how happy they would be . . . if they rambled away together, and lived in some lonely place, where there were none of these troubles'. Only the fool could be left at the end of the novel determined never to 'be tempted into London' again. The advance upon *The Old Curiosity Shop* was immense. Dickens was beginning to put ideas from Wordsworth to work.

Martin Chuzzlewit (1843–4) began the main sequence of twenty-part serials which was to give Dickens all his major successes but one. In the first experiment in this form, *Pickwick*, the preface had apologized that 'no artfully interwoven or ingeniously complicated plot can with reason be expected' in twenty such numbers, each 'to a certain extent complete in itself'. The preface to *Martin Chuzzlewit* claimed he had 'endeavoured . . . to resist the temptation of the current Monthly Number, and to keep a steadier eye upon the general purpose and design'.

The 'design' he announced 'of exhibiting, in various aspects, the commonest of all vices', self-centredness, allowed too much of the main action to be carried by simple figures from the comedy of humours—young Martin in his immaturity, old Martin as the suspicious benefactor, Mark Tapley with his cheerful persistence in believing that self-fulfilment is possible only in adversity. If such a design gave very little stimulus to the reader to find out what came next, it freed Dickens to develop those modes of the self's activity for which words were the supreme vehicle, in the rich confusion of mind of Mrs Gamp and the vertiginous heights of Pecksniff's moral pretensions. Here the delight of the performance depended less upon the moral point, though the reader was still to keep it in view, than upon the full folly of its transmutation—Pecksniff's selfish reasons in chapter 31 for dismissing Tom Pinch, brazened out with high-minded rhetoric, capped by quiet pseudo-sincerity: ' ... having discharged: I hope with tolerable firmness: the duty which I owed to society, I will now ... retire to shed a few tears in the back garden, as an humble individual'.

Yet Dickens's mind was now working with a steadier sense of the interconnections among the parts of his creation, the kind of control which he had been moving towards in *Oliver Twist* and *Barnaby Rudge* and had been ready to forfeit in *Nickleby* and *The Old Curiosity Shop*. Mrs Gamp's delicate orders for supper, before taking the patient's pillow for herself and descanting on the 'blessed thing it is—living in a wale—to be contented', are the funnier when we see that her 'Piljian's Projiss of a mortal wale' is Tapley's world of trial, in which he hopes to 'come out strong', while she is determined above all to survive in comfort. Her nice derangement of the language of Bunyan and the Bible may become for the reader almost an end in itself—'Rich folks may ride on camels but it an't so easy for 'em to see out of a needle's eye'—but, like Pecksniff's perversion of the language of conscience, it gives rise to a humour which is all the richer because the groundswell of the book makes us alive to the self-centred application of such language. Mr Pecksniff's cadenza on the beauties of his own digestive processes as a public service, Mrs Gamp's invention of a friend for purposes of self-praise and her noble inability to forgive an injury to that friend—'But the words she spoke of Mrs Harris, lambs could not forgive ... nor worms forget!'—are the extreme examples of the high fantastical delight with which Dickens contemplates the self-centredness revealed by affected altruism.

Where the language, in the American chapters, is that of contemporary catchwords—'in the van of human civilisation and moral purity', 'the Palladium of rational Liberty at home, sir, and the dread of Foreign oppression abroad'—its misapplication is for the most part mere satire, untouched by this rejoicing at the sheer range and intricacy of radical human self-deception. The American names—Lafayette Kettle, Hannibal Chollop, Mrs Hominy—are funnier than the people. In chapter 34 even the names are generalized into 'a Pogram or a Hominy, or any active principle to which we give those titles'; the assembled 'literary ladies' then hilariously dissolve them into the undifferentiated inane where 'Mind and matter . . . glide swiftly into the vortex of immensity. Howls the sublime, and softly sleeps the calm Ideal, in the whispering chambers of Imagination . . .' To show the American characters forfeiting individuality in their obsession with social and political institutions makes for strong contrast to the English chapters, crammed with individual idiosyncrasy, and that is the best Dickens can do, increasingly awake as he is to the analogies between such idiosyncrasies and the salient features of a whole society. It is amusing to see American localities equally characterless—'four cold white walls and ceiling', 'interminable whitewashed staircases, long whitewashed galleries up stairs and down stairs, scores of little whitewashed bedrooms'—but interest is forfeited with the loss of the accustomed specific detail. Dickens goes instead straight for the symbolism, for instance, of a fetid Eden. In a similar fashion he goes for the abstract point when everybody in a New York boarding-house eats 'his utmost in self-defence . . . to assert the first law of nature', in contrast to the complicated humour in Mrs Gamp's fastidious choice of the food, for herself, which 'does a world of good in a sick room', or Mr Pecksniff's disinterested praise of his digestion as 'one of the most wonderful works of nature'.

The English part of the book makes a steady contrast between town and country which reinforces the contrast between the assumed 'simplicity of innocence' of Pecksniff and the actual 'simplicity of cunning' in the London Chuzzlewits, credulous 'in all matters where a lively faith in knavery and meanness was required as the ground-work of belief'. The affectations of the first are exaggerated in speech and the actualities of the second in deeds, and Dickens works these out with detail which bears specific relation in each case to the setting and sometimes to particular American episodes as well. So Pecksniff pretends to the virtues of unfallen Adam as he takes up 'an ancient pursuit, gardening. Primitive, my dear sir; for, if I am not mistaken, Adam was the first of our calling.' The claim to innocence is as false as in the American parody in chapter 21 when General Choke, selling Martin land in a place called Eden, claims that 'man is in a more primeval state here, sir'. Unlike the Americans, however, Pecksniff reveals layer upon layer of self-deception; he approves of the state of nature where he himself is concerned, but where other people are concerned, his enemies note 'he always said of what was very bad, that it was very natural'.

The 'simplicity of cunning' in the city arises from taking 'knavery and meanness' as natural, in the precept 'Do others for they would do you.' This decisively breaks the comic mould of the book, substituting for Pecksniff's sanctimonious twaddling a terrible application of the precept, by the son against the father who had taught it. Here the play with what is natural and what is unnatural appears in the deeds not the talk; the father dies not from the unnatural attentions of his son but from natural causes. When the son, Jonas, falls victim to blackmail and plans murder to release himself, the crime is staged in the countryside, where the victim has been visiting Pecksniff, and this setting becomes a commentary upon the action. Jonas is judged by reference to a less disingenuous form of the pastoral piety which Pecksniff has steadily affected and which Dickens has affected ironically himself in chapter 30 when describing 'the worthy Pecksniff in tranquil meditation', the 'summer weather in his bosom . . . reflected in the breast of Nature', just before making his proposal of marriage, backed by threats against young Martin, to the girl whom young Martin loves. The account of Jonas's crime, though hectoring in Dickens's 'sledge-hammer' way, is in parts oddly moving; it reads like lavish echo and extension of Wordsworth's sonnet to Toussaint L'Ouverture. In the poem wickedness is opposed by

> . . . air, earth and skies;
> There's not a breathing of the common wind
> That will forget thee . . .

In the novel Dickens invokes, as the 'sentinels of God', trees, stars, wind, the night itself, which 'never winked' (recalling Rudge's obsession with the 'secret' of his guilt—'The stars had it in their twinkling, the . . . leaves in their rustling'—and Barnaby's fear of him as the man at whom the stars winked), and elaborates the contrast of the deed to the 'vistas of silence' which open in the woods, 'beginning with the likeness of an aisle, a cloister, or a ruin open to the sky'. The description, despite being over-emphatic and less original than that of 'Town and Todger's' in chapter 9, has power. It is gained not only from the obvious horror of the occasion but also from the persistence of pastoral allusion in the surrounding comedy which Dickens has developed from his initial laboured jocularity at the expense of the Chuzzlewits, 'undoubtedly descended in a direct line from Adam and Eve; and . . . closely connected with the agricultural interest'. With the imminence of the grim climax he even effects a modulation into comedy based on the loss of paradise, when Mrs Gamp, with her characteristic befuddling of what is in the author's mind, flatters the murderer's wife, immediately before he comes home to organize his alibi, with the wish that there might be many houses like hers 'which then this tearful walley would be changed into a flowering guardian'. Eventually the stature of this novel appears to depend not only upon the vitality of its major characters but also upon the consistency and force with which they activate motifs and metaphors drawn from the contemporary religious consciousness.

Although, according to Forster, *Dombey and Son* (1846–8) in its earliest plan was 'to do with Pride what its predecessor had done with Selfishness', it went as far beyond this mere abstract subject as its predecessor had done, albeit in a different direction. Writing at a time when the extension of railways was reviving the economy after the depression of the early forties, Dickens grounded the whole upon the sense of sudden economic change. In the plot which he devised once writing began, Dombey's obsession with a son to succeed to the proprietorship of the family firm led to much more than the exhibition of varieties of pride. For Dombey was not only proud but curiously passive in contrast to his Manager, a sycophant and a sybarite, determined to make his percentage on the firm's dealings yield more than its traditional operations would allow.

Counterpointing the main developments of this action against the vigorous development of the railways, in writing energized by his own fascination and fear, Dickens worked out his sense of triumph at economic advance and fear for its consequences. A pair of old-fashioned family firms, one in the plot, the other in the sub-plot, identified what was comforting in the state of being 'all behind the times'. Sentimentality here was at first tempered in the main plot by the mordant clear-sightedness of Carker the Manager. Although there was much of the stage villain about him, he went on to become the centre of the energy of an action in which Dombey's iciness, after being at first made comic, in the long run denied him all power of initiative. The personal and family aspect of this action accumulated pathos to rival that of Little Nell, as the economical and moving opening with the death of Mrs Dombey was followed by the less reticent treatment of the son's death and the daughter's disregarded affection for her father. Yet this aspect was turned to the reinforcement of a subject which looks at this distance curiously like that of *Culture and Anarchy*, twenty years later, in so far as it stresses the tendency 'to set doing above knowing'. Knowing was dramatized in the children and young women of the plot and doing in its businessmen, Dombey in his exclusive solicitude for his son's future career, Carker in what Arnold would call 'rough and coarse action, ill-calculated action, action with insufficient light'—his use of the firm's money for speculation (which Dickens would expect to be interpreted in 1848 as railway speculation) and his giving up of an intelligible plan to recommend himself to Dombey's daughter, and so become the heir, in favour of a much less rational plan to elope with Dombey's second wife. The author's forming hand showed as Carker was likened to a 'monster' swimming 'down below' Florence on her 'uneasy sea of doubt and hope' and keeping 'his shining eye upon her', and the same figure of a monster was applied to the lamp in the office strong-room in chapter 13, to the railway, in Dombey's tortured thoughts in chapter 20, and in chapter 33 to London itself, 'the monster, roaring in the distance', fascinating the victims it continually absorbed as 'food for the hospitals, the churchyards, the prisons, the river, fever, madness, vice, and death'. Finally, the 'red eye' seen in the office strong-room reappeared as that of the railway engine by which Carker was cut down.

Paul Dombey had judged with intuitive knowledge the power of finance, the monster in the strong-room that Carker released as disastrously for the firm as for the family—'Papa, what's money, after all?' In this he represented the class of persons, in this novel made up of children and young women, who know, but, compared with the characters who are adult and male, are granted only limited power of doing. Over this group played the associations of fairy-tale—Florence Dombey the sleeping princess, Paul as much a knowing gnome as a living child, Alice with her witch of a mother. Although these associations were developed with an archness that has dated badly, their power to make the reader pause and consider is not exhausted. We are prompted to notice that the women and children know, each of them, what has been done to them, whereas the adult males do not. We see their limited powers of action, borne down as they are by resistless parents, and their vain attempts to break out of such limitations. Alice, although the sexual victim of Carker, is even made to recognize that he is more than an object of hatred to her and to try and save him from Dombey's revenge. Nothing comes of the effort and that is the very point of it. In this novel those who know can apparently do little or nothing.

Dombey, having lost in the fall of the firm all power of public action, enters this group of the sentimentally saved by a change of heart the suddenness of which Dickens was still defending in the preface to the Charles Dickens edition of 1867. The effusive overwriting of the conclusion in the manuscript made the knowledge he had attained a knowledge of what the voices in the waves—already over-stressed, as some of the early reviewers saw—were speaking to him. Although the appeal to another dimension than that of action, to 'the invisible country far away' of chapter 1, as opposed to the all too visible country of restless change, was no doubt facile, it formed part of the coherent emotional structure which the author was trying, if rather too hard, to give to the whole book.

In the major novels of the fifties, a more sophisticated control of this kind of construction would come from drawing into it not only the pastoral vein of the novels before *Dombey* but also a subtler view of subjective experience developed in *David Copperfield* (1849–50), the last representative of the rather looser work, with a single figure at its centre, which characterized the first phase. This novel grew out of the interest in his own past stirred by Paul Dombey's childhood. Much of Dickens's own childhood appeared in Copperfield's 'Personal History', as we can see from the fragments of an autobiography which survive for comparison, in Forster's *Life*. Once, indeed, the hero ceased to be vulnerable as a child he ceased to be interesting as a character; but what survived, and that most impressively, was a curious and subtle concern with his processes of mind, the interaction of observing and recalling which made him successful as an artist. The business of growing up was somewhat laboriously moralized as the disciplining of the heart: Copperfield and most of the people who were of importance to him exhibited the calamities, both within marriage and without it, which led to this discipline or showed its lack. But another action was proceeding all the time, linked to it but far less simple, the history of the hero's powers of perception, his ability, for instance, to look at nothing but to see everything, or to see but not understand.

Dickens chose to write the whole as one person's recollection, and moreover with repeated dramatization, both comic and pathetic, of the act of recalling ('The shade of a young butcher arises ... Who is this ...?'). Such a method was natural after beginning to write his own autobiography, to 'go on the same shelf with the first volume of Holcroft's, he said— that is, with the first book of the life of a novelist and playwright, written by himself, in the first person. Dickens must have been struck by his own resemblance to this autobiographer, like himself 'prone to observation' and son of a father 'driven, by ... a brain too prone to sanguine expectation, into many absurdities'. (See Vol. XII of this history, Chap. 9.)

The method followed naturally, too, from the sort of story told in the first and last of the Christmas books of 1843–8. In these stories, moral regeneration is a matter of the central character's grasping the meaning of his past and present experience in order to move on to a different kind of future. In *A Christmas Carol* (1843), Scrooge, shown the shades which arise at the call of the Ghosts of Christmas Past, Present, and Yet to Come, resolves, 'I will live in the Past, the Present, and Future. The Spirits of all Three shall strive within me.' The reader may wish that there were more striving within Scrooge between what he recalls from the past and what he does in the present, instead of the mere over-mastering elation of his charity. The naïve dramatization of a miraculous reconciling of present with past, so that he accepts life as a gift and becomes a giver in turn, is potent nevertheless. After this, we are told, 'He had no further intercourse with spirits.' David Copperfield speaks of himself as one privileged, 'in consideration of the day and hour of my birth ... to see ghosts and spirits', but as not having done so 'yet', at the outset of his narrative. His narrative is precisely that process of haunting in which, as he says later, there arise the 'ghosts of many hopes, ... remembrances, ... errors, ...

sorrows'. Writing enables him to fulfil Scrooge's strange resolution to 'live in the Past, the Present and the Future'. Yet no more than Scrooge is he capable of any striving within him of the 'Spirits of all Three': his present and his future seem to be secured for him by external agency, the dishonouring of his first childhood love, Emily, and her removal to Australia, and the death of his second, the child-wife Dora. The very deficiencies of the man, however, develop the strength of the writer he is to become.

Dickens is not content to use narration in the first person as an easy way to the intensities of emotional identification between the central character and the writer or the reader. Rather he makes the method yield a steady sense of the limitations of the first-person viewpoint. This shows a clear contrast with the novel before. Paul Dombey had been, above all, the knowing child, aware of much more than his elders. That this was the author's aim is clear from a letter commenting on 'Mrs Pipchin's establishment' in Number III: 'I was there—I don't suppose I was eight years old; but I remember it all as well, and certainly understood it as well, as I do now. We should be devilish sharp in what we do to children.' The young David Copperfield has a similar sharpness on occasion: without ever having seen a coffin or heard one being made, he knows, when the tune of the hammer stops and a young fellow comes in with a hammer and a mouth full of little nails, 'what he had been doing', just as in the debtors' prison 'I divined (God knows how) that though the two girls . . . were Captain Hopkins's children, the dirty lady was not married to Captain Hopkins'. But the humour and pathos in the main plot repeatedly depend on his not understanding 'as well as I do now'. Spoken of as Brooks of Sheffield when his mother's second marriage is discussed in his presence, he almost sees through the device because, when 'Somebody' is said to be 'sharp', he knows who it must be, but his understanding at the time takes him no further. Later on, he notices the details which make Steerforth less than a hero, but his admiration quite hides their significance. The earliest episodes are full of such observations, the significance of which the child cannot know. Reasons for the 'metallic' inflexibility of Miss Murdstone appear, for instance, when the child innocently observes her 'constantly haunted by a suspicion that the servants had a man secreted somewhere on the premises'. The fears and tensions of the child's relation to a dead father and a doting mother are even more powerfully suggested in a sequence showing him greatly frightened by the story of Lazarus, but stilled by the sight of the churchyard 'with the dead all . . . at rest', and then recalling this with terror on hearing he has again 'got a Pa!' The tyranny of the stepfather and the resentment it arouses in the stepson are simpler matters than the force of the guilt the boy feels on taking his revenge, a force which the reader can see to have complicated origins in these early terrors and tensions.

The reader's awareness of such things is prompted by the constant interplay between what the hero knows and what he does not. His ignorance appears again and again where we do not expect it or where it seems exaggerated—

> 'You're the new boy?' . . . I supposed I was. I didn't know.
>
> If my grief were selfish, I did not know it to be so—

as well as, of course, in his repeated failure to understand, particularly to understand the Byronic selfishness of Steerforth, from which so much of the plot derives. His knowledge appears in the acuteness of immediate perception and in the poignancy of its re-creation in the memory long after its consequences are known. Yet perception may be paradoxical—'I looked at nothing, that I know of, but I saw everything'—and what memory creates with most immediacy in the narrator's own present time seems tinged not only with the sadness besetting 'the days that are no more' (Dickens was a friend and admirer of Tennyson, whose lyric containing these words was published in 1847), but also with doubt concerning their status as shadow or substance, as things lost in time or regained beyond it:

> I never hear the name, or read the name, of Yarmouth, but I am reminded of a certain Sunday morning on the beach, the bells ringing for church, little Em'ly leaning on my shoulder, Ham lazily dropping stones into the water, and the sun, away at sea, just breaking through the heavy mist, and showing us the ships, like their own shadows.

Common-sense reality seems called in doubt when it intermingles readily with dream: the boy drowsing to the sound of Mr Mell's flute is both indoors and out on a journey, and what in fact he sees, the demonstrative affection of old Mrs Mell, belongs to the dream, or to 'the middle state between sleeping and waking'.

Once more the little room ... fades before me ... The flute becomes inaudible, the wheels of the coach are heard instead, and I am on my journey ... I dreamed, I thought, that ... the old woman ... gave him an affectionate squeeze ...

Such passages give us the inner history of the growth of a novelist. Beneath the marital and extra-marital calamities of his plot, whether comical in the case of David's first marriage or over-pathetic, as with the seduction of Emily and her father's rescue of her with the help of a reclaimed prostitute, Dickens was absorbed in contemplating, as he did nowhere else so directly, the processes of perception and re-creation. When he wished to rise to the climax of his plot, the storm which would destroy both Steerforth and the man whom he had injured, it became the climax of the book's inner history, a climax of confused perception and distinct memory.

I had lost the clear arrangement of time and distance ... So to speak, there was in these respects a curious inattention in my mind. Yet it was busy, too, with all the remembrances the place naturally awakened; and they were particularly distinct and vivid ... Something within me, faintly answering to the storm without, tossed up the depths of my memory, and made a tumult in them.

At the catastrophe, memory dominated even in the outer action: the narrator was brought to the dead Steerforth by 'a fisherman, who had known me when Emily and I were children', with 'the old remembrance ... in his look'; Steerforth was found 'lying with his head upon his arm, as I had often seen him lie at school'. This is disturbing to the reader who recollects the unbridgeable gulf which David sets between himself and Emily in her actual misfortune. It is as if he discards her except as a character in the timeless charade of the imagination; that this is intended to be noticed appears in an early episode in chapter 10:

... a curious feeling came over me that made me pretend not to know her, and pass by as if I were looking at something a long way off. I have done such a thing since in later life, or I am mistaken.

Steerforth too is valued solely for his place in the narrator's recollections, though to a different effect:

I believe that if I had been brought face to face with him, I could not have uttered one reproach ... I should have held in so much tenderness the memory of my affection for him ...

Here, it may be suggested, is where the connection is strongest between the inner history of the developing novelist and the 'personal history'. For the recollecting David reveals, as inexorably as Proust's Marcel, the voracity of the artistic ego, drawing all things into itself. At any rate, realistic detail, closely observed, may repeatedly bear this interpretation. The culminating metaphor is afforded by his own face reflected in the glass of the window through which he attempts 'to look out' at the storm. The activity of his quiet egoism is plain in his first love of all, that for his mother, not only in the episodes already referred to, but also, when she dies, in his sincere grief, the 'importance' of which is 'a kind of satisfaction' to him—and he acts up to it with his schoolmates. When she is buried, she not only wings 'her way back to her calm untroubled youth' but becomes 'the mother of my infancy; the little creature in her arms, was myself, as I had once been, hushed for ever on her bosom'.

The corollary of his egoism is the fact that eventually he is known to the reader only by the people and things he has absorbed into himself in this way. It is odd in a 'personal history' to find any strong sense of the resulting personality absent. Even when he is 'steeped in Dora', the individuality in the situation is Dora's, not his. Here is the contrast with Thackeray's Esmond, who dwells equally upon the long perspectives of memory. Esmond's recollections are for the sake of the present; he is given a Wordsworthian sense that the passion he has 'once felt forms part of his whole being and . . . becomes a portion of the man of today'. Copperfield's memories seem to be there for him to get over them, with Dora dead and Emily in the antipodes, as Esmond never does.

Dickens is in fact interested in his central character less as a man than as an artist, so that to subordinate him as a man in this way to the things and people re-created in the memory is right, when the artist appears prepared to treat them as secondary to the re-creating process which constitutes his essential self. Dickens has already written in a letter to William Sandys, compiler of a collection of *Carols* which furnished hints for *A Christmas Carol*, of his satisfaction that 'so little is known' about the life of Shakespeare. 'It is a fine mystery . . .'. The full strangeness of his own extended treatment of the artist's life stands out if Copperfield is compared to Thackeray's Pendennis, whose 'History' was running at the same time throughout much of 1849–50. Pendennis also becomes a novelist; but the connection between his fiction and his own experience is merely a theme for further satire of his immaturity: the nearest that Thackeray gets to making the matter of novel-writing into fiction is in the discussion about 'selling his feelings for money' between Pen and Warrington, a sadder, wiser character very close to the author in age and personal predicament. Dickens persistently suggests or exemplifies, in the fiction itself, without argument or analysis, the oddities or dubieties which are inseparable from the novelist's acquiring and re-creating of experience.

This is supplemented, at a simpler level, by the ecstatic account of Copperfield's reading and story-telling. Smollett, Cervantes, Le Sage, Defoe, Goldsmith, and Fielding 'kept alive my fancy, and my hope of something beyond that time and place—they, and the *Arabian Nights* . . .'. When Copperfield tells stories regularly to his schoolfellows, they encourage in him 'whatever I had within me that was romantic and dreamy', recalling the dreaminess Dickens in person has spoken of in a letter of early 1849 as 'essential' to the 'effect' of *The Haunted Man* (the Christmas story, with its emphasis again upon memory, written just before this novel) and recalling too his own phrase for the writing of the instalments of *Dombey and Son*: 'I . . . fell into my month's dream . . .' In chapter 11 of *Copperfield* 'the manner in which I fitted my old books to my altered life and made stories for myself out of the streets' shows 'how some main points in the character I shall unconsciously develop, I suppose, in writing my life, were gradually forming all this while'. Only in the proofs did he add 'for myself' to 'stories', making plain the element of personal fancy which both distorts facts and gives them a new life.

The greatest of the characters for which *David Copperfield* is famous emerges from this interest in the artist. Mr Micawber has his fullest triumphs in the imagination, under the influence of words. Even emigration to the antipodes becomes 'merely crossing, merely crossing. The distance is quite imaginary.' His volatility of mood, his exaggeration of the feelings of the hour to the exclusion of all others, and his pleasure in making speeches and writing letters (as Dickens dreams in novels, Micawber 'dreams in letters') with the contents of which he may be affected to the point of being cast down completely, but in the making of which his satisfaction is undiminished, he is a joyous parody, amounting to a celebration, of the nature of the artist struggling to be free and in his own way succeeding. It is a pity that such a figure, realizing, as Chesterton has it, 'the inexhaustible opportunities offered by the liberty . . . of man', should be put to work for the plot and turn out to be 'a most untiring man when he works for other people', or that he should succeed in his work in Australia, as part of Dickens's Carlylean propaganda in favour of colonization. Micawber's joyful hope in the midst of his afflictions is a so much greater thing than his prosperity or his altruism. To see him, when in debt to the narrator, 'filling up the stamps with an expression of perfect joy', and 'with the relish of an artist', and then plunged in despair at the warning that he should 'abjure that occupation', is not only to see that promissory notes are the outward signs of an inextinguishable hope, even though he may stigmatize them as the serpents that have poisoned his life-blood; it is also to see portrayed, even in this most beautiful foolishness, the separation of the man who suffers from the artist who creates.

The prodigal freshness of a multitude of walk-on parts has all the spontaneity of Dickens's early fiction; the underlying coherence arises from a preoccupation with the dubieties of perception and re-creation and the paradoxical egoism of the artist. His strange, embarrassing, unequal treatment of the matter remains the novel of Dickens which stands beside *Pickwick* as known in some fashion to everyone.

II

Fortified by his confident exploration of David Copperfield's development from ignorance to knowledge, Dickens turned his interest to a parallel development with respect to social change. The resulting series of major novels was at first, except for *Great Expectations*, less appreciated than the work of the earlier phase, but has since come to be seen as its culmination. The four outstanding examples of this second phase offered a vision, selective and strongly distorted, of a whole society, clamant with individual demands, burgeoning with individual variety, but assimilated by the imagination as one. The focus was still upon London but now upon London as the nerve-centre of the whole country, the place where the results of its energies were concentrated and where, if anywhere, some sort of directing intelligence might be expected. As the railway boom of the 1840s launched the country upon a course of commercial prosperity extending over the rest of Dickens's lifetime, his sensibility to what was happening led to novels the strangeness of which still delights and disturbs. The same sensibility, too, was active in relation to his own life, moving him to make discomforting correlations between private experience and certain aspects of social change. At the same time he was a successful journalist, writer for and editor of two periodicals of his own in succession, *Household Words* and *All the Year Round*, with notable effect on the detail of his fictions.

The first result, *Bleak House* (1852–3), was a remarkable application of the technique of *David Copperfield* to a vision of the social organism. Part of the story was recalled, in the past tense, by a limited participant looking back after it was over. The rest had a narrator from outside the action, who commented at the same time, using the present tense and a vigorous, ironic voice, kindling an extravagant verbal blaze of indignation, humour, and pity. Contemporaries, notably the Brontës, had intricate means of viewing parts of a novel's action, or the whole of it, from the standpoint now of an onlooker, now of a participant; but no one else generated so much excitement from a steady play of the limited against the omniscient view.

The first-person narrator, Esther Summerson, was an orphan with a mystery attaching to her birth, brought up only to 'know what would be serviceable to her'. The omniscient narrator developed the hints which her narrative furnished of a connection between her position and a major case in the court of Chancery, so as to open up further mysteries and connect them with social evils which had come to be accepted as things of 'precedent and usage'—a dilatory court of Chancery, an aristocracy without function, a government preoccupied, like Chancery, with its own procedures rather than with the interests of the people they were meant to serve. In such a scheme, knowledge and ignorance were not limited to the mysterious details of the immediate story, but suggested the ignorance in which a whole nation was managing its affairs and a possible knowledge which might transform them. The topicality of the book, begun in the year of the Great Exhibition, touched more aspects of contemporary life than any Dickens had hitherto written, but he sought less to make particular recommendations than to induce a certain tendency of mind in the reader by the way the story was told, as a movement from ignorance to knowledge.

Persistently the omniscient narrator used his knowledge to suggest what lay beyond its limits—to contrast the 'world of fashion' first with 'this world of ours, which has its limits . . . when you . . . are come to the brink of the void beyond', and then with the 'rushing of the larger worlds . . . as they circle round the sun'. He delighted to give the Dedlock family a ghost-story going back to the English revolution of almost exactly two centuries before (the revolution which had made 'King Charles's head' the obsession of a character in *Copperfield*), but, beyond this, to imagine a prehistoric cataclysm as just over at the opening of the novel, so that 'it would not be wonderful to meet a Megalosaurus' on Holborn Hill. Out of such a catastrophe, some scientists still thought, came a new creation of living things: Dickens was drawing on the science popularized in *Household Words* (where the Megalosaurus had appeared on 16 August 1851), in order to suggest a background of limitless knowledge. Similar suggestions, but with reference to human conduct, were later to be carried by the conventions of pastoral.

The opening presented mere contiguous occurrences unmarshalled by the finite verbs necessary for narrative:

> LONDON. Michaelmas Term lately over, and the Lord Chancellor sitting in Lincoln's Inn Hall. Implacable November weather. . . . Smoke lowering down . . . Foot passengers, jostling . . . and losing their foot-hold . . . adding new deposits to the crust upon crust of mud, . . . Fog everywhere. . . . Gas looming through the fog . . .

By avoiding as much as possible the time-colouring of the kind of verb which changes in form to differentiate past from present, these opening paragraphs stood outside of both the present tense, unaltered throughout, in which the narrator satirized an ignorant society which refused to alter, and the past tense employed by Esther's story of the knowledge gained through the capacity to learn, like David Copperfield, from experience.

The knowledge of the impersonal narrator satirized ignorance almost as if on a graduated scale. The lowest point came to be represented by a crossing-sweeper, Jo, with his repeated 'I don't know nothink.' When employed by Lady Dedlock (in disguise) to lead her to the graveyard of 'Nemo', her lover and Esther's other parent, he became the means by which pleasurable mysteries in the plot might suggest the profounder ignorance of a whole society about itself. The answer to the question 'What connection can there be . . . ?' was found in their common devotion to the memory of the outcast lover, as well as in the smallpox Jo disseminated, from living as an 'outlaw' in a slum, Tom-all-Alone's, which was 'in Chancery, of course'. Next above Jo stood Krook, the rag and bottle man, who could not read but made a pretence to write and, in his power to amass paper, formed a tendentious analogy with the Lord Chancellor. And so the scale led up through the Chancery practitioners themselves, able to read but not to interpret for any benefit but their own, and through 'the fashionable world', its real status concealed from it by 'the fashionable intelligence', its characteristic members comically unintelligent compared with an Ironmaster, one of Carlyle's Captains of Industry, introduced showing scrupulous care for the education of his son's fiancée. The pinnacle of the scale was occupied by the gentry's legal protector, Mr Tulkinghorn, who on social occasions 'concerning which the fashionable intelligence is eloquent' remains 'speechless', receives salutations from 'half the Peerage . . . with gravity, and buries them along with the rest of his knowledge'. The calamities of the last third of the novel, as he confirmed his suspicions about Lady Dedlock, came from knowledge misused, ostensibly to protect a great family, but actually to feed a hatred which the narrator insinuated by unanswered questions, 'Whether he be . . . absorbed in love of power, . . . whether he in his heart despises the splendour of which he is a distant beam . . .'. The knowledge he sought had revolutionary overtones from the analogy between Lady Dedlock and the Lady in the story of the Ghost's Walk, caught between the opposing sides in a previous English revolution as Lady Dedlock was between two classes. When, in chapter 48, Tulkinghorn insisted she should remain with the class into which she had married, the omniscient narrator's triumph emerged and his melodrama was transcended in a conspectus of London and the country beyond, under the moon and stars which had earlier, in chapter 41, carried associations with an eternal omniscience transcending human life. The motives of the main actors in chapter 48, Lady Dedlock and Tulkinghorn, were, for the most part, left to be inferred; the reader's pleasure was in the panorama and the invigorating sense of seeing further than they could in it.

The personal narrator, Esther Summerson, who seemed most ignorant of all, for she did not even know who she was, appeared as the only character who could learn from experience, because she could forget herself and her own position in the contemplation of other selves and their relations to each other. Dickens had to endow her to a quite implausible degree with powers of perception and reflection which could only be his own. In the most famous instances indeed, like David Copperfield, she was imagining rather than perceiving—when Turveydrop was so 'pinched in and swelled out' by his clothing that, 'As he bowed to me in that tight state, I could almost believe I saw creases come into the whites of his eyes', or when Vholes the lawyer took his 'long thin shadow away. I thought of it on the outside of the coach, passing over all the sunny landscape between us and London, chilling the seed in the ground as it glided along.' In a book opposing knowledge to ignorance, here was where her powers of knowledge were strongest. Where they were weakest was in self-knowledge, a part of the whole matter which Dickens had had before him in *A Christmas Carol* (1843) and *The Haunted Man* (1848) and of which his treatment resembles Richardson's in *Pamela*. As Pamela's protestations of virtue had delightfully revealed how far she fell short of virtue, so Esther's claims to humility were undercut by their persistence. The limitations of her self-knowledge hardly made her a character to fascinate as Richardson's sanctimonious minx still does, and she was inferior in interest, too, to Thackeray's major women, Becky Sharp, or Beatrix—though not, perhaps, to Amelia or Laura. Self-knowledge was specifically linked to Esther's knowledge of her own position by the form of the disclaimers with which she began and ended her narrative. Her position was no doubt too much that of the domestic angel, but her ability to be utterly content with her immediate duties was justified by the clarity with which she saw how little other characters could do so, in a large gallery of satirical portraits from Mrs Jellyby with her eyes on a distant place, Africa, to Mr Turveydrop with his eyes on a distant time, the Regency, and a dead Prince whom he 'adored . . . on account of his Deportment' (the whole portrait making crisp and vivid Thackeray's satire of the 'moral man of fashion' in Yellowplush in the thirties). Esther's position gained overtones from the double narrative. The mutual recognition between mother and daughter in chapter 36 might be stagey in itself, but Esther's walk afterwards along the terrace of Chesney Wold, imagining her own footsteps to be those of the Ghost's Walk, generated more than a pleasurable Gothic shudder. By specifically relating Esther to the omniscient narrative in which the legend and the Lady had first been met, her terrace walk gave the reader a sudden intimate experience of what that narrative had stressed, the sense of a society divided into classes which tried to keep separate from each other but could not avoid being connected.

The mother's attempt to escape set the mystery and the moral pulling in opposite directions, since her thoughts and feelings had been kept too much concealed, in the interests of the mystery, for her flight to be fully comprehensible. She had had too much of the inflexibility of Edith Dombey, without Edith's suggestions of wild energy breaking out. But the sheer activity of her outcast state (in contrast with her weariness in 'society') and her aim in her wandering, to have sight of her daughter and then of the father's grave, even her note to the daughter, ending 'Forgive', were moving nevertheless, backed up as they were by the whole book. The pursuit might seem factitious in itself but the reader could not fail to see the significance of the pursuers—the detective, the character from the omniscient narrative now most eagerly in pursuit of knowledge, and the contrasting personal narrator who was to reach the culminating knowledge of how the extremes of society in fact met in herself.

Bleak House was among the earliest full-length novels to have a detective interest. Dickens had first-hand acquaintance, when preparing for articles in *Household Words*, with the plain-clothes Detective Force (organized in 1844 by the Home Secretary in Peel's cabinet) and drew upon this acquaintance for his astutely courteous Inspector Bucket. The murder mystery, however, was more important in its suspects than its criminal. It became, by the inclusion of Lady Dedlock among the suspects, part of the larger mystery of the Chancery suit—which only collapsed into laughter, telling the reader nothing of its possible connections with either Esther or her mother. The hints about such connections had in fact quite another function: to draw attention to the inadequacy, for those faced with the problems of the present, of all 'solutions' arising merely from knowledge of the past. Evidence from the past was accumulated in bags by the Chancery lawyers (and their hanger-on, the rag and bottle merchant) and stored in his own person by Sir Leicester Dedlock, a 'baronet with a head seven hundred years thick', 'bound to believe a pair of ears . . . handed down to him through such a family', but contemporary life crowded in resistlessly. Sir Leicester was able to face it because of his genuine attachment to Lady Dedlock. He suffered a stroke while being told of her past, but his love, at variance with his family's past, enabled him to instruct the detective: 'Full forgiveness. Find—'.

On the other hand, Miss Flite and Richard Carstone, unable to look beyond the outcome of the Chancery lawyers' slow investigation of those sacks of evidence, were marked, respectively, by madness and a progressive fatal apathy. Miss Flite's madness was moving because it seemed to release her from time altogether: she expected a judgment 'On the Day of Judgment'; she kept in cages birds whose names stored up, like Platonic forms, all the things deformed by Chancery, from Hope, Joy, Youth, and Peace to Words and Wigs, as well as the things Chancery fostered, like Folly and Despair. Richard Carstone was taken beyond time in his death, a death sentimentally tinted, no doubt, but well attuned, all the same, to his surroundings in the novel. For, after a long sequence of refusals to look decisively to his own future, he was made to 'begin the world' only in his fulfilment of the desire to 'go from this place, to that pleasant country where the old times are'—a country as much beyond his physical grasp as Miss Flite's Youth and Peace.

A pastoral land of lost content like this was repeatedly invoked by the language in both the personal and the impersonal narrative; it gave remarkable overtones to the satire of a whole society chained by its past, and inattentive to signs of the future. The narrator's celebration of 'a very quiet night', in chapter 48, when Tulkinghorn was moving to exert his power over Lady Dedlock, contrasted the freshness of the water meadows with Lincoln's Inn Fields 'where the shepherds play on Chancery pipes that have no stop, and keep their sheep in the fold by hook and by crook until they have shorn them exceeding close'. These had already appeared in chapter 42 as 'these pleasant fields, where the sheep are all made into parchment, the goats into wigs, and the pasture into chaff', and the reader could connect them with the ironical account in chapter 21 of the place inhabited by the Smallweeds, part of the neighbourhood called Mount Pleasant, 'in a little narrow street . . . where there yet lingers the stump of an old forest tree'. In each case the description suggested a range of life greater than the people in question could reach, a matter made explicit in the case of Tulkinghorn, 'so long used to make his cramped nest in holes and corners of human nature that he has forgotten its broader and better range'. Other satirized characters were reduced to sub-human forms of life too, the Smallweeds to 'a horney-skinned, two-legged, money-getting species of spider' and their children to 'old monkeys with something depressing on their minds'. Such obvious and extravagant types, which troubled sober contemporaries like Trollope, appeared to be denied any subtlety of inner life, though the unanswered questions about Tulkinghorn's motives suggested its possibility. Their essential being, however, emerged strongly from the coherence of metaphor and image in relation to the conventions of pastoral, and these steadily suggested a moral point of vantage for the criticism of a society that had lost its way. It was a mistake to bring this point of vantage literally into the first-person narrative, as a pastoral retreat, the Yorkshire Bleak House. Fortunately nothing could obliterate the city and the ignorant and damaged lives it housed, in both comfort and squalor, in this culminating work to which Dickens's imagination had been moving ever since *Oliver Twist* was conceived.

Hard Times (1854), written (as a weekly serial) on a scale less than half that of *Bleak House*, included as great a complication of contemporary issues: education, divorce, the relations of capital and labour, the monotony of life in the new industrial towns. But the single important character or predicament, needed to bring all these preoccupations into focus, was lacking. Gradgrind was an abstraction, Bounderby a puppet, and Louisa, daughter of the first and wife of the second, while she came nearest to the central role the book required, was rendered passive by its very plot. Her Gradgrind factual education, leaving her without spirit, left her without much vigour of presence for the reader. Her speech was too much that of the author and too much at the service of his own polemic: 'What are my heart's experiences? . . . What escape have I had from problems that could be demonstrated, and realities that could be grasped?' and later, 'Father if you had known . . . that there lingered in my breast, sensibilities, affections, weaknesses capable of being cherished into strength, defying all the calculations ever made by man, and no more known to his arithmetic than his Creator is ...'. These sensibilities were indeed enlivened in their metaphorical correlation with the domestic fires into which Louisa gazed while exercising the forbidden faculty of imagination, and with the Coketown furnaces which appeared dormant by day: 'There seems to be nothing there but languid and monotonous smoke. Yet when the night comes, fire bursts out, father!' But her predicament was dramatized with success only in two extraordinary sequences, the first when the girl's attachment to her brother, the only outlet for her inner fires, brought her to him (urging him to confess a felony) 'barefoot, unclothed, undistinguishable in darkness', and the second, captivating in its narrative impetus, when she escaped from her husband and his over-zealous spy, through a thunderstorm and by means of a railway, imagined, as in *Dombey*, as an explosion of energy, 'a fit of trembling, gradually deepening to a complaint of the heart'.

For most of the time, the thoroughly Wordsworthian championing of heart and imagination against the dominion of 'fact, fact, fact' was illustrated, rather than embodied. The case was made in mild tones when compared with the sharp percipience of Esther and the combative vigour of the omniscient voice in *Bleak House*, or with the terms in which readers could now find the case made in *The Prelude*. Gradgrind's phrase, 'destructive nonsense', for *The Arabian Nights* was mild beside Wordsworth's for the 'Forgers of daring tales':

> Imposters, drivellers, dotards, as the ape
> Philosophy will call you ...

Even Mr Sleary and his company, there to minister to the Coketown 'craving', in the manuscript, 'for some physical and mental entertainment', were presented in a somewhat doctrinaire fashion as examples, with their 'remarkable gentleness and childishness', their 'special inaptitude for any kind of sharp practice', their 'untiring readiness to help and pity one another'. Compared with the entertainers of earlier novels they were lay figures, and Sissy Jipe with them, adequate perhaps for allegory but falling below Dickens's own standards as fiction.

The part of it which concerned the employer Bounderby, or the trade-union orator opposing him, was sparkish enough in the superficial manner of the American chapters of *Martin Chuzzlewit*. There was a contrasting lugubrious realism in the presentation of Stephen Blackpool, the prematurely ageing employee, the 'man of perfect integrity' rejected by both employer and union. But the book was too short for the contrast of Blackpool's marital calamity with that of his employer, Bounderby, to be properly stressed; this left the employee's predicament a mere sketch, and obscured the motive he had for promising not to join a union. Space was, however, only part of the problem. Dickens's imagination, as distinct from his wish to do good as he saw it, had only been fitfully stirred. Taking a 'leap From hiding places ten years deep', his imagination eventually reaped the fruit of this particular encounter with industrial England only in *Our Mutual Friend*.

As if he was still learning from this experiment with an immediately topical action in a setting he had recently visited, his first plans for the next novel, *Little Dorrit* (1855–7), took it in that same direction. Its first title, *Nobody's Fault*, summarized much contemporary journalism about the mismanagement of the Crimean war; once a Select Committee of the House of Commons had reported, in mid-1855, that nobody in particular was to blame, the title had obvious marketing power. Yet it was rejected. Although Dickens showed his own opinions by joining the newly formed Administrative Reform Association and by expressing disgust, in letters, with 'our political aristocracy' and 'aristocratic red tape', and although the comic vehicle for this disgust, the Circumlocution Office, or at least its 'Waiting room', appeared among the first of the notes for Number I of the new novel, he chose less immediately topical ideas for its opening scenes. By the time of the notes for the second number, the notion of two 'Parallel imprisonments', the one actual, the other self-imposed, had hold of him and so the way was open for a development of the interest in inner life which the writing of *Copperfield* had stimulated. The culminating plan for the Dorrits, 'of overwhelming that family with wealth', gave Dickens the chance to link the Marshalsea prison of his own childhood to scenes which had recently impressed him in Switzerland and Italy, as well as to widen the novel's social range.

In earlier days this would have given him all the plot he needed; his second manner demanded that there should be knowledge withheld, in an elaborate mystery, as well. Although, as in *Bleak House*, it revealed the actions of the past only that they might be forgiven, and although it sporadically quickened the lagging pulse of the novel with a sense of strange connections among the cast, its solution was peripheral, important chiefly for remaining at the end unknown to the central male character, Clennam.

Once the female lead was taken by a young woman with 'Little' as part of her name and, like Nell, with an elderly dependant (though with a smaller disparity of age), idealization became a danger Dickens found hard to avoid. At first the constraints upon Little Dorrit's freedom were fully realized in the fiction, particularly in her family and in the shabby staleness of the Marshalsea prison to which its male members accommodated themselves only too completely. Her own nostalgia for light and air, too, was rendered by reference to substantial things, showing her imagination at work upon her surroundings. She not only longed for the open fields but walked on the Iron Bridge, 'to see the river, and so much sky . . . and such change and motion', in contrast to 'that cramped place'. All the same, her moral status was sentimentally exaggerated and, what was more, it failed to survive in the face of a new direct realism in the presentation of inner life, exemplified in her father. Dorrit, in chapter 19, humiliating himself by suggesting that, on his account, she should 'lead on' the turnkey's son, was less grotesque and more embarrassingly realistic than similar self-imprisoned figures in the novels before *Copperfield*. His punishing of himself by refusing his dinner, and his lavish self-pity, were so intimately observed as to make the praise of Little Dorrit ('the devoted child . . . whose love alone had saved him to be even what he was') seem ironic. 'He only requires to be understood' expressed her love, no doubt; but the understanding of him shown in the actual presentation was so much greater than hers that it threw into relief how his condition of mind had been aggravated by the comfort she gave. Even the humour underlined this, for instance when he 'became grave on the subject of his cravat, and promised her that when she could afford it, she should buy him a new one'.

In the characterization of Arthur Clennam which formed the pivot of the plot there was at first no such clash between the old methods and the new. Even when the old sledge-hammer technique marked the grim Sunday of his return to London, the accent fell on his flaccid resignation and on an outward quietness of speech and gesture which appeared as a triumph over inward promptings to resentment and self-pity. A childhood glimpsed from the inside, in the manner of *Copperfield*, suggested how a man could be left without sufficient self to constitute a prison, despite the neurotic tendency of the vestige that remained; the plot suggested that dubious results might follow from such a deficiency, for his main attempts at altruism were calamitous. One led to the discovery of the Dorrits' wealth (which made a decisive difference only to the daughters and that, as Dickens presented it, for the worse), and a second led him to ruin the firm whose money he managed, by investing it in Merdle enterprises which crashed. Insisting on imprisonment in person for the debt, Clennam suffered the stifling oppression of the Marshalsea, in chapter 29, at first with all the old exaggerations; but when he fell into the 'desolate calm' of 'a low, slow fever', a more specific language enabled his state to be felt from the inside as the nadir of a life without personal will, a contrast even to the half-life of the prison.

> A blurred circle of yellow haze had risen up in the sky in lieu of sun and he had watched the patch it put upon his wall, like a bit of the prison's raggedness. He had heard the gates open; and the badly shod feet that waited outside shuffle in; and the sweeping, and pumping, and moving about, begin, which commenced the prison morning.

To learn in such a place that Little Dorrit loved him was to learn the dangers of the state he had been in for a long time. His own heart's 'curious enquiry' was a good example of the kind of introspection that had now become subtler: behind Clennam's concealment of the truth of his feelings by self-effacement lay the further possibility that even such restraint might be used by the self for its own comfort, as it 'whispered . . . that the time had gone by him, and he was too saddened and old'. Little Dorrit, brought back to the prison to visit him, brought with her the tendency to idealization. Images of light which had been capable of subtle modulation to give the feel of inner experience, like that blurred circle of yellow haze, now became subject to the grosser purposes of allegory.

> There was one bright star shining in the sky. She looked up at it as she spoke, as if it were the fervent purpose of her own heart shining above her.

Moreover, Clennam was drawn into competition with her, refusing her money as a 'noble offer' which he might have accepted 'if, . . . through my reserve and self-mistrust, I had discerned a light that I see brightly now that it has passed far away, and my weak footsteps can never overtake it . . .'. With the young woman actually standing before him, it was all too much.

Such crudity stood out all the more when the activity of the senses counted for so much in this novel. The reader's senses had been stirred so as to increase both the unity of the whole and the diversity of personal vision. The sense of smell formed a strong ingredient of this unity, in a vein of rather selfconscious humour, from the Barnacle house 'like a sort of bottle filled with a strong distillation of mews' out of which the footman, opening the door, 'seemed to take the stopper', to 'the prevailing Venetian odor of bilge water and an ebb tide on a weedy shore'. Such things, though developing out of Dickens's interest in sanitary reform, went far beyond it to the idea of disease which was recurrent in the metaphors and the action—in the financier Merdle's 'complaint' which was 'simply, Forgery and Robbery', or in the actual illnesses of the Clennams in 'close, stale' interiors.

The senses gave, too, sudden vivid glimpses of people's lives from the inside, whether by means of metaphor—a pauper's days out of the workhouse became 'those flecks of light in his flat vista of pollard old men'—or direct perception, as when Little Dorrit would look 'up at the sky through the barred window, until bars of light would arise, when she turned her eyes away, between her and her friend'. No doubt the author's design upon the reader was plain, to draw attention to freedom as a state existing in the mind, but the instances were so odd in themselves as to amount to an exploration of the matter. The only person, for instance, to see Merdle in anything like his true light just before his suicide was Little Dorrit's sister Fanny.

> Fanny passed into the balcony for a breath of air. Waters of vexation filled her eyes; and they had the effect of making the famous Mr Merdle, in going down the street, appear to leap, and waltz, and gyrate, as if he were possessed by several Devils.

Her own bondage had come about, in some of the liveliest sequences of the book, through seeking only the freedom to revenge herself upon a patronizing mother-in-law; her odd perception here suggested her awareness of the empty man about to be repossessed by demons, in the parable; like him she had certainly wandered 'through dry places, seeking rest' and finding none. The strange episode, echoing at the end of a chapter late in the book, brought out the analogy between the mind's creation of a state of bondage and the mind's contribution to what was physically perceived.

Detail of this kind showed how a subtler concern with experience seen from the inside could refine the novel's significance. Even the old emphatic methods had a stronger inward reference. The two opening set pieces, contrasting the glare of Marseilles, 'burning in the sun', with the gloom of its prison, were made of iterations which critics then and later found to be damnable, but the art of the opening lay in the transition from the one passage to the other, where the common element between them consisted of the people 'lounging and lying wherever shade was'. The reader, strongly aware of the discomforts of first the light and then the shade, was led to direct experience of a question, something like 'Which of the two *is* hostile to the lounging men?' Even if the question was not articulated, it was none the less present within the experience of reading the two passages.

It may be surmised, indeed, that it was this question, arising as he read the opening over, that led Dickens to the idea of overwhelming the Dorrits with wealth. Of course, if the main action was to move into the prison and poverty, the opening contained the possibility that it might move out again into riches and foreign travel. From the external evidence, however, it is quite clear that the possibility became distinct only in the course of writing the third number. The example suggests that Dickens's apparently naïve showmanship in passages like the opening of *Little Dorrit* has caused their exploratory nature to be overlooked. As much as performing for the admiration of readers, he was trying out, on his own sensibilities as well as theirs, the words which would stir sensations, in order to see how these sensations might clarify and give form to the nascent idea of the book. The driving force of such work was the relation he hoped to establish with his readers in a strong shared experience, but he was conscious of his audience less in order to separate himself from them as the showman than in order to enter a close kinship with them in the show. The evidence in this respect of the prefaces, and indeed of the very adoption of the serial mode, is supported by the public readings and other activities like those of the amateur actor-manager. Working in this latter way on *The Frozen Deep*, before *Little Dorrit* was finished, he had been enabled, he said in a letter, 'as it were, *to write a book in company*, instead of in my own solitary room'. His aims in his professional work are clearly illuminated when it becomes, like this, an analogy for the most intense of his amateur activities; in the same way even his labouring of his effects in his 'solitary room', in *Little Dorrit*, resulted from a determination that they, his readers, should be at one with him.

Shared sense experience was the substratum. The opening scenes of these later novels showed it being laid down. What followed, in *Little Dorrit*, had less of mockery than in *Bleak House*, and more of sombre denunciation. Nevertheless the mocking voice came forward on occasion with comic force—in a phrase like 'these Arcadian objects' for London's poisoned river and 'miles of close wells and pits of houses'. It focused a significance already present and opened the way for the elaborate humour which worked Arcadian suggestions into the contrast between make-believe and what Little Dorrit called 'the plain truth'. Again pastoral carried the satire, but this time pastoral itself was made laughable. In this vein the financier Merdle's wife could claim a primitive tenderness of heart by means of a fake nostalgia for Savage life in Tropical Seas—'most delightful life and perfect climate I am told'. Clennam's superannuated sweetheart was called Flora, as if for the goddess appropriate to this pastoral ideal, however faded and suburban its form (her dwelling looked on to a 'wilderness patched with unfruitful gardens and pimpled with eruptive summer-houses'). Her language set her beside the great soliloquizers of the early novels, even though, in harmony with the tendency of the book, she was brought delightfully out from her romantic pretence to acknowledge the real world; that world had indeed been implicit all along, since her humour depended not only on the mixture of romantic cliché with fact but also upon the way her imagination worked with sensuous exactitude upon everyday analogies: they made Little Dorrit's name resemble 'something from a seed-shop to be put in a garden or a flower-pot and come up speckled'.

At the same time the tendentious championing of 'the plain truth' was itself exposed to mockery: Flora was accompanied repeatedly by Mr F's Aunt, firing off unconnected facts; and Clennam in the Marshalsea, facing that 'curious enquiry of his own heart's', was visited by the Plornishes, shopkeepers who had decorated their parlour to look like 'the Golden Age revived', and whose half-hold upon reality was beautifully dramatized in Mr Plornish's conversation:

It was in wain to ask why ups why downs; there they was, you know. He had heerd it given for a truth that accordin' as the world went round, which round it did rewolve undoubted, even the best of gentlemen must take his turn of standing with his ed upside down and all his air a flying the wrong way into what you might call Space. Wery well then, what Mr Plornish said was very well then. That gentleman's ed would come up'ards when his turn come . . .

Picturesque grammar drew attention to the few statements which were true—'which round it did rewolve undoubted'; the repeated 'Wery well then' caught the very accents of vacuous 'philosophical' consolation; even though the humour connoted charity, it threw its light on Clennam's attempt to be resigned in the face of his new awareness of the facts. It all suggested that to live in the real world was to make a complicated inner adjustment. Clennam might seem inadequate to embody this idea. But Dickens, though still over-explicit at times, was now, after *Dombey and Son*, relying also upon implication; in Clennam's case the process of adjustment was suggestively left incomplete, through his ignorance of his true parentage until some point after Mrs Clennam's death itself, outside the main action.

Such indirect methods were certainly more successful in dealing with the difficulties of inner life than the direct attack upon them in 'The History of a Self-Tormentor', which only showed how transparently mistaken Miss Wade, the Self-Tormentor, was in believing she could not be loved. Moreover, the absence from her story of the sensuous particulars in which also he was subtler than in direct statement further deadened his treatment. Only for an aberrant sexuality, comparable with Louisa's in *Hard Times*, was there a strong correlative object; with a friend, Miss Wade felt often 'as if, rather than suffer so, I could so hold her in my arms and plunge to the bottom of a river—where I would still hold her, after we were both dead'. This in turn reflected strangely upon the power she gained, in the main action, over another girl, Tattycoram, herself a vivid sketch of resentment which the reader might recognize as justified. Another abnormal state, that of religious hypochondria, ambitiously attempted in Mrs Clennam, cumbered much of the book with an automaton. She was enlivened (apart from Flora's exact remarks—'person . . . with no limbs and wheels instead') only at the beginning by her delicate appetite and at the end by her ability, under profound stress, to recover the power, all the greater for a puppet, of motion. The first showed clearly the subterfuges of the self in directing the machinery of self-denial; the conclusion uncovered the 'half-hope' that Clennam, although not her own son, might come to love her, an unexpected insight.

Despite its unevenness, the novel was one to grow up round the reader as a distinct and self-consistent world, its strong, sensuous particularity leading him to experience directly the questions of freedom and bondage which it raised. What was dramatic in it tended to be factitious, generated as it was by the mystery part of the story, but the essentials of the novel belonged to an inner dimension which, although it was affected, of course, by the events of the plot, opened long vistas to readers who valued the humorous, pathetic, and, at its best, realistic illumination of the secret self.

Dickens attempted an opposite kind of novel in *A Tale of Two Cities* (1859), 'a story of incident', he called it, 'a picturesque story, . . . with characters . . . whom the story should express more than they should express themselves by dialogue'. This was to deny himself a great deal and to forget how much his most successful characters had been 'expressed' by their physical setting as well as their speech. It is clear that he attempted to compensate as well as to abbreviate by treating the story in terms of an opposition of death to rebirth which was made 'picturesque' as an opposition of darkness to light. He deepened the shadows by repetition, so that the mere relief from them would constitute an experience to suggest rebirth. He could be seen again and again overworking the words 'dark' and 'shadow' as a shorthand for what, with more space, he would have attempted to realize more fully: 'the dark deference of fear and slavery' paid to a pre-revolutionary Monseigneur, 'the dark part of my life' of Dr Manette's long imprisonment in the Bastille, 'the fatal darkness' of 'the cloud of caring for nothing' in Sydney Carton's unregenerate state, the 'threatening and dark' shadow of Mme Defarge and her vengeance, 'the air around . . . thick and dark' at the time of the Terror, and the 'black malice' of the law of the Suspect. Where the darkness and death were associated with imprisonment, the result, when compared with *Little Dorrit*, was allusive, almost allegorical. Darnay, committed to La Force ('a gloomy prison, dark and filthy, and with a horrible smell of foul sleep in it'), stood in the presence of 'Ghosts . . . all waiting their dismissal from the desolate shore, all turning on him eyes that were changed by the death they had died in coming there'. Such passages had power, but it seemed to come from English Virgil read by candlelight, when compared with the specific presence of the Marshalsea prison.

The prisoner, Manette, 'buried alive' in the dark and released to touch his daughter's golden hair and the 'golden thread' of the generations following him, was equally allegorical. It was true that his reversion, in moments of stress, to the shoemaking which had soothed his imprisonment was a moving development of Mr Dorrit's return in mind to the Marshalsea when he was physically free of it; but now, to carry out the idea, Dickens had only an aged actor of limited range. 'Recalled to life' was startling as a summary of his part in the plot, and suggestive as a message delivered to a traveller at night in the dead of winter; but the dramatization of this new life was trite, in Dr Manette's 'resolute face and calm strong look' as he worked for the release of his son-in-law from La Force. At the end Sydney Carton, dying to save his successful rival in love, had a similarly allegorical 'strong face', 'sublime and prophetic'.

Dickens worked hard to enliven the novel's leading idea, rebirth, by reference to successive generations or to the seasons. He had in fact foreshadowed this in the first jottings of titles, before *Little Dorrit* was begun: 'TIME! THE LEAVES OF THE FOREST. . . . TWO GENERATIONS'. The justly admired scenes of rustling leaves and late summer lightning in a quiet corner of London, close to the 'forest trees' then growing 'north of the Oxford-road', carried both the threat and the promise of the seasonal imagery into the subject of public events disturbing private lives. It was a pity that the private lives of Darnay and his family were so slightly realized that the promise of rebirth in successive generations became little more than a gesture. In the plot and in the imagery of seed and fruit, the excesses of the French Revolution were presented as a 'natural birth' from the evil of a 'previous time', but Carton's final 'prophetic' thoughts were mere platform oratory, good enough for a politician but not for a novelist.

That Dickens wished the novel to be read, like *Little Dorrit*, with contemporary politics in mind was suggested by its dedication to Lord John Russell. When rising prices after the Crimean War had set back living-standards and caused renewed agitation for parliamentary reform, Lord John had been expected to join with Bright and lead the democracy against the aristocracy, discredited by its management of the war. (The anti-aristocratic tendency of *Little Dorrit* was still there in the *Tale*, typifying the *ancien régime* by 'military officers destitute of military knowledge; ... civil officers without a notion of affairs; ... all nearly or remotely of the order of Monseigneur, and therefore foisted on all public employments . . .') Just before the first instalment appeared, Russell brought the Conservatives down with a motion in the House 'that no re-adjustment of the Franchise will satisfy this House or the Country which does not provide for a greater extension of the Suffrage in Cities and Boroughs'. By the time the completed novel was dedicated to Russell, at the end of 1859, Palmerston (satirized in *Little Dorrit* as Lord Decimus), the arch-opponent of an extended franchise, was Prime Minister, with Russell at the Foreign Office. The dedication hinted that Russell's reforming aims should not be forgotten. The novel's emphasis on death and rebirth, seed and fruit, and (following Carlyle, to whom Dickens made acknowledgement in the preface) fire and flood, in the presentation of a 'terrible time' past, was a warning of forces which Dickens believed it foolish to disregard. Although he feared them, he seemed to be asking, as Russell in his *Memoirs of the Affairs of Europe* had asked, of revolutions, 'On whom lies the blame of their excesses? ... On those who have long and patiently borne what no man ought to inflict; or on those who have inflicted what no man ought to suffer?' But if such views were to be transmuted into art, all of Dickens's habitual methods were needed, in particular a specific Paris instead of a collection of stage properties, so that details from the physical setting might elaborate and elucidate the inner life of his people. Famous words about resurrection from the service for the Burial of the Dead were no substitute, not even when grotesquely parodied in the figure of a resurrection-man, robber of graves. Dickens had made a good skeleton for the novel but left off the flesh.

Great Expectations (1860–1), on the other hand, had substance to match its structure. There was not only a strong central idea, but Kent and the London of Dickens's childhood to embody it with sensuous particularity and to enliven the characters when physical objects became the index of inner states. Moreover, instead of developing, as *A Tale of Two Cities* had done, the topical and political part of *Little Dorrit*, *Great Expectations* developed the part concerned with perennial questions of freedom and its opposite. It began with Pip bound and oppressed in the country and took him to apparent freedom in London: the two first sections might indeed have been called, like the two divisions of *Little Dorrit*, 'Poverty' and 'Riches'. Such titles, however, would now have appeared crudely ironic, so extreme was the contrast between Pip's human riches in the country, so long as he 'believed in the forge as the glowing road to manhood and independence', and the relative emptiness of his idle life in London. London had been, in *Dombey and Son*, the 'monster, roaring in the distance', towards which, in hopes of bettering themselves, people 'seemed impelled by a desperate fascination'. Pip accepted a similar fate without question, believing it to be the way to freedom, though the end of the second division of the novel made plain his different view when he looked back, as his own narrator and judge. There the image of the iron ring and stone slab slowly and inexorably prepared for the destruction of the flushed conqueror in 'the Eastern Story' (which as a child Dickens had turned into a play) was an image for the facts, inexorably accumulating and waiting to be discovered—that the money for the fulfilment of his snobbish dreams came from a convict, himself a snob. From the bondage of that hopeless treadmill it was the business of the final part of the novel to make him free.

Behind the elaborate mysteries of Little Dorrit was a maimed recluse mockingly called 'Fate in a go-cart'. Fate, to Pip, had at first appeared to assume a similar form, in Miss Havisham, a woman in white, aroused from her slumbering origins in Dickens's own memories by the title of Wilkie Collins's major success, which had followed *A Tale of Two Cities* in the same periodical, *All the Year Round*. Miss Havisham might be thought, with some reason, to be the source of Pip's monetary expectations, but in fact she made him meet his fate only in the sense of the cliché, by bringing him acquainted with Estella. The similar name Stella, Dickens could have found, in capitals, in the sonnets of Sidney discussed in the *Essays of Elia* (which he was given to quoting in letters), and moreover in suggestive collocation with

> . . . ambition's rage,
> Scourge of itself, still climbing slippery place,

and holding the star-lover's 'young brain captiv'd in golden cage'. In Dickens it was the lady who stirred ambition, ambition to be a 'gentleman'. Through her, in spite of her deficiencies as a fictional character, the subject of snobbery became a matter of feelings which were, as Elia said of Sidney's poems, 'full, material, and circumstantiated'.

Despite her lady's 'air of completeness and superiority', Estella's family connections were discovered to be with criminals, one of them Pip's benefactor. Her story suggested that it was no simple matter to decide what sort of guardian might best serve a child's interests. A criminal lawyer like Jaggers was predisposed by his profession to simplify the problem, for his profession was part of the process by which society, for its very preservation, must attempt to separate the guilty from the innocent. The court scene near the end of the novel, however, made explicit the difference between such judgements and 'the greater Judgement that . . . cannot err', and so reinforced the implication of the narrative that actual guilt is less easily assigned than guilt in the eyes of society. Moreover, society, which awarded blame so decisively itself, bore some responsibility for the creation of criminals, as Magwitch's own story in chapter 42 made clear. It was by understanding this, through what he was told and through the guilt he had himself incurred in allowing himself to be drawn into London (the centre of what Thackeray had called, in *The Book of Snobs*, 'Man-and-Mammon-worship, instituted by command of law'), that Pip gained his freedom. The gradual revelation of what had preceded his 'expectations'—amorous, financial, and social—was skilful compared with the last-minute explanations in *Little Dorrit*. No doubt Dickens's reading of *The Woman in White* 'with great care and attention' (as he had written on 7 January 1859) had something to do with this. But his object was not narrative suspense but rather the revelation of the causes of the situation in which Pip's 'blind' reaction had been almost unavoidable.

The persistence of the prison taint in the story, even when the plot did not demand it, emphasized the preoccupation with bondage and freedom, both actual and apparent. Reinforcing this, the sense of place was particularly strong to underline Pip's vulnerability in the adult world which bound him. The marvellously economical opening scene isolated him as a child in a churchyard near the river and the sea, on Christmas Eve. With characteristic exaggeration of contrast, he was identified first as a 'bundle of shivers' in 'that flat dark wilderness', under threat from 'the distant savage lair from which the wind was rushing', and then as a boy of implausible precision of speech when actually threatened by a prisoner on the run. Impressive in itself, the scene also held the essentials of the whole novel—the country setting, the boy a little too sharp for his humble place in it, the outcast claiming by terror a share in the feast which Christmas, taken seriously, would have allowed him.

The irony used by the later Pip, in the early stages of his story, to convey and at the same time criticize the intensity of the boy's feelings, and his disregard of those of his protectors, was a development from 'The History of a Self-Tormentor' and from *Copperfield*. The latter Dickens actually reread, 'to be quite sure I had fallen into no unconscious repetitions'. He risked echoing it when Pip, near the end, planned to go back and offer an 'errant heart' to Biddy, the country girl from whom it had 'strayed'. But now the critique of this heart was carried on into his second choice. There was none of the high-minded self-congratulation with which David Copperfield had returned to Agnes: Pip's plan showed him up relentlessly—he would ask to be accepted in marriage 'with all my . . . disappointments on my head', when the chief of these was the failure to gain Estella, his obsession with whom he had already confided to Biddy in a memorable episode of ironic humour in his first 'stage'. Having now understood his fate, he was trying to undo it: not only to return to the country life which had earlier failed to satisfy him, but to return to childhood and a mother he had never known. He looked to Biddy to receive him 'like a forgiven child' in 'need of a hushing voice and a soothing hand', in contrast to his actual harsh upbringing, 'by hand', by his sister. The plot made it plain that all this was mere nostalgia: Biddy was in fact about to marry the blacksmith whose service Pip had been keen to leave (albeit for reasons the reader could understand).

In striking contrast to *Copperfield*, too, the conclusion to *Great Expectations* as first published left Pip, far from having conquered the first obsession of his 'errant heart', hand in hand with Estella, seeing 'in all the broad expanse of tranquil light . . . the shadow of no parting from her'. In the conclusion as originally written the pair met in sympathy but decisively separated, Pip 'sure and certain' he did not 'fret for her' and Estella supposing him to be married. This had left him in his final, sad freedom. For the last edition in his lifetime Dickens edged back towards his first ending and wrote, 'I saw no shadow of another parting', as if there might be another to come, but, like Thackeray at the end of *The Newcomes* or Charlotte Brontë in *Villette*, he was leaving the matter to the reader. Did he hope the reader might recall chapter 29, in which Estella was inseparable from 'all those ill-regulated aspirations that first made me ashamed of home and Joe', inseparable from what had been, in the second 'stage' of his 'expectations', 'the innermost life of my life'? From this fate Pip, in the original version, had now fought free, acknowledging that Joe the blacksmith, not himself, 'bred to no calling, and . . . fit for nothing', was the gentleman. Now, following Joe, Pip was to cultivate the self-forgetfulness of which he had become capable in his concern for the convict, to face things as they were, and to 'work' for his 'profits'. Things as they were included Estella's connection not only with the convict but also with the accumulated snobberies of the story.

The prize at the end was not the girl, ghostly and inscrutable, but the state of mind in which he refused to fret for her, forgiving her guardian and infusing their story with a humour which repeatedly went against himself. Such a state of mind was presented as a kind of rebirth after the (otherwise quite unnecessary) episode of Pip's narrow escape from being murdered by a man who claimed to be confronting him with his own image, accusing him of things of which Pip had earlier, in his victimized state, falsely accused himself. In his rebirth into freedom lay the point of the conclusion, the achievement, that is, of the state of being which had enabled the whole to be written.

Our Mutual Friend (1864–5), the last of the complete major novels, was an even more grotesque rehandling of matters that had haunted him since *Little Dorrit* and *A Tale of Two Cities*—determinism and freedom, death and rebirth. But now the time of the action was closer to the time of writing, while its localization was, to an even greater extent than in *Great Expectations*, both geographically precise and metaphorically suggestive. Especially was this so in the opening, where a corpse was recovered by Hexam and his daughter on a carefully identified reach of the Thames, 'between Southwark Bridge which is of iron and London Bridge which is of stone'. From this scene grew a comic action centred on a freakish evasion of fate and a tragic action centred on a wilful submission to it. From the exact localizing of the first scene Dickens developed a heavily sombre social background; from the climax of that scene in the finding of a corpse identifiable only from its clothing, he developed a series of strange, fictional images of death and the transcendence of death. It was not the least remarkable of his mature constructions and certainly the most elaborate in its symmetries.

The comic action arose from the mistaking of the drowned man for John Harmon. In consequence, he was free to decide to be 'reborn' as Rokesmith and so to test, on his own terms, the people with whom his relationships as Harmon had been predetermined by his father's will. The plentiful contrivances of this part of the action, in which further wills were discovered but the legatee under the latest of them, Boffin, refused to inherit at Harmon's expense, all formed a semi-farcical extravaganza of human freedom defying legal (which was also, as in *Bleak House*, social) predetermination.

A heavy sense of submission to fate hung over the contrasting action concerned with rivalry for the love of a working girl, Lizzie Hexam, between two men—one, Wrayburn, from a class above hers, the other, Headstone, from a class below hers, since he had begun life as a pauper. The two bridges at the beginning, one of iron and one of stone, gave this part of the action overtones which were to be fully elucidated only in the final number. 'Southwark Bridge which is of iron' had been the place of Little Dorrit's visions of freedom, but now, after the iron chains, the iron ring, and stone slab of *Great Expectations*, its associations were with a sense of fate, which was a source of great danger to those who, like

Pip for two-thirds of that novel, made no attempt to resist it. Iron and stone came together again in the last number of *Our Mutual Friend* when Headstone dealt with his blackmailer by gripping him and drowning with him, 'girdled' in Headstone's grip, an 'iron ring, and the rivets of the iron ring held tight'. By that time the associations had accumulated, especially from the striking episode at the very centre of the book, its tenth instalment, in which Headstone walked round the perimeter of a graveyard in 'the Leadenhall Street region' near that stone bridge, declaring his hopeless love. From the coping of the burial enclosure he loosened a stone to which he returned again and again, finally bringing his clenched fist down upon it 'with a force that laid the knuckles raw and bleeding'. Before this he had resisted his fate as a pauper; now he was helpless in the power of the forces which he had suppressed in order to succeed, that substratum within him 'of what was animal, and of what was fiery': he believed that there was no escape from this 'tremendous attraction' that could draw him to any 'disgrace' or 'to any good—to every good, with equal force'.

To this simplification was opposed another, his rival Wrayburn, a drifter 'incapable of designs' until shaken out of his 'cursed carelessness' by almost dying after Headstone made an attack upon him. The moving dramatization of his resistance to being borne away by delirium, in his near-fatal illness, made the reader share in the awakening of a hitherto flaccid will.

Unlike the men, the main female characters, as in *Dombey and Son* and *Bleak House*, were presented as having a vantage-point outside the processes in which they were involved. But now they could act as well as being acted upon, resist as well as being impelled. In the comic action, Bella Wilfer's resistance took the form first of honestly understanding her own temptation to over-value money and then of seeing how small a part of her nature it was. In the more sombre part of the novel, Lizzie Hexam was presented as rising preternaturally above her fate. She could accept education for herself (discreetly shown in the gradual alteration of her homely—if implausibly far from vulgar—manner of speech) without, like Headstone, suffering a divorce of head from heart; loving Wrayburn, she could hide herself away to keep him from falling into Headstone's hands, and, when discovered, reject his offer of marriage—in the interests of them both, as she believed, because of their difference of social class. It was all too extreme, and yet the resonance of the strong chords of determinism and free will infused a kind of life even into a character as much the creation of the author's wishes as this. With Bella Wilfer, Dickens came nearer to success by finding a form of speech that would individualize her, somewhat in the way that a stronger idiom of wilfulness had individualized Fanny Dorrit.

Behind the foreground characters Dickens's obsession with fate and freedom sharpened his sense of a current social malaise. For the third time he made financial speculation of major importance in his plot. On each occasion, in *Dombey and Son*, *Little Dorrit*, and *Our Mutual Friend*, he was writing in the years leading up to one of the three big banking crises in which Peel's Bank Charter Act of 1844 had to be suspended—1847, 1857, and 1866. The particular reach of the river in the first chapter was chosen in order to begin the action in the City, the financial centre of England. The Royal Exchange and 'the mighty Bank' stood 500 yards back from the midway point between the two named bridges, in the very locality of Dombey and Son's offices. This was in Dickens's imagination a centre both of life and of death. Even in the early novel the adjacent Bank was identified by 'its vaults of gold and silver "down among the dead men" underground'. In each novel there were two lines of economic force radiating from this centre, the one of actual achievement, the other consisting of the more dubious dealings of finance. Dickens's image for the actual achievement, uncomprehended and uncontrolled by his characters, was again a railway and the district adjoining it, this time the line to the south starting (then) from London Bridge Station and bestriding 'that district of the flat country tending to the Thames, where Kent and Surrey meet'. It was here that Headstone had his school, 'newly built', like so many all over the country,

> in a neighbourhood which looked like a toy neighbourhood taken in blocks out of a box by a child of particularly incoherent mind, and set up anyhow; here, one side of a new street; . . . here, another unfinished street already in ruins; . . . a medley of . . . rank field, . . . viaduct, . . . canal . . . As if the child had given the table a kick and gone to sleep.

Like Headstone, this changing society was neither completely free nor completely bound; if it failed to recognize the limits of its freedom and of its bondage it would destroy itself as he did.

From the first book of the four into which the novel was divided Dickens's imagination seized the shadow of impending calamity in comic form in allusions to Gibbon's *Decline and Fall*, as well as in metaphor and simile drawn from popular science. Boffin, newly enriched and, like three other characters, seeking education, employed a half-literate ballad-seller, Wegg, to read Gibbon to him. The resulting garbled names, like Commodious or Vittle-us, comically foreshadowed a 'Decline and Fall off' for financiers and sharpers like the Podsnaps and Veneerings who, as the Romans in Gibbon's first chapter did, 'enjoyed and abused the advantages of wealth and luxury'. Contemplating their insular and philistine ignorance, Dickens punctuated the novel with the mock-innocent amazement of an onlooker, indignant that people should be judged by reference to their reputed incomes and their traffic in the 'mighty shares' which were accepted as substitutes for all the things that should have marked the nerve-centre of the country—'established character, ... cultivation, ... ideas, ... principles'. Such passages functioned like the omniscient chapters of *Bleak House*. Although fewer and less varied in tone, they made up for this by their consistent texture of metaphor and simile derived from popular science and ancient history. A servant carrying dubious drinks became 'a gloomy Analytical Chemist'; a bony matron a specimen for Professor Owen, the comparative anatomist; while the 'friends' of undiscriminating hosts were 'like astronomical distances ... only to be spoken of in the very largest figures'. Popular science as from *All the Year Round* (where those astronomical distances had appeared at the end of 1864, six months earlier than the reference to them in this fifteenth number of the novel) drew attention to a known system of cause and effect which, like that of history, the satirized characters were ignoring at their peril. Mordant reference to Antinous, the Colossus at Rhodes, or the sociable baths of Rome reinforced the point.

The one character from Gibbon to be named but not ridiculed in the action was 'the amiable Pertinax', who, as we see in the fourth chapter of the *Decline and Fall*, considered 'the pure and genuine sources of wealth' to be 'economy and industry'. Such sources of wealth came into view in the non-farcical half of the novel with the model paper-mill, beyond London (like the paper-mill at Dartford in Kent in *Household Words* on 31 August 1850), where Lizzie Hexam, in hiding from her lover, found employment. It was identified specifically as a Jewish enterprise, to underline a contrast with the misuse of money by Gentiles in the City. The mill-workers, released to country pleasures beside a 'silver river', contrasted with those going home from the City like 'a set of prisoners departing from gaol'. The country setting suggested earth raised to heaven—'the sky appeared to meet the earth, as if there were no immensity of space between mankind and Heaven'—whereas the City appeared to drag heaven down, 'the towers and steeples of the many house-encompassed churches, dark and dingy as the sky that seems descending on them'. Each of these heavily contrasted places was the setting for the rejected proposal of one of the rivals—Wrayburn's, in the country, the declaration of a drifter only now finding resolution, Headstone's, in the City, made under apparently inexorable pressure from energies his ambitious will had disregarded. Over-emphatic the contrast might be, but it did link suggestively the personal and the social subjects of the novel so as to stir the reader to take pleasure in reflecting upon them. Both subjects suggested the lack at the centre, in London, of the kind of understanding from which might come some degree of guidance and direction of vital energies.

The contrived escapade of Betty Higden, resolved to 'walk and work' to keep out of the workhouse, focused these suggestions upon an immediate issue. Podsnap had voiced the complacency of people at the centre about 'a country ... where so noble a provision is made for the poor', while at the same time he had rejected central responsibility for the efficient control of this 'provision': 'Centralization. No. Never with my consent. Not English.' The inequitable financing of provision for the poor had been prominent in *Punch* and in *The Times* earlier in 1864. Indeed, prejudice against a uniform, country-wide recasting of the 1834 Poor Law survived the specific recommendations about it by the 1909 Royal Commission and was not overcome until the reforming government of 1945–51. Dickens spoke too for those to whom the continued punitive character of the Poor Law (despite some softening in its application) was a scandal in conditions of unexampled commercial prosperity.

The origins of this prosperity in the railway boom of the eighteen-forties had stirred his imagination in *Dombey and Son*. In his second phase, each of the major novels, whatever the period of the plot, showed a London which refused to understand its position as the nerve-centre and to develop the institutions required by the role—legal and political in *Bleak House*, bureaucratic and financial in *Little Dorrit* and *Our Mutual Friend*; only by such a development could the energies which were transforming the country be mastered instead of becoming the masters themselves.

That submission to fate in this form could lead only to calamity was suggested by the correlation of Headstone's uncomprehended resources of 'what was animal and what was fiery' with the economic energies transforming the country, rather like the correlation in *Hard Times* of the factory fires with Louisa's 'sensibilities and affections'. In the plot and its associated metaphors, this calamity took the form of death—in Headstone's case, murder and suicide at once. The threat to society was reiterated in Harmon's dust heaps, as the emblem not of the creation of wealth but of the accumulation of it, and of the apparently uncontrolled power it conferred—extending in this case beyond the grave, by means of a sequence of wills. The monthly green wrapper showed a skeleton-sexton climbing a mound with spade and bell beneath the label 'Dust Contractor', and his associations darkened melodramatically the successive sombre descriptions of the financial centre of Britain as 'gritty', its evening 'gray dusty withered', its spring wind choking the passengers with sawdust and blowing about the rubbish, 'that mysterious paper currency' of London, while its 'closed warehouses have an air of death about them, and the national dread of colour has an air of mourning'. The resonance of the metaphor of dust had increased greatly since its casual use in *Hard Times* to describe Gradgrind's parliamentary duties in 'the national cinder-heap . . . and his throwing of the dust about into the eyes of other people . . .'. As in *Hard Times*, too, the moral blight upon society took the stock form of the mercenary marriage—commanded by old Harmon's will, recommended by Wrayburn's father, narrowly escaped by Miss Podsnap, and actually contracted by the Lammles in their mutual deception, with the Veneerings' encouragement. Such a threat could apparently only be evaded by some sort of rebirth. Young Harmon could evade it only by being reborn as Rokesmith. Wrayburn could evade it only after his return from delirious wanderings in 'endless places' in the very shadow of death, his will at last awakened. Marriages which were the opposite of mercenary seemed again to be the nearest Dickens's imagination could get to that rebirth of a society which had been the content of the 'prophetic' thoughts of Sydney Carton. The resolution of the matter of Harmon's accumulated riches seemed facile when they were redeemed by the new purpose to which they were applied, to furnish an ideal nursery for John Harmon's child. The archness obscured the simple facts that both parents had been able to see the folly of treating money as an end in itself and that they had inherited only after Boffin had refused to be bound by the will. Nevertheless, for all Dickens's description of Bella's marriage as if it were a death and a rebirth ('the church-porch, having swallowed up Bella Wilfer for ever and ever, had it not in its power to relinquish that young woman, but slid into the happy sunlight Mrs John Rokesmith instead'), the whole passage was greatly inferior to a comic-realistic escape from death—that of Rogue Riderhood, impudently unaffected, still thinking as a scavenger and calling for his lost cap:

> 'And warn't there no honest man to pick it up? Of course there was though, and to cut off with it arterwards. You are a rare lot, all on you.'

There was a quite different, allegorical variation on the theme of death and rebirth in a crippled girl's vision of freedom, at the apparent centre of bondage, on the roof of a money-lending firm, just off Leadenhall Street. Her invitation, 'Come up and be dead! Come up and be dead!' was reversed at Headstone's catastrophe, out in the country, in the final number, when he answered Riderhood's claim that 'the man as has come through drownings can never be drowned' with

> 'I can be! ... I am resolved to be. I'll hold you living and I'll hold you dead. Come down!'

Such neatness of correspondence and contrast was characteristic of this most intricate of the novels of Dickens's later phase. It made a brilliant show, over-carefully elaborated, the bones of its main design too salient, its dialogue too patterned, its symmetries of plot and description and character too much insisted on, but alive with the power to surprise, and unified by a strong sense of place and the visionary overtones to which it gave rise.

The Mystery of Edwin Drood concludes the second phase with a surprising departure. Here Dickens does not develop the implications of its single action so that they concern the whole country; although the book is full of echoes of the preceding novels which embody this concern above all, the echoes seem chosen so as to emphasize his avoidance of it. Overtones are developed which transcend time altogether, while matters of free will and its opposite, which were oppressive in the previous novels of the second phase, now seem to have their burden lightened.

Edwin Drood, like John Harmon, has been betrothed 'by will and bequest', and in the early stages of the novel Rosa Bud's guardian is believed to be 'bound down to bestow her'. Actually, the marriage 'was not bound upon' either of them, but was a mere 'wish' or 'friendly project' of the parents, although the reaction of Drood's uncle, John Jasper, on hearing that the marriage is not to take place emphasizes that he at any rate has believed it to be inevitable. As there is not much doubt that Jasper is guilty (his exploration of the cathedral is said, in the working notes to chapter 12, to be laying 'the ground for the manner of the Murder, to come out at last'), there appears to be a parallel with Headstone in the previous novel: both exaggerate the element of the unavoidable in the future, and their crimes follow from this mistake. What is inevitable is the discovery of the culprit: the echo of *Great Expectations* in the author's commentary ('the mighty store of wonderful chains that are for ever forging, day and night, in the vast iron-works of time and circumstance') refers to the ring which Drood neither gives nor mentions to his betrothed and which is to lead to the unravelling of the mystery.

Other references to stone, to iron, and to dust touch with comedy even what is inevitable beyond doubt, as for instance in the small but substantial part allocated to the stonemason 'chiefly in the gravestone, tomb, and monument way', Durdles. Even the associations of inexorable Time itself are lightened before the fragment finishes. In chapter 9, the city's cathedral, the most impressive of its 'vestiges of old time and decay', is a haunt of damps and shadows, but in chapter 19, on a 'brilliant morning' in summer, 'a soft glow seems to shine from within' these same walls, and the compulsion exerted by time upon all things becomes beneficent. Eventually, in the last chapter Dickens wrote, the works of Time in its seasonal turning not only transfigure 'the old city', but 'penetrate into the Cathedral, subdue its earthly odour, and preach the Resurrection and the Life'.

One cannot argue from this paradoxical account of the work of Time to some happier ending to the mystery than murder, for the radiance is itself an echo of the contrast between sunlight and Sikes's culminating crime in *Oliver Twist*, which Dickens was reading on the public stage right up to March 1870, while working on *Drood*. He had underlined in his acting copy of *Sikes and Nancy* the 'brilliant light' falling 'through costly-coloured glass ... through cathedral dome' and into 'the room where the murdered woman lay'. Other words in the chapter from which he selected for the reading, for instance, 'the bright sun, that brings back, not light alone, but new life, and hope' make the echo even more pronounced. It would be consonant with what we know of Dickens's methods, however, that he should be working towards a parallel between the Cathedral, smelling 'earthy' but becoming the place for an affirmation by a rejuvenated earth, and the chief of the choristers, Jasper, to whom the cathedral services are a mere 'daily drudging round' but who might eventually come to find in them a meaning he had overlooked. Jasper speaks of the service as sounding 'quite devilish', mocking his drudgery: 'What shall I do? Must I take to carving [demons] out of my own heart?' He seems, in the fragment we have, to sing really well only when carving such demons—before the opium dream in which he rehearses his crime, or as he goes home on the eve of the murder and pulls off 'that great black scarf' which looks like the weapon. It is emphasized that he is oblivious to the sense of what he sings—and even in private it is specifically 'choir music' that he chants in a 'beautiful voice', for instance, before the expedition which will 'lay the ground for the manner of the Murder'. This seems the mark of a double nature, which must have formed at least some part of the 'very curious and new idea' which Dickens told Forster he had for the book. Forster believed it was to lead to the eliciting from Jasper of his own crimes 'as if told of another'—much, one may add, as Carker in *Dombey and Son* had 'no more influence' over his senses, rehearsing his 'treachery' and humiliation, 'than if they had been another man's'.

Unlike Carker, however, Jasper may have been allowed something more than remorse, if indeed there was to be a parallel between the Cathedral and its chief chorister. For it is noticeable that the cathedral singing has a force which is greater than that of the singers and that this force is steadily associated with 'new life'. In chapter 9 the 'cracked monotonous mutter' of the single priestly voice has its life beaten out by 'a sea of music' from 'the organ and the choir' which 'lashed the roof, and surged among the arches, and pierced the heights of the great tower'. The choir on emerging are called 'the living waters', and the cleansing associations of water and the sea appear in a reference to *Macbeth* as compendious as that to St John (though less accurate), when Mr Crisparkle, the contrasting minor canon to Jasper, is 'as confident in the sweetening powers of Cloisterham Weir and a wholesome mind, as Lady Macbeth was hopeless of those of all the seas that roll'. Even without the echo of 'the dark and unknown sea that rolls round all the world' in *Dombey and Son*, the sea holds associations, too, with eternity, present when Jasper looks down in chapter 12 on the 'sanctuaries of the dead', the 'houses of the living', and the river, in the manuscript 'already lashing and heaving with the knowledge of its approach towards the sea'. The associations make it another instance, underlined by the echo of chapter 9 in 'lashing', of Jasper encountering what he fails to understand.

In the first words of the novel, Cloisterham and its cathedral are seen in Jasper's opium dream. As he goes about in his everyday persona the city has an excitement for the reader which comes from the suspected presence of evil as well as from the satire upon the willingness of the inhabitants 'to suppose . . . that all its changes lie behind it, and that there are no more to come'. Nevertheless the place emerges, in what would have been the central chapters of the novel if it had been finished, as quite transformed by the light and the season. The parallel transformation of the cathedral music into an engulfing sea, and even, in the context of 'living waters', into something more, suggests a similar exposure of Jasper's life to the light, in the sense not only of discovery and confession as Forster affirms, but also of repentance implied by the references to Shakespeare, Bible, and liturgy. How this might have been accomplished is lost to us. All we have is the author's word in an aside, chapter 20, referring directly to Jasper, about 'the criminal intellect, which its own professed students perpetually misread, because they persist in trying to reconcile it with the average intellect of average men, instead of identifying it as a horrible wonder apart'. Villains he had presented before, and from the inside, with a vehemence and empathy which put them among his most remarkable creations. But none of them appeared in the extreme paradox of Jasper's position, making a drudging routine out of the pleas and affirmations which were relevant to his state and to which the author obviously attached great importance.

The heterodox importance attached to the climactic affirmation in the fragment, in which 'changes of glorious light from moving boughs, songs of birds, scents from gardens, woods, and fields . . . preach the Resurrection and the Life', recalls two other well-known instances. Whether or not Dickens himself knew of the passage from his friend C. R. Leslie's *Life of Constable* in which the painter found 'that sublime expression in Scripture "I am the Resurrection and the Life" ' verified out of doors in spring and remembered Wordsworth's having done so too, the words used by two earlier artists who devoted themselves to celebrating what Dickens celebrates here bring to mind the triumph over time which the celebration itself constitutes. So even the fragment we have may afford suggestions that the end of the second phase of Dickens's career was to sound the note of the triumph over time as strongly as the end of the first. Such a suggestion would be appropriate for a body of work which gives every sign of bearing it out: Dickens's readers continue to grow in number; articles and books show the still increasing interest he has for those who wish to understand what he accomplished and how. But although his literary sophistication becomes more and more apparent, his appeal remains undiminished for those who are satisfied to immerse themselves in his over-mastering illusion. In both respects he resembles Shakespeare.

5. The Brontës

The fiction of Anne, Emily, and Charlotte Brontë[1] is related less to that of their contemporaries, with the exception of Thackeray, than to the literature of the early part of the century, to Scott, Wordsworth, and the popular ballads they revalued, and to Byron. What such poets have to say about human feeling and its mysterious harmony with the non-human part of creation is channelled into fiction by these novelists above all others, for a large number of readers. For one reader who responds to Wordsworth's 'ennobling interchange Of action from without and from within', there are hundreds who know of the interplay between human passion and storm, sunset, or harvest moon in these novels. For one who is acquainted with the rhetoric masking a restless sense of insufficiency in the heroes of Byron's eastern tales, there are hundreds who rise to a similar bait in Mr Rochester or Heathcliff. The Brontës' continuing hold over a very great variety of readers depends, however, not only on their access to the same springs of experience as had fed major writing early in the century, but also upon the remarkable powers of construction which they developed as they learned from each other how to master that experience in fiction.

[1] Charlotte, 1816–55, Patrick Branwell, 1817–48, Emily Jane, 1818–48, and Anne, 1820–49, were born in Yorkshire to an Irish Anglican clergyman, Patrick Brontë (originally Brunty) and his Cornish wife, who died in 1821. An important part of their education was gained from discussing the news with their father and from reading his books and those borrowed from the Keighley Mechanics Institute and perhaps from Ponden House, not far from the parsonage at Haworth where they all spent most of their lives. For a time in 1824, Charlotte, Emily, and two elder sisters, Maria and Elizabeth, who died in 1825, attended a school for daughters of the clergy at Cowan Bridge near Kirkby Lonsdale, Westmoreland. Charlotte went to a boarding school at Roe Head, near Dewsbury, 1831–2, and returned to teach in 1835, when Emily and Anne had, each in turn, a few months there as pupil. Charlotte and Emily also studied in Brussels in 1842, at the Pensionnat Héger, where Charlotte returned to teach throughout 1843. Back in England Charlotte and Anne were each briefly unhappy as a resident governess. The three sisters' hopes of taking pupils at home proved fruitless, but their poems were published in London, at their own expense, as 'by Currer, Ellis and Acton Bell', the pseudonyms under which their first novels appeared. After the death of her brother and her remaining two sisters, Charlotte made the acquaintance of her future biographer, Mrs Gaskell, was befriended by her publisher, George Smith, and married, 1854, her father's curate, the Revd A. B. Nicholls (1817–1907).

The three sisters and their brother, Branwell, were precocious readers of Renaissance and Romantic poetry (including poetic drama), as well as of the few periodicals which came their way, particularly *Blackwood's Magazine,* which Branwell claimed in 1835 had been 'read and reread' in childhood, and *Fraser's*, to which their aunt, who lived at the parsonage, subscribed. They were also, from an early age, precocious in self-expression, as the authors of articles and stories written as if in print and bound to form minute books and periodicals. Out of their naming and sharing of a set of wooden toy soldiers in 1826 grew heady chronicles of love and war among the aristocratic rulers of a confederacy of English-speaking states on the Guinea Coast of Africa. They elected one of their number, the Duke of Wellington, as their king and built their capital at Glass Town, that is, Verreopolis, later Verdopolis. At first the 'Guardians' of the land were the four Chief Genii, Brannii, Tallii, Emmii and Annii, but by 1834 the stories had divided into those of Emily and Anne centred upon Gondal, 'a large island in the north Pacific', and those of Charlotte and Branwell about Angria, the African kingdom conquered and ruled by the Duke of Wellington's elder son, Arthur Wellesley, Duke of Zamorna. Some eloquent verses are all that remains of Gondal. Of the prose of Charlotte and Branwell we have perhaps too much. Although it is heavily melodramatic, its headstrong plotting is not out of touch with real life—Verdopolis is a commercial and manufacturing centre, with mill-owners and rebellious operatives—and Charlotte experiments with narrators who mock its emotions and criticize its simplified characters, sometimes with the full deflating power of common speech. A phrase for Zamorna, 'the great blethering King of Angria', is the rough talk of his brother, Charles Townshend, a frequent, if less than consistent, narrator. The authors themselves or the habits of other authors may be mocked: Captain Bird, who tells the story of *The Green Dwarf*, looks back with amusement to 'the Genii's Inn at Verdopolis, the almost exclusive resort of wayfarers . . . in spite of the equivocal character of the host and hostesses being the four Chief Genii, Tallii, Brannii, Emmii and Annii'; his heroine, the Lady Emily, is one who 'instantly did what will not be thought very becoming in the heroine of a novel, viz: coined a little lie'; when she meets a 'ghost' its language is less than terrifying ('It's a rare thing for me to be in this dog hole at night without a candle.

That last pint made my hand rather unsteady, and I can't see to find the keyhole'), and she is not above pretending to be a ghost herself. Though the selfconsciousness is immature in its expression compared with early Thackeray, the developing awareness of the diverse ways in which a story may be written is not unlike his. As in his case, it is all far from solemn; but mockery does not inhibit the Byronic extravagance—Zamorna's 'dark large eye burning bright with a spark from the depths of Gehenna' and striking Caroline Vernon 'with a thrill of nameless dread'—or the preoccupation with the knowledge of passion which comes to those who are made unhappy by it. This is apparent in the earliest of Charlotte's love stories, written at the age of 14, in which Albion, separated from Marina, hears her calling to him at the moment of her death. It is equally apparent seven years later, when Mina Laury, in love with Zamorna, recognizes that 'She had lost her identity; her very life was swallowed up in that of another', so that 'she could no more feel alienation from him than from herself'.

By the end of the 1830s Charlotte had formally recognized her need to get away from over-heated fantasy:

> . . . I long to quit for a while that burning clime where we have sojourned too long—its skies flame—the glow of sunset is always upon it—the mind would cease from excitement and turn now to a cooler region where the dawn breaks grey and sober, and the coming day for a time at least is subdued by clouds.

Something like the realistic presentation of ordinary life can be seen at the beginning of *Mina Laury* in 1837, and it takes over at the end of the second chapter of the unfinished *Ashworth* of 1840–1. She made a determined effort to bring it into contrast with the heady attractions of the 'infernal world' of romance in *The Professor* (1857), which was unsuccessfully submitted to publishers in 1846; in the preface written when it seemed likely that it would be published, in 1851, she spoke of the effort by which 'I had got over any such taste as I might once have had for ornamented and redundant composition'. The actual story showed the difficulty she had in recognizing what was ornamental or redundant and how slowly she came to see what constituted its central material and method. About a fifth of the book was taken up with an account of the narrator's reasons for going abroad, which had little to do with his profession of schoolmaster once there, or his feelings on the way. Details, like the stillness of the winter day when he left home, or St Paul's bell heard in a London lodging, were portentously announced:

> There was a great stillness near and far; . . . A sound of full-flowing water alone pervaded the air . . .
> . . . I first heard the great bell of St. Paul's telling London it was midnight; and well do I recall the deep, deliberate tones, so full charged with colossal phlegm and force.

It was only the injunction, 'Treasure them, Memory', not the later content of the narrative, that justified the portentous overtones; the injunction seemed addressed to the author herself thinking of her own experience, rather than to the imagined narrator. And indeed the metaphor of treasure became important in relation to the character modelled on the author, Mlle Henri, one of the professor's pupils, the 'treasure', 'the lost jewel' he sought throughout Brussels when she abruptly left the school. His relationship with this woman of slight figure and 'a certain anxious and preoccupied expression of face', neither beautiful nor plain, was quietly and movingly unfolded. It was set in a context which seemed to mix impressive observation with Angrian excess. The girls' school where the narrator taught was realistically present, but the narrator's account of his repelling of advances from its directress, Mlle Reuter, brought Zamorna ludicrously back on stage: 'When she stole about me with the soft step of a slave, I felt at once barbarous and sensual as a pasha.' The tale of her 'singular effect', eliciting 'all that was noxious in my nature', opened the possibility, once she had married, of 'romantic domestic treachery', drawing the narrator into participation in 'a practical modern French novel'. On first acquaintance with Mlle Reuter he had asked, 'Is she like the women of novelists and romancers?' Failing, at first, to find her 'made up of sentiment, either for good or bad', like 'the female character as depicted in poetry and fiction', he finished by suspecting she had the moral ambiguity of a 'modern French novel'. He implied a total contrast when presenting his own marriage, begun 'with the notion that we were working people' and leading to the alternation of separate work and domestic companionship, remembered as 'a long string of rubies', each 'unvaried' but none the less 'brilliant and burning', the treasure now multiplied. The author could not yet control the emphasis or sustain the interest of this ambitious construction, made to contrast the rejection of romantic fiction and the acceptance of solid work-a-day felicity. She was to return to techniques of contrast once her reading of the completed work of her sisters had suggested ways in which contrasts might be more lucidly managed.

In *Agnes Grey* (1847), the novel which Anne sent on its rounds of the publishers in 1846 at the same time as her sisters submitted *The Professor* and *Wuthering Heights* respectively, her own past experience as a governess supplied much of the material but she was better able than Charlotte to select and shape. Everyday detail was deftly mingled with the feeling of the characters towards it, to form the fluctuations of a small plot. The cottager's cat was an instance: kicked across the room by the rector in his eagerness to pursue the squire's daughter, it was saved from a gamekeeper by the curate, who, when returning it, learned that Miss Grey read to the cottager and never saw her employer, the squire, at all 'to speak to'. This was at the exact centre of the story, following her confession to herself of extreme loneliness in the great house. Her narrative of the family's behaviour there, which she hoped would 'benefit those whom it might concern', had in fact unobtrusively established her own character so as to engage the reader as a partisan in the simple vicissitudes of her love. Yet, as George Moore saw in his *Conversations in Ebury Street*, 'Anne's eyes were always upon the story itself and not upon her readers': she was making something, and her attention, like Emily's in her verse, was all upon the relations of its parts to each other. That the curate should be substantial enough only for his personal compatibility with Agnes Grey, and his contrast with the rector, mattered less than his placing in the sequence of the heroine's despair and hope. It was a sequence not only of external but of internal events, and presented with some subtlety: 'I did not believe half of what she [Hope] told me: I pretended to laugh at it all; but I was far more credulous than I myself supposed.' The natural setting was given prominence for its harmony with these hopes or its quiet elucidation of them—quite obviously when the high rookery was reduced at sunset 'to the sombre, work-a-day hue of the lower world or of my world within', less obviously at the climax, where the sea dashed against 'a little mossy promontory with no prodigious force, for the swell was broken by the tangled seaweed and the unseen rocks beneath'; there the description summed up with exactitude the moderate scale of the personal drama even though its groundswell flowed from a greater sea.

Wuthering Heights (1847), on which Emily was at work at this time, combined Anne's simplicity of subject with a structural sophistication from which both her sisters learned a great deal. It was praised by Sydney Dobell, in a notice in the *Palladium*, September 1850, for its 'art', and 'thinking out', as 'the masterpiece of a poet rather than the hybrid creation of the novelist'. Her verse, which is all that survives of her early work, had already shown ability of a different order from her sisters'. The management of rhythm, for instance, enabled her to create without relying on extravagance of incident or character, however much these may have stimulated her in the first place.

The night is darkening round me
The wild winds coldly blow;
But a tyrant spell has bound me
And I cannot, cannot go.

The giant trees are bending
Their bare boughs weighed with snow,
And the storm is fast descending
And yet I cannot go.

Clouds upon clouds above me,
Wastes upon wastes below;
But nothing drear can move me;
I will not, cannot go.

In such work, instead of evoking ready-made emotions with conventional counters ('Gehenna . . . thrill . . . nameless dread'), the rhythms were made to contain a more particular sense of inner constraint. The words became active themselves; out of them the poet clearly saw the possibility of forming solid artefacts, to be tested and changed by reference not to her own wishes but to the internal relations between the poem's actual constituents.

This kind of ability was evident in the 'thinking out' of

Wuthering Heights. Emily Brontë started with 'bare and simple details', such as (according to Mrs Gaskell) were narrated to the family by their servant Tabby, of 'family tragedies, and dark and superstitious dooms', the material of the ballads, long lost to literature before the eighteenth and early nineteenth centuries and even then in varying degrees obscured by the writers who revived them, whether Percy, 'Ossian', Scott, or Wordsworth. Out of this ballad material she constructed not only a story but the story of its telling. Its variations upon a limited range of names and motives suggested strongly the inbred intensity of a remote region where people lived 'more in earnest, more in themselves', showing 'the essential passions of the heart' which Wordsworth had claimed, in the preface to *Lyrical Ballads*, were 'less under restraint' in such places and spoke a 'plainer and more emphatic language'. Although the story dealt in ultimate intensities of love and death which, to the simplest as well as to the most sophisticated contemplation, are mysterious, it was straightforward enough, beginning with Heathcliff's disturbance of the Earnshaw family by receiving, as an interloper, the favours and affection of a father and a daughter which would otherwise have belonged to the son of the house, Hindley; the feud between the two men was eventually extended by Heathcliff to the Linton family into which the daughter, Catherine, had married, believing, despite her attachment to him, that 'it would degrade me to marry Heathcliff'; finally he yoked together as instruments of his revenge the son who issued from his own marriage to a Linton, the daughter born to Catherine and her husband, and Hindley's son Hareton. For all his selfconscious hatred, Heathcliff was made the agent of the opposite result to his intention, starting with his involuntary saving of the life of Hareton, and finishing with the failure of his own will to take revenge just at the climax of his power to do so and just when two members of the next generation, brought together by his own actions, were ready, through a reconciling marriage, to undo all his work.

This modified saga of revenge was influenced by major poetry—by *Paradise Lost* (where all Satan's 'malice serv'd but to bring forth Infinite goodness') as much as by the echoes of Satan's rhetoric in the heroes of Byron's dramas and eastern tales. But it was written in such a way as to withhold from Heathcliff the admiration allowed to Byron and his avatars and to their descendants in the stories of Angria and no doubt of Gondal. Even in Angria there were, as we have seen, critical voices. Now such voices were more steadily heard. Their curious relationship to the story they told set up strange reverberations, adding to those inherent in it.

It was a story of an overwhelming attraction between foster-siblings, in relation to which its two narrators each claimed a privileged position. The secondary narrator, Ellen Dean, claimed some knowledge of the relation of foster-siblings, having been nursed with Hindley and with him scampered the moors, like Catherine with Heathcliff; this narrator even longed, like Catherine, to resume the earlier relationship after the man's deterioration, and was given a vision of him as a child which matched the primary narrator's vision, at the opening of the novel, of Catherine as a child trying to get back into the house now it was Heathcliff's. The primary narrator, Lockwood, claimed some knowledge of the relation of lovers, for he was a seeker after solitude, fleeing from his own tentative skirmish with a girl in love with him. But his misanthropy was mere pretence—since the whole narrative arose from his visiting his landlord and, when illness put an end to visits, begging his housekeeper to talk to him—and his claim to be misjudged as 'deliberately heartless' in relation to the girl had to be measured against his obvious lack of self-knowledge (for he seemed in that episode either not yet grown up or already set in too close an attachment to 'my dear mother'). He was given the mincing language of the fashionable novel in matters of love: 'the favoured possessor of the beneficent fairy', 'a most fascinating creature: a real goddess'.

All the same, Lockwood had the ability to be emotionally stirred and he was subject to strange undercurrents of unacknowledged fear which generated dreams to stir the reader. The character seems to be based on observation of Branwell Brontë, who is known to have fled, like Lockwood, once the girl he was eyeing had acknowledged her love for him. Lockwood now had to face a broken attachment belonging, apparently, to quite another world. He was led into it by a comical nightmare fantasia suggesting revenge, forgiveness, and his own guilt, followed by the dream of a waif begging entrance with an importunity which made him cruel; on waking he was himself mistaken for a ghost, like Lady Emily in *The Green Dwarf*. The reader was with him in his scepticism and his fear, but against him when he called Heathcliff's plea ('Come in! come in! . . . hear me *this* time— Catherine, at last!') a mere 'piece of superstition . . . which belied, oddly, his apparent sense'. The reader was, eventually, to be with him too when, apparently moved to take the other world more seriously, he was given the last word about it and the last word in the novel.

Lockwood, associated as he was with the city and with what one of the first reviewers (perhaps John Forster) called 'the affectation and effeminate frippery, which is but too frequent in the modern novel', was moved to this final wonder by Nelly Dean, the local observer and participant. She could explain the two main sequences of the story to him, after he had seen their results at two crucial points, in 1801 and 1802 respectively, just before and just after Heathcliff's death. Though Lockwood's dreams suggested that his fears gave him a better chance than she had of understanding the first sequence, her story showed that she had more good sense than he and could in some degree overcome her obvious limitations (as almost an Earnshaw herself, and so prejudiced against Heathcliff). But she was afraid of dreams. She accepted the country superstitions; she held to an undogmatic country religion which made her equally at home at the kirk or at 'the Methodists' or Baptists' place (I can't say which it is)'; but she feared to 'shape a prophecy' from the telling of a dream. Emily Brontë seems to have been using the distinction De Quincey made, when writing on 'Modern Superstition' in *Blackwood's* in 1840, between omens, those 'agencies of the supernatural which . . . become terrific by being personal', and the Ovidian form of the supernatural which connects 'the unseen powers moving in nature with human sympathies of love and reverence'. For if the dream which Catherine did not tell may have belonged to the first kind, the one she told, of being expelled from heaven, certainly belonged to the second. Nelly's dislike of being told was in fact at the very centre of the novel's chinese-box construction: what she in her turn could tell might have been rejected by Lockwood if he had been the solitary misanthrope he pretended to be, but, once heard, it could be turned into an omen for himself, in his fear of the consequences of love. Nelly's objection to hearing Catherine's dream was of a piece with her 'sensible' disapproval of Catherine's marriage to Linton. The attitudes of the two narrators enabled the author to present Catherine's own much less sensible account of herself without appearing openly to demand sympathy or approval for it. Catherine was trying to understand how she could be 'convinced I'm wrong' to marry Linton. The dream in which she was thrown out from the heaven of angels 'into the middle of the heath, where I woke sobbing for joy' was part of her sense of being founded upon the 'rock' (her own extravagant figure) of an inescapable attachment to Heathcliff. So the reader received what are remembered as her famous affirmations—'I *am* Heathcliff', 'If all else perished, and *he* remained, I should still continue to be'—through the bewilderment of a girl and through the story told by a countrywoman, sensible but far from impartial, to a sentimental alien from the city. Catherine's marriage, not to Heathcliff but to Linton, was like others in contemporary fiction, like Hope's in *Deerbrook*, for instance. But it was unlike these in being dramatized only in brief, suggestive hints of wilfulness and calm, while it was contrasted, in this strange but unforgettable rhetoric, with a fierce, paradoxical fidelity to Heathcliff, and all by means of the narrative of sceptics. So the reader was brought to contemplate the central relationship in varying lights—on the one hand to see Catherine's perversity and confused motives, but to go on himself to try to do what Nelly would not, that is, to 'make sense' of Catherine's 'nonsense', and on the other hand to hate what Heathcliff did in revenge, but to try to understand him, despite the self-centred defiance of his faded rhetoric.

The techniques all seemed to have come from Thackeray; the pastiche, the narrators who only half understood the story, even the attitude towards a heroine, as in Thackeray's *Catherine*, whose erring was explained without being justified, could all be found in his fiction from 1837 to 1844 in *Fraser's*. Such techniques were applied to an extreme situation of 'nature erring from itself', within which stirred a strange undertow of interest in the four last things, death, judgement, heaven, and hell. The dreams, brief but startling, brought these things into the immediate present time of the story and made them interpenetrate it like Catherine's own dreams which, she said, had 'gone through and through me, like wine through water'. The dream that the heaven of angels was not her home gave a dramatic context to Emily Brontë's poem of 17 July 1841:

We would not leave our native home
For *any* world beyond the Tomb
No—rather on thy kindly breast
Let us be laid in lasting rest;
Or waken but to share with thee
A mutual immortality.

Such a dream was to be set beside the waking imaginations of heaven by the children of Catherine and Heathcliff respectively, and these in turn beside the conviction which eventually possessed Heathcliff, in a kind of trance: 'I have nearly attained *my* heaven . . .'.

At this final point both narrators contemplated a peaceful hereafter for the villain-hero. The narrator of Catherine's death in chapter 16 had asked Lockwood whether he shared her assurance of an 'endless and shadowless hereafter' for 'such people', but he merely found her question 'something heterodox'. At Heathcliff's death, when the orthodox judgement came from the half-comical bigot, Joseph, 'Th' divil's harried off his soul', Nelly Dean held to her view that 'the dead are at peace'; this time Lockwood saw only Hareton and the second Catherine—'*They* are afraid of nothing . . . Together they would brave Satan and all his legions'—and was then moved to visit the graves of the previous generation and wonder 'how any one could ever imagine unquiet slumbers for the sleepers in that quiet earth'. The dying fall of that final paragraph has seemed to many readers at the most a resolution of their own emotions. But it rested upon concern, common to all three Brontë sisters, with Universal Salvation. In her verse, Emily characteristically tried out this idea and others opposed to it, by means of dramatic monologues, now seeing, like Milton, eternal consequences following from unshaken pride—

Shed no tears o'er that tomb . . .

The time of grace is past
And mercy scorned and tried
Forsakes to utter wrath at last
The soul so steeled by pride—

now reasoning, from the intuition of a God of love, that even

A God of *hate* could hardly bear
To watch through all eternity
His own creation's dread despair!

The whole novel was a construction in which characters in their known situations might express such views while the resolution of them, if any, was only suggested. It was constructed to pursue intuitions which had appeared in the verse, to set them in contexts where the reader might apply a logic of impressions which was not alien to them and so did not murder to dissect. In showing, without comment from the author, schemes of revenge deflected into reconciliation, it developed the glimpses the poems had afforded of a goodness immanent within the natural order. A poem of 1842–3, for instance, had turned away from contemplating broken love and damnation—

> And is this she for whom he died:
> For whom his spirit, unforgiven,
> Wanders unsheltered, shut from heaven—
> An outcast for eternity—

with a confidence which, though born of mere moonshine, was accepted as all-sufficient:

> O'er wood and wold, o'er flood and fell
> O'er flashing lake and gleaming dell,
> The harvest moon looks down;
> And when heaven smiles with love and light . . .
> Earth's children should not frown.

That incomplete intuition was pursued in the novel by making the same harvest moon present at the crucial points of the action. It was harvest time when Earnshaw went on the journey from which he returned bringing the child Heathcliff to be accepted, he said, 'as a gift of God, though it's as dark almost as if it came from the devil'; the harvest moon shone at Heathcliff's own return in chapter 10 on hearing of Catherine's marriage (when her welcome turned his thoughts from suicide and, for the time, from revenge); it shone again at his entry, in chapter 29, as master of the Grange eighteen years later; and finally, it shone when Lockwood came back on 'a sudden impulse' in chapter 32, and was moved, by the end, tentatively to share the final conviction of Nelly Dean about the peace of the dead. The moon was no mere sentimental reach-me-down but appeared among the ordinary business of life with a Wordsworthian matter-of-factness—at Linton's death, for instance: 'I put the jug on the banister and hastened to admit him myself. The harvest moon shone clear outside. It was not the attorney.' In a novel where the melodramatic action suggested the subservience of hatred to love, and moreover related this to a countrywoman's undogmatic piety by reconciling Hareton and the younger Catherine on Easter Monday, the setting of place and season carried, for the author, overtones of the power in *Adonais* which 'wields the world with never-wearying love'—at least, if the Shelleyan associations are to be trusted in the poem which perhaps comes closest in time and in mood to the novel. Dated 2 January 1846, it addresses the 'God within my breast':

> With wide-embracing love
> Thy spirit animates eternal years,
> Pervades and broods above,
> Changes, sustains, dissolves, creates and rears.

In the novel such suggestions were the most disturbing of all, for this power had apparently been working through Heathcliff.

With such a subject, the author apparently did not care how tawdry was the presentation of Heathcliff's evil, for he had only a limited rein. She would not at any rate allow him the magniloquence of even Byron's heroes but only a turgid echo of the boasting of Scott's *Black Dwarf*. No doubt there was conscious archaism here, for the action was thrust back into an earlier age, one in fact before her father had made his influence felt in the district: its conclusion was given the date at which he first set foot in England, 1802, and its first dive back into that past, in chapter 3, could be dated to the year of his birth, 1777. Heathcliff's language of hatred in the second half of the novel was as much a parody as Lockwood's finical phrasing in the first half was a parody of the language of love. Only a knotty vein of tortured logic in Heathcliff's talk seemed to give close acquaintance with the nerve that was in him—'"I love *my* murderer—but *yours* ! How can I?"'—and to make him seem like a character from Jacobean drama.

What the author was attempting belonged with the strangely reasoning and unreasoning figures of Middleton, Tourneur, and Ford, the presentation of a love which, when thwarted, became selfish, 'wolfish', melodramatically devoid of compassion, and which yet deserved to be contemplated: even Heathcliff, 'an arid wilderness of furze and whinstone' like his name, could be touched by some power greater than his own simple will-to-power. The result has, from the time of its very first readers, been disquieting; hardly any two seem to have read quite the same book, for it was so arranged as to balance against one another impressions of the events and characters which might suggest ways of interpreting it, but left that interpretation to the reader as he penetrated further into the construction of the whole. Criticism has not yet done with it.

The greater unity, indeed the perhaps over-zealous construction, of *Jane Eyre* (1847) bore the marks of a close study of both *Agnes Grey* and *Wuthering Heights*. It was a story told by a governess who was much more fully known to both reader and author than the Crimsworth of *The Professor*. Her loneliness, as at the centre of *Agnes Grey*, was intensified and multiplied, starting with Jane's situation as an orphan and subjecting her to separation from each of the individuals in turn to whom she had become attached—from Helen Burns by death, from Miss Temple who married a clergyman, from Rochester by his (and her own) desire for love without marriage, from St John Rivers by his wish for marriage without love. There was obviously some debt to the fruitful, implausible repetitions in the action of *Wuthering Heights*. Contrast of locality, such as had been observed in one of the very first notices of Emily's novel, had its counterpart too, and, as with the heroine's loneliness, the simple contrast was multiplied, in particular when the wooded warmth and luxury at Rochester's Thornfield Hall contrasted with environments which were bleak and hostile: with the earlier winter of discontent at Gateshead and at Lowood School, and with the subsequent region of heath and mountain where Jane was for the first time made quite destitute, before being led to the Rivers family. The place-names, like Emily's, had metaphorical overtones—Gateshead, the start of the heroine's self-discovery; Thornfield and Ferndean, where respectively she would lose and gain Rochester; Marsh-End, where her destitute wanderings ceased; and Moor-House, the same place seen from the aspect which would link it with Morton, where she would achieve a brisk and bracing independence. Such contrasts were accentuated, in the manner of Dickens, by strong and repeated sensory oppositions of warmth and cold, running through the impressions which discriminated between people as well as places, and by a relatively unsubtle but insistently suggestive use of weather and season, particularly the behaviour of the moon.

Above all, *Wuthering Heights* had shown how such impressions, when distilled into dream and reverie, might open a line of communication between the world of Angria and the 'real world' in the study of which, according to *The Professor*, chapter 19, 'Novelists should never allow themselves to weary'; out of this intercommunication might come answers to the romantic question posed by Crimsworth in chapter 1, 'Shall I . . . feel free to show something of my real nature, or—'.

To frame an answer matching the complexity of the question and risking the incomprehension behind that 'or—' with reference to her heroine, Emily had needed a structure which to its first readers verged on the sibylline, 'a dark tale darkly told'; the originality of *Jane Eyre*, on the other hand, was found in the heroine's simple freedom to show her 'real nature' from the very first page. To G. H. Lewes this made the novel 'an autobiography . . . in actual suffering and experience, . . . an utterance from the depths of a . . . much-enduring spirit: *suspiria de profundis*'. The Brontës would have read De Quincey's *Suspiria*, which had begun in *Blackwood's* in 1845 with 'The Affliction of Childhood' (later transferred to the *Autobiographical Sketches*). Its piercing visionary recollections had an important place in the development of that power to understand the world as children understand it which was first suggested to many people by Wordsworth. At the very time that Charlotte was beginning the book, if Mrs Gaskell's account is correct (though as late as 14 October 1846 Charlotte still says, 'I am doing nothing'), Dickens too was interesting himself in what could be 'expressed in the child's own feelings—not otherwise described' (according to his notes for *Dombey and Son*, which began to appear that October). To this ability Charlotte added another, which she could have found in Thackeray's first-person narratives in *Fraser's*, the ability to make the reader aware of more than the first-person narrator could be, carrying out as a novelist a feat such as De Quincey had described in a footnote in *Blackwood's*: ' . . . though a child's feelings are spoken of, it is not a child who speaks. I decipher what the child only felt in cipher.' Jane Eyre was imagined as writing ten years after the end of the story and so giving repeatedly the benefit of a critical view, as in Emily's novel, to a narrative which was a stronger, more emphatic form of Anne's story, of a girl whose attempts to find the great world and be independent in it led her at first only into greater solitude and greater dependence. Jane was obviously more passionate and this was so presented as to appear both in the good and the bad senses of the word. She was entitled to sympathy, but so too were those to whom she was an inconvenience in herself and in her clamorous self-assertion. The continuity of that self, learning to recognize the demands of others and to modify the expression of her own but never to give them up, was the nerve of the novel.

Jane was known from within by literary means which not only made the instant impressions stab sharply, but presented and marshalled them for contemplation as experience and not as abstractions. It was less a matter of self-analysis, though her habit of pausing to review what had happened to her obviously made the reader aware of her mind and nature, than of the grouping of the parts of the action and their concentration into metaphor. She was made interesting by not being beautiful and by being set, like the solitary governess in *Deerbrook*, in contrast to more favoured women who might have made heroines in more conventional novels. In the successive waves of the action she was thrown on to her own resources by the loss, one after the other, of two female mentors and two male. Through all this, the metaphorical contrasts of fire and ice gathered and grouped her developing experiences. Fire, from the 'lighted heath' of her first successful rebellion onward, subsumed her innate freedom and preserved the sense of its wavering activity. The metaphor was obviously related to Charlotte's and Branwell's term for the uncontrolled wishes of Angria, 'the burning clime'. This freedom was called into its fullest activity by the answering fires of Rochester, with his background of life in the tropics and his Angrian amorous history. The opposing metaphors from ice, generated by the literal wintry settings of the first two sections of the book at Gateshead and Lowood, gathered together the restraints of the 'real world'. Startlingly physical, these restraints were also, as Blake would say, 'mind-forged', willed by named characters, Reed or Brocklehurst, whose natures were grasped by what they had willed, though the anger of the young Jane's immediate repudiation was tinged with the comedy seen by the mature narrator. The results were then carried over into later parts of the novel when, for instance, the same metaphor of a 'black column' was applied both to Brocklehurst, whose power over the heroine was limited and comic, and to St John Rivers, who spoke with a less limited moral authority and a less insignificant egotism, but one which would 'force' her 'to keep the fire of my nature continually low, to compel it to burn inwardly', an 'imprisoned flame'.

As the true subject of the novel moved into view with 'you are too impulsive, too vehement' from Helen Burns, the first mentor Jane could love, literary analogies multiplied to define it. Helen Burns's favourite book, *Rasselas*, with its firm emphasis upon the discipline of reality, had 'nothing about genii' in it—meaning the freedom of fantasy in *The Arabian Nights*, but no doubt also in the work of those 'four Chief Genii' of the juvenilia. Helen's epitaph, 'Resurgam', stood similarly in antithesis to the motto of Angria, 'Arise!', which had implied that to contemplate such a kingdom of fantasy, its free flag a rising sun on a crimson field, was sufficient for the fulfilment of the ambitious command. In *Jane Eyre*, the way to a life worthy of resurrection was, however, not through the bleak subjection of the world within to the world without, but, in a fashion well understood by Wordsworth, by 'an interchange of action' from both worlds. It was as important that Jane should love her mentor as that she should accept the tranquil force of her rebuke. Throughout, the stress fell on Jane's uncoerced adjudging of what she received in this way. The only coercion she could accept came, finally, from her own independence itself. Feeling this threatened, once she had engaged herself to Rochester, she wrote to her West Indian relatives, so setting in motion the discovery to her of his prior marriage. This separated her from him by the 'hand of fiery iron' in the 'vitals', her own resolute independence: having found even the position of a wife unacceptable unless she was allowed to retain her employment as a governess, she could hardly be expected to tolerate an even greater submission as mistress. She believed it would be against 'the law given by God', but her words before this were '*I* care for myself.' The offer of marriage without love, from St John Rivers, also undervalued the self, as if it were, in her own satirical phrase, merely 'the husk and shell to the kernel' of her missionary energies.

It was in relation to the demands of the heroine's nature that the two men were discovered for what they were, with a definition and a melodrama as persuasive as anything in *Wuthering Heights*. But the decisive judgement was exercised by the woman who found herself between them, not by an observer. The defects of the one man went with the natural warmth which might make it possible for one whole self to meet another. The defects of the other were in the 'hardness and despotism' to be found beneath his heroic self-sacrifice. As in Angria, the more obviously flawed character was the one to succeed in love. In the novel there might be fuller reasons given for this, but little more change in their characters was exacted by the story than in the juvenile romances. Rochester was required to accept, as a penitent and blinded Samson, the dependence he had tried to force upon Jane, Rivers to recognize his solitary state and his supernatural marriage; the heroine's view of each man became less limited, that was all. She, for her part, was kept before the reader less in her changes—though these appeared, impressively, when she visited Mrs Reed and forgave her without fuss—than in her consistency, as more and more of 'the jewel within' was disclosed. She spoke, for instance, with the same astringent honesty as a child to Mrs Reed as to Rochester the lover and Rivers the 'comrade' (and dared, moreover, with the same honesty, to give the latter the last word, even though it might seem to be against herself).

As much as in behaviour or speech, the disclosure to the reader of the heroine's inner self was dramatized in dream and in related activities mediating between the inner and outer worlds. They offered at one extreme the uncontrolled emotions of Angria—the 'dreams many-coloured, agitated, full of the ideal, the stirring, the stormy —. . . unusual scenes, charged with adventure, with agitating risk and romantic chance'—into which she 'used to rush' in her separation (permanent, as she believed) from Rochester. At the other extreme there were the equally vivid, uneasily waking visions in which, when imagination caught a glimpse of 'the hills of Beulah', 'Sense would resist delirium: judgement would warn passion.' Sometimes there seemed to be plainly visible a development of detail taken from *Wuthering Heights*—the shivering child, for instance, in the dream before Jane's marriage, who clung to her in terror as Jane bent forward from the top of the thin wall of Thornfield in ruins, to take a last look at Rochester departing, and who, once the marriage had been stopped, was seen to be the love 'which he had created; it shivered in my heart, like a suffering child in a cold cradle'. Sometimes dream took the form, as for an author awake now to the mediating power of art, of vision externalized in the dreadful pictures she made at Lowood, with life still unexplored—the shipwrecked owner of a fair arm, the female forehead amongst the stars, the icy, half-concealed face of Brocklehurst, and the crowned spirit of Helen Burns, 'a ring of white flame'. Like the pictures of John Martin which the author admired, the dreams were rather too stagey to go through the book 'like wine through water'. Yet they were points in a romantic progression of inward activities running from the subjugation of the inner self by what Wordsworth called 'outward sense' all the way to the transforming power of which the imagination was capable when stirred by love.

As in *The Ancient Mariner* and *Wuthering Heights*, the intermediate steps in this progression depended on a developing awareness of a subliminal kinship between the self and 'Nature'. Jane's childhood experience of being friendless was made one with her intense awareness of carking cold weather; the vicissitudes of her somewhat schematic human relationships—particularly her first meeting with Rochester, her proposal of marriage to him ('You glowed in the cold moonlight . . . when you mutinied against fate, and claimed your rank as my equal. Janet . . . it was you who made me the offer') and her final reception of his telepathic summons—all took place under the moon, the prompter of the imagination and guarantor of what it discerned. Events at the time of Jane's 'offer', like the fluttering of night moths as in the serene last paragraph of *Wuthering Heights*, or the behaviour of a tree, groaning and riven by lightning, like the tree split in the storm when Heathcliff fled from the Heights, carried the sense of a mysterious kinship between external nature and the inner life. Characteristically, Charlotte went on, like Wordsworth in his *Ode to Duty*, to a vision which gave a similar source to the sense of inexorable moral obligation, a source both within and beyond the order of the visible world: at the first crisis of her threatened independence, Jane, dreaming of the terror of childhood punishment in the shadows of the red room, saw the one gleam of light there expand into the light of the moon itself before it became an admonitory form with 'glorious brow': 'My daughter, flee temptation.'

The attempt to make the sense of providence vivid by the methods of *Wuthering Heights* could be seen when the same harvest-time as had marked the main parts of the action in Emily's novel marked Jane's and Rochester's first plighting of their troth and also the burning of Thornfield, in which Rochester's failure to save his wife—and, it was implied, his sacrifice in trying—made marriage to Jane possible. In *Wuthering Heights*, a harder and more commonplace kind of goodness had been directly shown in the younger Catherine's developing ability to put away resentment and forgive. In *Jane Eyre*, readers were required to make more of a leap to believe in the connection suggested between human action and a goodness immanent in the created order, at the moment of Jane's summons to Rochester's side by an inner feeling 'as sharp' as 'an electric shock', when she was resisting Rivers's overwhelming appeal, in a room 'full of moonlight'. But there is no doubt that, as Coleridge would say, a 'human interest' was transferred to such a shadow of the imagination, from 'our inward nature'.

How well Anne Brontë had understood the work of her sisters appeared in her second novel, *The Tenant of Wildfell Hall* (1848). There was a double action: one part, like the Rochester plot of *Jane Eyre*, connected a mystery with an offer of marriage to a character who was already married; the other part, inset within the first, developed a version of the situation in *Wuthering Heights*, and this at two similar points quite late in the action's development, in 1827 just before and in 1828 just after it was resolved by the death of the 'villain', Huntingdon. This accomplished conflation of two difficult structures was a failure only if compared with its models. Its departure from them was the cause of its main weaknesses. The narration of the two actions, for instance, was given, in turn, to Gilbert Markham and Helen Huntingdon, whose marriage was to heal the wounds inflicted in both stories. Markham also passed the second one on to the reader and so was the teller of the tale as a whole; but he had none of the status in relation to the second story which went with that position in *Wuthering Heights*, for he had no kind of interaction with the heroine's story which he passed on, and he was not imagined with sufficient force to form a contrast to Huntingdon in his view of what the latter did or in what he did himself. In large part this was because, in his own section, he quite missed the immediacy of *Jane Eyre*; like her, he was writing it all up years later, but unlike her, he was doing so in letters to a friend who stood completely apart from the story and needed confidential explanations which were of no importance beyond showing realistically why it came to be told. To explain this was itself a source of weakness and prevented an electrifying jump, like Jane's, *in medias res*. The instinct of whoever lopped the novel for the Parlour Library reprint in 1854 by leaving out the unnumbered introductory epistle was sound, however reprehensible textually. (As he was followed by Smith Elder in their reprint of 1859, the omission extends to many in the twentieth century.) It was even a help when 'you', the first word of chapter 1, seemed to refer to the reader rather than to the 'crusty old fellow' of the introduction. The same difficulty, of having more constituents in her work than were justified by the relations between them, affected the flashback which Helen gave by means of a journal of events leading up to late 1827, and subsequently by letters between then and 1828. The journal explained the mystery all right, but it was so lengthy—no fewer than twenty-nine chapters out of the fifty-three—that the reader forgot that a mystery was being unravelled for a would-be suitor, and became absorbed in the heroine's sufferings. Subsequently her letters threw into relief the sufferings of her husband before his death and the faith which the author wished to express in the ultimate salvation of even the most wicked. But neither the narrator who passed all this on nor his friend who received it had any relation to the fears of the dying man or the faith of his wife, out of which might come sympathy with the one or corroboration or questioning of the other.

The talents of the author for direct narrative were left, then, to fill out a scheme which was more complicated than they needed. Domestic comedy admirably conveyed the orderly, intensely discussed happenings in the circle of a prosperous farmer. The comedy stood in contrast to sharp domestic suffering in the heroine's higher range of society, imagined with considerable—if not quite equal—skill. But there was no stronger link between the two than this contrast. The linear interest was well maintained in each; in the second of the two stories, the permutations of encounter were fearlessly presented, once a liaison had been discovered between the husband of one character and the wife of another; at crucial points the natural setting gave moving emphasis to the emotions; but the impress of a unique vision was lacking in the book as a whole. In this *Agnes Grey* had succeeded, from its very ordinariness and directness. By challenging her sisters on their own more imaginative ground with a story and a mode of telling it that resembled, respectively, each of theirs, Anne had been inveigled from her own vein, a vein which it took the author of *Esther Waters* to estimate at its full value.

The family competition and collaboration which lay behind the achievements of the three sisters in their mature fiction, just as much as in the juvenilia, was at an end before Charlotte finished her second published novel, *Shirley* (1849). She had been looking for a subject that would not limit her to private life. While anxious about the reception of *Jane Eyre*, she feared, in October 1847, that 'a mere domestic novel will . . . seem trivial to men of large views, and solid attainments'. First among such men, it was clear from the preface she wrote in December to the second edition of *Jane Eyre*, was Thackeray, whom she regarded 'as the first social regenerator of the day—as the very master of that working corps who would restore to rectitude the warped system of things'. It is a strange tribute to the Equilibrist and Tightrope dancer, but it shows clearly the bent of her own mind at a time when she saw signs of the 'warped system of things' in economic depression, unemployment, and the threatened violence of Chartist agitation, particularly in the north. When *Mary Barton* (1848) appeared Charlotte recognized that Mrs Gaskell had anticipated her in 'subject and incident', even though her own work was concerned with the clash of employer and employed in the Luddite insurgency of more than thirty-five years before. This period was connected, like that of *Wuthering Heights*, with her father's career, though more obviously. As a curate near Halifax at the time of the attack upon Cartwright's mill in 1811, the model for the attack upon Moore's in *Shirley*, he had been in some personal danger and in some anxiety about his own duties; the experience lay behind what he had written of the repelling of an armed attack upon a house, in his own novel, *The Maid of Killarney . . . a Modern Tale* (1818).

It was also the period when the writing of Scott and Byron and the deeds of Nelson and of Wellington had formed the daughters' ideas about masterful men and women. Both Nelson and Shirley were called 'Titans' in the book, and Byron himself had been associated with its action, since his maiden speech in parliament, quoted in Moore's *Life*, which Charlotte knew, had opposed a bill to make the Luddites' frame-breaking a capital offence. So the subject was attractive in many ways. In addition, it could give to the factors common to 1811 and 1848 a certain objectivity, resting on the suggestion that this was how, after a similar lapse of time, the issues of 1848 would appear (a technique which both George Eliot and Meredith were to use). Although, then, the plot turned upon the loss of markets consequent on the Orders in Council of January 1807, and upon popular resentment at the resulting unemployment, the action proper concerned the attitude towards the unemployed of clerical Tory, well-to-do radical, and innovating mill-owner, in relation to the impartial compassion of the female characters. Of these, Shirley had many of the qualities of Emily Brontë, while Caroline came increasingly to resemble Anne, suggesting that, for the author, the position of the daughters of a clerical family who could think for themselves was a position of vantage.

Much of the novel was devoted to the sources and development of this power to think for themselves—Caroline was just emerging from the 'marvellous fiction' of life before 18 into 'the school of Experience'; for Shirley, who had already emerged, her inner development in youth was recalled by her intolerable tutor. As in *Jane Eyre*, the inner life was dramatized in dream and vision, while close reference to literature made more precise its forms and its goals. The action then applied the inner resources so developed to the outer life of social relationships. Caroline was shown pointing out to Robert Moore how literature might stir him to 'new sensations ... to make you feel your life strongly, not only your virtues, but your vicious, perverse points', so that he might avoid being 'proud to ... working people ... under the general and insulting name of "the mob"'. Shirley, the mature woman, was allowed a vision at moonrise of the earth as an Eden, the joy of which gave her 'experience of a genii-life' and the power to join the angels on Jacob's ladder, while her tutor told how, as a girl, she had imagined 'a Seraph, on earth, named Genius' who, married to Humanity, fought the Serpent, 'rectified the perverted impulse, detected the lurking venom, ... purified, justified, watched, and withstood'. Recalling, it must be confessed, the inflated manner of Disraeli, such visions, stirred by natural beauty under the moon, were intended to identify the source of the independence out of which arose Shirley's passionate harangues, her wonder that 'people cannot judge more fairly of each other and themselves', her doubt 'whether men exist clement, reasonable, and just enough to be entrusted with the task of reform'.

In so far as the 'manners' presented were 'not essentially

different from those of the present day', the novel resembled the greater one set in a similar period, which was appearing while Charlotte was at work, *Vanity Fair*. The mode of treatment was similar to Thackeray's, a social panorama displayed with satire and sympathy by an omniscient 'I'. The touch, however, was much heavier; the author tended to assume an intimacy with the reader and then to hector and shout; the sense of having arrived and of daring to follow 'such writers as Dickens and Thackeray', which appeared in a letter of early 1848, when the book was making its 'slow progress', made the language frequently pretentious. Yet there was a steady Thackerayan tendency to expose to criticism, even to mockery, the opposing attitudes which formed the mainstay of the action. The church, which the author believed should be pre-eminent in that 'corps who would restore to rectitude the warped system', was presented in the vigorous farce at the expense of the curates and in much comedy where the criticism was only implied, as with the cleric whose wife 'held an opinion that, when her lord dropped asleep after a good dinner, his face became as the face of an angel'. The whole sequence in which a 'priest-led and woman-officered company' put to flight a similar force of Dissenters, while the soldiers were glimpsed on patrol in expectation of trouble at the mill, shaded into the actual attack; comedy, which in Thackeray's novel had moved the reader to notice vanity in a variety of senses, here gave a vision of the natural state of society as a state of disunity, arising from almost unavoidable limitation of view in the members of its main groupings. The group which the author knew best, the established church, led her to exclaim, 'God save it! God also reform it!' That that group contained some enlightened members was stressed in the narrative; that other groups might, too, was implied. Their leaders, the fat dealer in spirits at the head of the Dissenters or the even more broadly ridiculed spokesman of the insurgents, were dismissed with little more than lampoon, but their members, inadequately represented no doubt in the 'good' workman Farren, were to be seen as not wicked, but misled.

It was crucial that the female characters at the centre of the novel should be among neither the defenders nor the insurgents and that the night attack upon the mill should be heard through their ears and glimpsed with their imperfect sight. This brought the matter within the author's self-imposed limits—'Details ...

which I . . . cannot personally inspect I would not for the world meddle with, lest I should make even a more ridiculous mess of the matter than Mrs Trollope did in her "Factory Boy",' she wrote in January 1848—while dramatizing the limitations of the women's view. It was a more impressive presentation of the strength, and the weakness, of their position than the pretentious dreams of the central character.

The uncertain control of the materials of romance and reality in *Shirley* was remedied in *Villette* (1853). With a single, 'fiery' heroine (the adjective is found in the earliest reviews), who herself narrated the whole, it applied the method of *Jane Eyre* to material, from the author's experience in Brussels, such as had formed the basis of *The Professor*; the principle of construction came from the contrast, present but imperfectly developed in the first novel, between romantic and realistic fiction. This contrast emerged distinctly when, after the heroine's return to her family in chapter 4, the calamity which left her quite alone was hustled through with strange mockery of the reader of conventional romance.

> It will be conjectured that I was of course glad to return to the bosom of my kindred. Well! the amiable conjecture does no harm, and . . . I will permit the reader to picture me, for the next eight years, as a bark slumbering through halcyon weather . . .'

The reality was given in a brusque opposing metaphor which would become important at the end of the novel: 'In fine, the ship was lost, the crew perished.' Once, however, she was driven overseas by 'unutterable loathing of a desolate existence past', the narrative increasingly took the form of romance itself, as people from the past she had rejected reappeared, in particular the first of the book's two heroes. This young Englishman, 'handsome as a vision', who rescued her in her first tussle with foreigners, was the son of her godmother, though the reader was not at first told that this was so or that the heroine had recognized him. When, losing her way, she happened upon the very 'Pensionnat de Demoiselles' which she had heard mentioned on the voyage, and obtained employment, the same princely young man attended the young ladies there as doctor, and appeared to have a romantic attachment to one of them, or to the portresse. When the heroine became lost a second time and fainted in the street, the same Doctor John took her home to what seemed, when she woke, to be the house she had known when visiting his family in England. The natural development of her regard for the fairy prince of these marvels was parallel to a tendency, quite independent of the marvels, for 'the life of thought' to take the form of an 'infatuated resignation', 'unstirred by the impulses of practical ambition', and 'nourished with . . . the strange necromantic joys of fancy'. Out of this she was shaken by the chance to become a teacher, a profession which would obviously give the life of 'thought' more commerce with the life of 'reality'.

The second hero, M. Paul Emanuel, appeared as her teacher and colleague, small, dark, choleric, even 'malign', contrasting in every way with the princely Dr John. When the interest became transferred to this M. Paul and his associates, and the power of the novel increased enormously to define his nature, the publisher protested. The author appeared to agree, in a letter of 6 December 1852:

> I must pronounce you right . . . in your complaint of the transfer of interest . . . from one set of characters to another. It is not pleasant, and it will probably be found as unwelcome to the reader as it was, in a sense, compulsory upon the writer. The spirit of romance would have indicated another course, far more flowery and inviting; it would have fashioned a paramount hero, kept faithfully with him, and made him supremely worshipful . . . but this would have been unlike real life . . .

She had in fact held back the main interest till the later part of the book so that real life, and the criticism of real life proper to a novel, might stand out in contrast to the pleasures of romance. The second hero, equally important in both the life of 'thought' and the life of 'reality', was one whom the reader could see for himself and not only through the heroine's reactions to him. Her reaction tended at first to be conventional repulsion, while the reader could see further, could see his touchiness, for instance, as one form taken by a growing interest in herself. The characteristic scenes became full of the surprises of character proper to a novel, instead of the simpler wonders of romance, or the oddities of his 'character part' up till then. M. Paul was expected, for instance, to explode with egotistical irritation at the interruption of his class and still more when the heroine, interrupting, broke his glasses; but instead he proved to be 'overflowing' with forgiveness for 'the real injury'. By the third volume the reader was made aware of a symmetry in the interplay of the two literary forms. Dr John, the very agent of romantic rescue, and (to all appearance) of amorous intrigue, had become more and more realistically viewed until, by the end of the second volume, in a passage immediately preceding the scene of M. Paul's spectacles, he appeared not only beyond the heroine's reach but personally incompatible with her, obtusely urging her to play 'the part of officious soubrette in a love drama' between himself and another girl, and above all showing that 'no tyrant-passion dragged him back; no enthusiasms, no foibles encumbered his way'. By contrast, once M. Paul was realistically established in the present, his past was discovered as another romantic sequence: first his lost love, then his celibacy in the service of those who had kept her from him, and finally the grotesque age and deformity of the bedizened dwarf who was their only living representative. The mirror image constituted by this movement from reality to romance was completed when the elements from romance, which had in the first half of the novel been beneficent, or at most puzzling like the nun's ghost, seemed now to move at the command of the wicked fairy 'Malevola' and to act out the heroine's conviction that fate was her foe.

This conviction was at variance with the romantic coincidences which had rather made her seem, to the reader, uniquely favoured. It was explained as part of her hypochondria. The author claimed in a letter that 'anybody living her life would necessarily become morbid'. Hypochondria, active without a cause in *The Professor*, was here related to the calamity, all the greater for being hidden from the reader, at her home in chapter 4. Solitary, 'living my own life in my own still, shadow world', she tried 'studiously' to deaden all hope, to hold, she said, 'in catalepsy and a dead trance ... the quick of my nature'; in her solitude in summer, the result was ungovernable depression: like those on holiday, she 'saw those harvest moons', but, unlike them, she 'could not live in their light'. Beneficent echoes of her own and Emily's earlier fiction, strong in that image, became engulfed in a headier rhetoric of which the master was De Quincey.

Sleep . . . brought with her an avenging dream. By the clock . . . that dream remained scarce fifteen minutes—a brief space, but sufficing to wring my whole frame with an unknown anguish; to confer a nameless experience that had the hue, the mien, the terror, the very tone of a visitation from eternity. Between twelve and one that night a cup was forced to my lips, black, strong, strange, drawn from no well, but filled up seething from a bottomless and boundless sea.

On the other hand, in chapter 26, under the moon as under the aurora twelve months before, she felt 'strong with reinforced strength'. Thereafter, the growing relationship with M. Paul forced her to hope, but at the same time entangled her in its attendant experiences from the world of romance, and now she could not disregard them. The culmination of this movement came in a remarkable scene in which the heroine was given a drug which caused her to be addressed in the motto of Angria, 'Rise!' The result was the overwhelming necromancy of her wandering, like De Quincey under the spell of opium on a Saturday evening, through the park of Villette on a 'festal night'. There she took for the last time her position of an unseen observer in the shadows: by a happy stroke, the music from the park had already been connected in chapter 13 with 'my own still shadow world'; by an even happier, the only person to notice her in the park was Dr John, the fairy prince. Here she succeeded in misinterpreting everything that was of importance to herself, to the accompaniment of comment which opposed everyday detail to its fanciful transfiguration: 'Ah! when imagination once runs riot where do we stop? What winter tree so bare and branchless—what wayside, hedge-munching animal so humble, that Fancy . . . will not . . . make of it a phantom?' It constituted a climax of the most persuasive type of romance, in which the source of the deception is within the character; this was the kind of insight De Quincey had added to the potions and poisonings of Mrs Radcliffe and of Scott in *Kenilworth*. After this, the unmasking of that lesser phantom the nun, from the Gothic stratum of the book's construction, was a prelude to the discovery of Lucy's mistakes about M. Paul himself. So she was brought decisively out from the dominion of the 'necromantic' life of 'Fancy' into the professional independence in which, with his real care for her interests, he had set her up.

Anything more than this independence was seen as belonging to the romance, not to the novel. It would have been the culmination of the romantic intrigue that M. Paul, after sailing to the tropics to complete the last act of his quixotic 'revenge of purest charity' upon the relatives of his first love, should return to marry his second, the 'heretic' they had been scheming against. His tropical destination, like the background of Mr Rochester's troubles, had the Angrian associations of 'the burning clime'; his ship was the *Paul et Virginie* with its associations of French 'sentimentalizing' (her own phrase when the priest had made his 'little romantic narrative' of M. Paul's past, in chapter 34). Readers who wished for the high-flown heroism of Bernadin de St Pierre's famous romance or for a happy ending, like those who had earlier been mocked for surrounding the heroine's return home with the conventional associations of comfort and security, were allowed to believe as they list: 'Trouble no quiet, kind heart; leave sunny imaginations hope. . . . Let them picture union and a happy succeeding life.' But in fact her lot was to be solitary independence. The point was in the heroine's resilience in the face of the kind of experiences which made up conventional romance—Gothic 'horror', amorous intrigue, 'agony piled . . . high' (the phrase she quoted in a letter querying 'whether the regular novel-reader will consider' this 'sufficiently high'), even the affectations of the fashionable novel, in the story Ginevra Fanshawe told in chapter 21 of her own cruelty to Dr John.

At the same time as the first half of the novel used inferior kinds of fiction to illuminate its subject by contrast, it also gave piercing glimpses of the narrator's mixed attitude towards the profounder kind she was to use in the second half. Recurring, in chapter 17, to the image of the story which had twice (first metaphorically in chapter 4, then literally in chapter 6) afforded a contrast between 'reality' and the expectations of conventional romance, Lucy spoke of herself as 'the half-drowned life-boat man' who 'keeps his own counsel and spins no yarns'. For his 'yarns' could only be of 'danger and death'—of a kind, that is, which people 'cruising safe in smooth seas', like Dr John's mother, 'could not conceive'. In chapter 21, concerned whether she should reply to a letter from Dr John, the narrator staged a debate between Reason and Imagination which had a similar oblique reference to the whole business of writing this very novel, in which 'danger and death' were to be faced.

'But if I feel, may I *never* express?'
'*Never!* ' declared Reason.

An extravagant paean followed, to Imagination, Reason's 'bright foe, *our* sweet Help, *our* divine Hope', gilding her dreams. From these she awoke to rain and a dying lamp. Yet the whole novel was clearly an expression of feeling from this very narrator, giving more than 'a truant hour' to Imagination. When chapter 23 showed Vashti, the great actress, not only representing her grief on stage but attacking and 'worrying it down', she became the very pattern of a destructive way in which art might deal with 'what hurts':

> Before calamity she is a tigress; she rends her woes, shivers them in convulsed abhorrence. . . . on sickness, on death itself, she looks with the eye of a rebel.

The whole passage, from that point where 'Suffering had struck that stage empress' on to her final exile remote from Heaven, was quoted by Meredith, over thirty years later, when the *Fortnightly Review* asked him to choose some 'fine passages in verse and prose'; the only other prose he selected was Hamlet's address to the players, another comment upon his art made by the writer himself at the centre of one of his works. Had Meredith seen the force of Charlotte Brontë's placing of this passage at the centre of her second volume? He does not mention her at all in his letters, but *The Ordeal of Richard Feverel* in 1859 undoubtedly used similar methods to those which can be discerned in *Villette*: Richard's self-indulgent, romantic heroism and the inadequate book about life which his father had written became the central features of a construction exemplifying the misguided modes of living which made a victim of another Lucy. There seems to be as little doubt in *Villette* that the story was told in this way to stir thought about the dangers of self-expression in art and its relation to experience. A parallel and a possible source of the speculation and the method may be found in Thackeray's argument, in the central, fortieth chapter (February 1850) of *Pendennis*, about the ignominy, and worse, of 'selling one's feelings for money' in fiction and the 'aptitude for that kind of truth' which alone can justify the risk. In the two later novels a debate of such a kind became the principle of construction of the whole.

The result which in *Villette* vindicated the heroine's independence and solitude also vindicated the independence and originality of the writer. Had she known of the just opinion (and the future achievements) of George Eliot she would have been satisfied. On first reading, George Eliot wrote to a friend, in February 1853:

> I am only just returned to a sense of the real world about me for I have been reading *Villette*, a still more wonderful book than *Jane Eyre*. There is something preternatural in its power.

Three years later, and fewer years before beginning her own career as a novelist, George Eliot confirmed her opinion when praising Meredith's romance, *The Shaving of Shagpat*, in the *Westminster Review*, as an imitation of the Arabian Nights 'from genuine love and mental affinity':

> Of course the great mass of fictions are imitations more or less slavish and mechanical—imitations of Scott, of Balzac, of Dickens, of Currer Bell, and the rest of the real 'makers'.

At a greater distance it can be seen that her sisters must be associated with Charlotte Brontë. They are a remarkable trio, not only in the sheer emotional power which, for each in her degree, still holds the common reader, but also in the constructive power which Matthew Arnold was recommending in the year the last of their works appeared, 'that power of execution, which creates, forms, and constitutes'.

6. Later Minor Novelists

I

The historical novel appeared to have just as much prestige in and after the middle years of the century as immediately after the death of Scott. A wide range of novelists, from Eliza Lynn Linton in *Azeth the Egyptian* (1847) to George Eliot in *Romola* (1862–3), still attempted at least one example. A line of critical justification, for instance in *Blackwood's* for September 1845, had taken the high ground that historical fiction 'founds the ideal [that is, the imaginary] upon its only solid and durable basis—the real'. The theory implied an over-simple view of 'the real' and a doubt as strong as Harriet Martineau's about the capabilities of unaided imagination. Major examples like *Sylvia's Lovers* suggested, however, that imagination came first, to search out the matters of history which it could then interpret, often with reference to issues which belonged to the time of writing. But critics appeared less ready to see this than to praise the connections of the fiction with historical scholarship. Wilkie Collins's showy first novel *Antonina* (1850) was taken seriously as 'a clear and distinct picture . . . of the causes of the final ruin of the Roman empire' (*Gentleman's Magazine*), although the story it told was lumbering and imperfectly seen, especially at its climax. Collins's preface might insist on the 'exact truth in respect to time, place and circumstance' of the historical events narrated, but they were interpreted so as to appeal to the contemporary reader—the Goths in AD 408 bringing hope to an oppressed Roman of 'the "middle class", despised in his day', the priests continuing 'to confuse that which both in their Gospel and their Church had once been simple'. There was no doubt, however, that to call it *Antonina, or the Fall of Rome* gave to this tale of revenge and attempted seduction an air of grandiloquent misfortune, like Bulwer's trio of novels bearing 'the Last of' in title or subtitle.

The steadiest contributors to the sub-genre in the second half of the century, apart from Ainsworth, whose flow continued until

1881, were Charlotte M. Yonge and Anne Manning.[1] Some historical fiction by Miss Yonge appeared in every decade from 1850 to 1899 with the exception of the seventies, the decade of most of her general historical work—not only textbooks and volumes of stories from a wide range of history, but also the editing and translating of memoirs, prefatory writing for such work by others, and her own *Life of John Coleridge Patteson* (1873). Her attitude to history is that of a serious popularizer, capable of facing some of the problems of actual historiography. Her prefaces to books like *The Lances of Lynwood* (1853–5), *The Dove in the Eagle's Nest* (1865–6), *The Chaplet of Pearls* (1867–8), or *The Armourer's 'Prentices* (1884) do not take 'the real' to be instantly and easily available in history, but speak of the variety of the records, in verse and prose, which sometimes afford no more than 'hints' for the novelist. Where there is disagreement amongst the sources, the 'story-teller' claims licence 'to follow whatever is most suitable to the purpose'; and the purpose comes from the imagination, for the story-teller's activity is 'the shaping of the conceptions which the imagination must necessarily form when dwelling on the records'. The result is 'not a presentment of the times themselves, but of my notion of them', a picture which 'cannot be exact, and is sometimes distorted', though the action must 'keep within the bounds of historical verity', forgoing the freedom which Scott might allow.

Within those bounds Miss Yonge seizes the opportunities to feed generous admiration and hope—whether from chivalry in Aquitaine under the Black Prince, or the progress of the rule of law in fifteenth-century Germany. The skill with which leading roles are found for extremely youthful characters—for it is 'the young' who, it is hoped, will be enabled 'to realize history vividly'—leads to variations on the recurrent subject of the non-historical novels, the initiation into responsibility. In the strange circumstances of history resulting from child-marriage, the sudden accession to military command, or the unexpected loss of a parent in a feud with neighbours, the initiation is more decisive and picturesque;

[1] Charlotte Mary Yonge, 1823–1901, lived most of her life at Otterbourne, near Winchester. Keble, whom she called 'my master . . . in every way', was vicar of Hursley, to which Otterbourne was joined. Anne Manning, 1807–79, daughter of a Lloyd's insurance broker, made her name with fictionalized biography, such as *Mary Powell, . . . Mistress Milton* (1849) and *The Household of Sir Thomas More* (1851).

but the same imagination is at work upon it. The contrast may appear more of a routine when pairs of young men come forward, one of slight build and studious habits, the other more muscular and aggressive, the one 'sensitive and speculative', the other 'simple and practical', according as the plot stresses the ability of the soldier or the artisan; there may be even less doubt than in the non-historical fiction about how these young men will act; but, as in those other novels, the reactions of the young women are watched more closely by the author and are capable of surprising the reader. The greater range of activity allowed by the circumstances of history to both men and women means a rather more obvious moral range. The outright villains tend to be caricatures, like Narcisse de Ribaumont in *The Chaplet of Pearls*, while, at the other extreme, there is more scope for idealization such as Eustacie, the child-bride of that novel, suffers once she has ceased to be the much more interesting spoilt child of the court.

The fact that the action arises from the author's contemplation of a tract of history means that it comes with the ideas for its interpretation built in, so that it is not, as the non-historical fiction tends to be, all foreground. The foreground is specifically imagined—the converted stable where a printer, fellow-countryman of Erasmus, lives and works, in *The Armourer's 'Prentices*, the sufferings of the heroine of *A Dove in the Eagle's Nest* when forced to leave her bourgeois foster-parents for a freebooter's castle—but it is set in a firm interpretive framework, however simplified and tendentious. The distant attempts of the Swabian League to end hitherto authorized feuding give point to the latter heroine's steady and effective moral opposition to it. Here is the main difference from the slighter historical fictions of Anne Manning, where, although the historical circumstances, the Great Plague of 1665 or the earthquakes of 1750, affect the main action, they do not afford parallels by which to interpret it. Miss Manning's foreground plot, though attractive from its very artlessness, and certainly developing from the nature of the personages, uncomplicated though they are, tends to be of the kind that might occur at any period, so that, for instance, two of the most popular examples, *Cherry and Violet: A Tale of the Great Plague* (1853) and *The Old Chelsea Bun-House: A Tale of the Last Century* (1855), each centre upon the same story.

Charlotte Yonge is more ambitious, in following Scott and Manzoni. As much as anything in history it is Manzoni's building of his novel upon the separation of the betrothed in *I promessi sposi* that has furnished the 'hint' for the action of *The Chaplet of Pearls*. The lawless background of *The Dove in the Eagle's Nest*, too, may well have been suggested by Manzoni, as its title almost certainly was. (Manzoni's villain overlooks all possible approaches to his castle 'come l'aquila dal suo nido'.) Nevertheless, Miss Yonge is also skilled in using specific, often homely, detail to bring her historically significant action to the bar of ordinary life and keep it clear of the pitfalls and pretensions of allegory.

It is quite otherwise with Charles Kingsley, who was seen at the time as above all the 'speculatist', Mrs Oliphant's term for him when writing on 'Modern Novelists' in *Blackwood's* for May 1855. Compared with his, the tendentiousness of Charlotte Yonge is mild, even discreet. He came to history, in addition, in search of the materials of a lucid plot such as he had failed hitherto to invent. *Hypatia* (1852–3) was his answer to Carlyle's criticism of his previous novel, *Alton Locke*, that it gave 'the impression . . . of fervid creation still left half chaotic'. In the history of Alexandria in the fifth century, Kingsley found a coherent action and actors who could be differentiated as characters, and at the same time given a certain contemporary relevance, by their contrasted opinions. The result was a spirited charade of 'New Foes', sceptical, Emersonian, Tractarian, 'under an Old Face'. From his own combative armoury, Kingsley added opinions in favour of sexual love and the Germanic peoples and against celibacy and 'a world drained and tainted by the influence of Rome'. There was excitement in the political intrigue, the attempt of Orestes the prefect to secure the virtual independence of Africa, culminating in the murder of his ally, the female philosopher Hypatia, by a mob whose emotions had been fired by the preaching and counter-plotting of the partiarch Cyril. The stereotyped eroticism arising from Kingsley's objection to celibacy, however, sent a good deal of this action swerving towards the ludicrous: differences of opinion tended quickly to become questions of love, and love viewed with no sign of a smile, no slightest awareness of possible self-deception. Hypatia, a 'cold-blooded fanatical archangel' to the male characters fluttering round her, was herself 'irresistibly' attracted to a sceptical Jew, Raphael Aben-Ezra, at the same time attracted to a sceptical Jew, Raphael Aben-Ezra, at the same time as she sought to make him 'her instrument for turning back the stream of human error'. Summoning another possible proselyte, the monk Philammon, to correct his 'youthful ignorance', and finding he was 'Beautiful as Antinous! . . . as the young Phoebus himself,' she concluded, 'That is no plebian by birth; . . . it shows out in . . . every motion of the hand and lip. . . . By . . . the instinct of my own heart, that young monk might be the instrument . . . for carrying out all my dreams.' Kingsley went on to bring the pair of them closer together in a set-piece staged by a sorceress, so that Hypatia, after being made ready, in her despair, to worship a copy of the Apollo Belvedere, might see instead the living youth, after he in his turn had been stimulated by dancing girls and drugged wine. When people complained about Hypatia's demise, 'naked, snow-white against the dusky mass around', Kingsley could justifiably claim to be following his source; he overlooked the dubious flavour arising from his handling of the antecedent events. He had controlled the plot but had little control over the tone, which depended not on 'creeds and theories', to quote a letter to him from Froude the historian, but on 'ordinary passions and human influences'. Racial theories were similarly made ridiculous when the Goths were praised for the physique which enabled them to knock the Egyptians about and 'make a poor little woman feel like a gazelle in the lion's paw'.

In *Westward Ho!* (1855) the 'passions and human influences' were reduced to the level of a schoolboy combativeness and the tone suffered proportionately. The issues between England and Spain in the sixteenth century became simplified out of recognition when 'the half-century after the Reformation in England' was seen as 'one not merely of new intellectual freedom, but of immense animal good spirits'. The hero, Amyas Leigh, was another huge Goth given to boxing the ears of lesser breeds. His activities were suggested by Froude's research into those of Elizabethan seamen, but the narration reduced them much below Hakluyt and not much above Ainsworth. Admittedly Leigh's pursuit of revenge left him physically blinded and ready to acknowledge he had been 'swollen with cruelty and pride'. For most of the novel, however, he had the author's approval. The many readers who approved too made this his most popular novel, now that Great Britain had drifted into the Crimean War and the penultimate chapter could link 'Russian Despotisms' with 'Byzantine Empires' and 'Spanish Armadas' as examples of 'the devil's work' for which 'A day of judgement has come, which has divided the light from the darkness, and the sheep from the goats . . .'. Kingsley admired Carlyle too much to indulge in mere fiction-making. It must always move towards allegory, saying one thing while meaning another, and that with an application to the reader's immediate circumstances. The exception was the part—and here it was a very large part of the book—which was lively with the pleasures of the open air. Reviewers who made comparisons with Scott were to this degree justified, for Scott could communicate that pleasure as fully. It was no mean ability, supported in *Westward Ho!* by occasional happy quotation, in the epigraphs, from the poets who had shown the same ability (while doing much more with it) during Kingsley's lifetime; it still leaves the book with some power of enchantment wherever the simplified emotions do not prove intolerable.

A final novel, written when he had become Regius Professor of History at Cambridge, *Hereward the Wake* (1865), showed a sensible choice of subject, for his way was mapped out by the written legends, and his limited control of tone mattered less when his task was to enliven an eventful external history of battle, disguise, and witchcraft. But the picturesque simplicity which did well enough for action to elevate Hereward into 'a hero of heroes . . . beside Beowulf, Frotho, Ragnar Lodbrog . . .' required much more skill if it was to be adequate for his marriage to Torfrida or the movement of his affection away from her afterwards. That Hereward, rescuing Alftruda, 'forgot Torfrida a second time' was worth saying; it was clownish to go on, 'But there was no time for evil thoughts even had any crossed his mind.' Repeatedly comment was at this level, supplemented on occasion with falsetto heroics in the dialogue, and fictional cliché in the account of what was done: Torfrida 'swept past them all and flung open the bower where Alftruda was with the dead hero. "Out, siren . . ."'. The vividness of landscape seen only accentuated the inadequacy of the actors—though it must be said that, for the quieter moments of marital dissension and the drudgery of the wife of a hero in hiding, there was an unexpectedly keen eye. Such things, however, were rare. History more often took the author away from 'the real' by offering easy substitutes for the activity of imagining.

The life of the historical fiction of R. D. Blackmore[2] was similarly in his appreciation of the rural outdoors, in this case regionally identified by subtitles: *Lorna Doone: A Romance of Exmoor, Alice Lorraine: A Tale of the South Downs.* Thomas Hardy in 1875 praised *Lorna Doone* (1869) for 'little phases of nature which I thought nobody had noticed but myself'. There was certainly much close observation in Blackmore's descriptions—the imminence of frost felt in 'the way the dead leaves hung, and the worm-casts prickling like women's combs'. But any sense, such as Hardy gave, of a vast system whose disciplines men could not evade was dissipated in sentimental anthropomorphism—the spring's 'over-eager children (who had started forth to meet her, through the frost and shower of sleet), catkin'd hazel, gold-gloved withy . . .'. The imagined action set all this at the service of a plot which was as much in the line of the sensation novel as Blackmore's earlier books—beginning with manslaughter, continuing in mystery about both Lorna and the Doones, and reaching a climax in murder and revenge. It was a well-sustained crime story, the naïveté both contrasting with the subtleties of the natural description, and sustained by them in so far as the most important transactions took place out of doors.

The connections of this plot with history intensified the excitement of some of its violence and mystery and, as it was told in the first person, gave opportunities for discreet archaism which served well Blackmore's selfconscious prose rhythms. At the same time, the skill of the romance-writer was plain in the management of the narrative so as to bring the reader's feelings in behind the hero's exercise of his legendary strength. At the climax the innocence of the victim and the strong emotions connected with an interrupted marriage (which Blackmore could by now have noticed in more than one of Charles Reade's plots) seemed to license in the reader the indulgence of cheering on the husband's revenge, especially as, having spared his enemy's father, he then spared his son. The unarmed pursuit of an armed man, the tearing of a branch from an oak to bludgeon down horse and rider, and the readiness to be satisfied with a mere victory at wrestling, were of a piece with the antecedent simplicities 'handed down, to weaker ages'. It was all

[2] Richard Doddridge Blackmore, 1825–1900, was born in Berkshire, educated at Blundell's School and Exeter College, Oxford, and called to the bar, but his occupations apart from writing were, successively, schoolteaching and fruit-growing.

make-believe, compared with Reade's obsessive frenzy, but, in its unsubtle lines of action, emotion, and morality, it was as clear as a good Western.

Blackmore wrote four further historical fictions, without matching the lucidity of this first one. Sensational materials—predictions fulfilled in the heroine's attempted suicide in *Alice Lorraine* (1874–5), smuggling, murder, and disputed inheritance in *Mary Anerley* (1879–80), the preparations in *Springhaven* (1886–7) to betray a village to the French invader—failed to lead to the simple and satisfying antagonisms he needed. The 'trouble to keep clear' which he acknowledged to his publisher in the first of these was equally apparent in others. Selfconscious care with his prose seemed to hinder his control over the larger rhythms of romance. And romance was all he could manage. The preparations for a tragic action in *Alice Lorraine* were not adequate to the serious interpretation of life which tragedy would have required and Blackmore rightly gave it up. The comparison with *Sylvia's Lovers* which was invited by the period and setting of *Mary Anerley* only showed up the episodic and wish-fulfilling magazine entertainment he offered. His insight into non-human appearances and behaviour was matched in the human world only in the episodes—few, and often irrelevant—where country people could speak their own language.

Other comparisons, however, might set Blackmore in a more favourable light. The historical spy thrillers of Watts Phillips,[3] for instance, offered glib narrative of adventure and surprise, but nothing more. Phillips was a dramatist turned sensation novelist, trying to assume the respectability which might be conferred by history and vague social criticism. He might quote *The Times* to support the claim that the 'corruption of manners arising out of the spy system' was as prevalent at 'the present day . . . in France' as in the time of Fouché in *The Hooded Snake* (1860). But his work remained caught in the commonplaces of life (the love which 'breaks down the barriers of worldly prudence') and of narrative (they 'now knew each other as foes—it was diamond cut diamond . . .'). Even as early sensation novels, his were much inferior to Blackmore's.

[3] Watts Phillips, 1825–74, dramatist and cartoonist as well as writer of *feuilletons*.

II

The social novel developed more strongly in the second half of the century. Mrs Gaskell, the most considerable new talent it engaged, must have a separate chapter along with the work of her 'hero', Charles Kingsley, in this vein. The main period of mid-Victorian prosperity (from 1850 to 1874) was beginning. Carlyle had long been urging the necessity of improving 'the condition of the people' and now the possibility of doing so—by means of education, colonization, or even trade unions—seemed to be opening out.

Novelists tended to preach rather than to investigate and, in such an immense field, they found it hard to concentrate their attention upon particulars. Catherine Sinclair,[4] for instance, could announce, in the preface of *Cross Purposes* (1855): 'In free and happy England there are four kinds of slavery', that of 'overdone education' (she meant education of the mind and memory rather than the imagination), 'of overworked needlewomen', 'of intemperance', and, 'worse than all united, of Romanism'. Not surprisingly, her 'attempt to weave a story in which the evil of all these heavy yokes might be warningly portrayed' foundered as fiction, whatever its effectiveness as admonition. When she had approached an apparently more determinate social subject in *Edward Graham, or the Railway Speculators* (1849) it had turned out to be merely peripheral to a tale of bigamy and crime.

Dinah Maria Mulock's[5] romance about the social status to be achieved by self-dependence and a willingness to work, *John Halifax, Gentleman* (1856), did better by centring all upon one idealized figure, even though the varied action was untroubled by dull details of the work upon which his success depended. Instead it dealt in his ability to curb food rioters by his 'firm, indomitable will', his love for a 'gentlewoman' and the attractive coincidences that furthered it, his refusal to allow the notorious Emma Hamilton to visit his house and his later opening of it to her friend, Lady Caroline, in her dishonoured old age. In the process the more

[4] Catherine Sinclair, 1800–64, was born in Edinburgh, became secretary to her father, Sir John Sinclair, MP for Caithness, and carried on his philanthropic work after his death in 1835.

[5] Dinah Maria Mulock, 1826–87, was born at Stoke-on-Trent and came to London in 1846. Her greatest popularity as a novelist lasted from 1850 (*Olive*) to 1865 (*Christian's Mistake*). In 1864 she married G. L. Craik, a partner in Macmillan & Co.

spectacular social phenomena of the earlier part of the century—speculation mania, bank failure, the lessening of the distance between master and men, emigration as the way to prosperity for the penniless, and the emancipation of the slaves—all came under the eye or the influence of Halifax as representative of 'the best men', who 'ought to govern, and will govern, one day, whether their patent of nobility be birth and titles, or only honesty and brains'. It was characteristic of the mild meliorism of the book that the son of a discredited nobleman should be shown as able to gain a place in this world by his own efforts. For all the triteness and sentimentality, this was the fairy-tale of the middle classes at a time of maximum confidence, looking back over what had been accomplished. Nothing else of the author's achieved the status of this novel, which became a part of received mythology.

Thomas Hughes[6] was equally without artistic aims. 'My whole object in writing at all was to get the chance of preaching,' says the preface he wrote in 1868 for the sixth edition of *Tom Brown's Schooldays* (1857). At the time, the preaching was no deterrent to popularity, any more than it was in the case of Dean Farrar's lugubrious and widely read *Eric, or Little by Little* (1858). Each developed an action from the headings of a sermon rather than from the connection of the events, but, where Farrar used the dangerous weapon of vague fear, Hughes appealed to the eagerness to live. Each book was well received at a time of developing interest in the kind of education needed to administer a nation of augmented powers and responsibilities, an interest exemplified in the 1861 commission to investigate nine public schools.

Part 1 of *Tom Brown's Schooldays* survives well as fiction, even though almost half of that part is devoted to life in the country in the thirties, contrasted with the time of writing, when 'gentlefolk and farmers have . . . forgotten the poor'. The notion that schools exist to make 'good future citizens' stirred in Hughes an interest in the kinds of people who forward that work, 'the most important part' of which 'must be done, or not done, out of school hours'. Part 1, in which Brown seemed to be going to the bad, had unusual

[6] Thomas Hughes, 1822–96, an equity lawyer who became a QC in 1869, was educated at Rugby under Dr Arnold and at Oriel College, Oxford. Active in the Christian Socialist Society for Promoting Working Men's Associations, and in many of the societies so promoted, he was examiner and instructor at the Working Men's College, London, 1854–61, both critic and defender of trade unions, and Liberal MP, 1865–74.

vivacity; the more didactic Part 2, culminating with Brown in tears at the tomb of Dr Arnold, the idealized headmaster, was more of a tract. The instant success of the book spread Arnold's ideas widely, though in a simplified form. Its message to the countryside, with apt reference to the contrasting, less idealized vision of *Yeast*, was developed briefly in *The Scouring of the White Horse* (1859) and at length in *Tom Brown at Oxford* (1859–61). Both had hardly a trace of the earlier liveliness of characterization. The second of these recorded Oxford life in the 1840s as faithfully as was possible with a set of human types, but little action of any force was derived from them. Brown's opinions were left relatively vague except for his excitement at reading Carlyle. The 'mixture of self-conceit in his relish of the name "Chartist Brown"' was well observed, but only in passing. His tepid amorous troubles supplied a tenuous thread while the book drifted on. Like most of those who tried to make fiction out of social problems, Hughes had insufficient of the novelist's concern to select and shape.

Selection, at least, came more easily to Eliza Lynn Linton,[7] a professional journalist and author of the notorious attack, in the *Saturday Review* for 14 March, 1868, upon 'The Girl of the Period' for her 'bold talk and general fastness ... dissatisfaction with the monotony of ordinary life, horror of all useful work'. Although she was by 1854 opposed to 'the Emancipated Woman', with her 'absurd advocacy of exaggeration' in pursuit of 'the glittering honours of public life' (as she put it in an article in *Household Words*), she displayed, in novels with titles like *Realities* (1851), *Grasp your Nettle* (1865), *Sowing the Wind* (1867), the ability of women in careers and in crime, and the difficulties of their position when marriages go wrong. There was little more power to stir the reader here than in the historical romances with which she began, except when feelings like her own were expressed—for instance, in *Sowing the Wind*, the pleasure of 'work—a man's work—work that influences the world—work that is power'. But her energy and integrity could not compensate for

[7] Eliza Lynn Linton, 1822–98, free thinker and critical feminist, was the daughter of the Revd James Lynn. She was employed to write articles by the *Morning Chronicle*, 1849–51, by the *Saturday Review*, 1866–77, and, as Paris correspondent, by the *Leader*, the shareholders of which included Charles Kingsley, G. H. Lewes, and the engraver W. J. Linton, whom she married in 1858.

the obvious difficulty she had in imagining people who were different from herself.

One way round this was to make the interest depend less upon the persons than upon their ideas, as in *The True History of Joshua Davidson* (1872). This son of David, 'beautified at all times' by a 'look of peace and love', ostracized for consorting with harlot and thief, and celebrated in the narrative of a disciple called John, had less of character than the central figure of the gospels, though hardly less of symbolic importance, to judge by the repeated reprinting of the book on into the early twentieth century. Davidson sought 'the meaning of Christ' by stripping away 'all the mythology' to reveal a political programme: 'the equalization of classes' and the elimination of poverty and ignorance. Although Davidson's saintly energy and forgiveness to seventy times seven hardly belonged to a credible character, the account of his charitable and political activity was far from completely naïve: people he tried to reform found 'the monotony of virtue tired them'; and he himself, too ready 'to distrust one-sided partisans', 'never got the ear of the International'. Neither people nor circumstances, however, were imagined with the fullness of realistic fiction.

A curious intensity in the account of the relations between women marked her other novels—the heroine in the one she thought her best, *The Atonement of Leam Dundas* (1875–6), devoted to her dead mother to the extent of murdering her stepmother; Virginia in *Under which Lord?* (1879), whose devotion to the sister of her vicar had passionate overtones; or Perdita in *The Rebel of the Family* (1880), taken up by a woman who believed that 'Men ... have demoralized us, denied us freedom and education that they might govern us more easily through our follies and weaknesses.' *The Autobiography of Christopher Kirkland* (1885) showed the same intensity in the central character's 'strange deifying reverence' for the wife of an effeminate husband, or the jealous attachment to Kirkland of an older woman who kept a 'strange unrest and impatience ... within the outer envelope of her habitual tenderness'. Since most of Kirkland's other experiences belonged to the author herself it was reasonable to infer that the transposition of the narrator's sex obscured the nature of these relationships. But here as elsewhere, the originality evidenced in the materials did not extend to the handling of them.

III

The most striking development in minor fiction after the mid-point of the century is that of the novel dealing in sensation in the newspaper sense and in a contemporary setting. Like its modern descendants as the dominant form of light reading, the sensation novel is romance for the unromantic—for what Keats called the 'consecutive man', not for the man who, as he said elsewhere, 'is capable of being in uncertainties, mysteries, doubts, without any irritable reaching after fact and reason'. Essentially it contains a mystery or secret to be resolved with facts and reasons. Not only are the setting and, often, the mystery and its resolution of a kind likely to be found in contemporary life; they may often, too, raise contemporary issues, or use, or suggest, contemporary events; hence this sub-genre merges easily, in the hands of all its chief writers, into the social thesis novel. It is fiction for the reader who wants realism, hence its close connections with contemporary drama, at that time the most simply realistic of all the arts—there *are* the people, they walk and talk, there *is* the room, the garden, or even, in a famous example, Boucicault's *The Colleen Bawn* in Dublin in 1861, enough water for the hero to dive into.

In the *Spectator* for 5 December 1863 a review of Mrs Henry Wood's latest novel claimed that the contemporary realism of such works, and indeed the preponderance of good novels over works in other forms, resulted from 'the great progress of science, and of a scientific mode of thought'. Two years later, the young Henry James spoke of Wilkie Collins as working in a genre peculiarly 'adapted to the wants of a sternly prosaic age', producing 'not so much works of art as works of science'. Whatever such claims amount to—and James was clearly speaking of scholarship in general and not merely experimental science—it is suggestive that the sensation novel should have become prominent in the years immediately following the discussion caused by T. H. Huxley's defence of Darwin's *The Origin of Species* (1859) and by Herbert Spencer's claims for science as the answer to the question he asked in his 1855 lectures at the Royal Institute, 'What education is of most worth?'

Equally, of course, these were the years when the results of the repeal in 1855 of the stamp tax charged per newspaper sheet, and the abolition in 1861 of the duty on paper, began to appear in the falling price and the rising circulation of newspapers and in the foundation of new periodicals carrying fiction, though some of them not only fiction—periodicals from the *Halfpenny Journal* in 1861 to the shilling *Belgravia* in 1867, edited by M. E. Braddon, and the *Argosy*, also from 1867, edited by Mrs Henry Wood. Long before, at the turn of the century, Wordsworth had claimed there was a connection between 'frantic novels' and 'a craving for extraordinary incident which the rapid communication of intelligence hourly gratifies'. The very use of 'sensation' as an adjective may indeed derive from the application of the noun, since the late eighteenth century, to public events which evoked strong feelings of admiration or revulsion, whence the noun passed easily into the vocabulary of literary criticism, as in the review of Anthony Trollope's second novel in *Jerrold's Weekly Newspaper* in 1848, recommending it to readers 'who dislike to be wrought up to a pitch of strong sensation'.

It was not that a sensational press was new. What was new was a more or less respectable daily press at a penny and selling in large numbers. With the foundation of the *Daily Telegraph* in 1855, it began its long haul to overtake the circulation of the scandal-mongering Sunday papers which had never been a threat to reputable fiction. By 1869 R. H. Hutton feared that 'among the more cultivated section of society, . . . as the mechanical appliances of communication improve, all kinds of light reading will be swallowed up by the most sensational of all, the hourly history of the world'. Such a view overlooked the novelist's control of what was divulged and his power to prolong the pleasures of desiring more and more to be revealed (making the reader more vulnerable to surprise). Nevertheless, in developing such techniques, novelists were not only indebted to stage melodrama and to other novelists, especially Dickens and Bulwer; they were also making entertainment out of what more and more of their readers were becoming accustomed to in the new penny press. Charles Reade was a forerunner of the sensation novelists in this, relying as he did upon newspaper and other sources of 'matter-of-fact romance'. But although he shared their aim of producing sensation in the newspaper sense of the sudden excitement evoked by contemporary events or revelations, the excitement was too important to him to be made matter for a game of concealment, so that he did not develop their techniques of mystery.

In the fifties those who dealt in mystery tended not to resolve it with 'consecutive' care. Anna Harriet Drury's plot, for instance, in *Eastbury* (1851) certainly had secrets involving murder, and a confusion of identity of the kind Wilkie Collins eventually used. But the secrets, instead of being gradually penetrated by 'clues', were suddenly dispelled by confession. Nevertheless the contemporary setting was of importance—as, later, in Collins, to leave London was to leave 'the last skeleton row of houses—the last pile of bricks'— while the connection between the woman on whom the mystery centred and another who was prominent in the civil wars of the seventeenth century anticipated *Bleak House*, itself to become a quarry for the sensation novelists.

The centre of the mystery in a more notorious example, Caroline Clive's[8] *Paul Ferroll* (1855), was a murder committed in the present time of the story and not, as in *Eastbury*, one long antecedent, but it too was resolved by confession rather than detection. Unlike the fully developed sensation novel, *Paul Ferroll* wasted the effect of the murder in events unconnected with it, the knife-wielding of a mad butler and a further killing by the murderer, this time of his friend, who was leading a riot. In so far as the novel had a focus it was rather upon the Byronic character of the hero-villain: his close relationship with his daughter, his indifference when arrested, his high spirits in conversation with a friend acting for the prosecution, and his eventual confession to save an innocent person. A sequel, *Why Paul Ferroll Killed his Wife* (1860), missed an important principle of construction of the emerging sub-genre, the detailed correspondence between strange events in the present and facts and reasons from the past, for the new revelations had little to do with the multifarious characters and events of the primary plot.

Stories which Sheridan Le Fanu[9] published, in the *Dublin University Magazine* from as early as 1838, and in London in *Ghost Stories and Tales of Mystery* (1851), appeared to make the connection between his sensation novels and Gothic romance. The terror they fostered, however, was less an end in itself than a

[8] Caroline Clive, née Meysey-Wigley, 1801–73, was born in London and became well known for her verse as well as her fiction.

[9] Joseph Sheridan Le Fanu, 1814–73, was educated at Trinity College, Dublin, and called to the bar there in 1839. He became a newspaper proprietor, then writer for and, from 1869, editor of the *Dublin University Magazine*.

means to dramatize the sufferings of a troubled conscience in 'The Watcher' and 'The Evil Guest'. Fact and evidence in the latter story might increase mystery, from the apparent use of two murder weapons when one had succeeded; but even the grossest physical horrors took second place to the stranger evidence of guilt and apparent demonic possession. The stories are part of the prehistory of the sensation sub-genre. Although their action was placed back in the past its criminal ingredients were those which became popular in the sixties.

The florid, emotional manner of narration that was to characterize M. E. Braddon's fiction may already be found in G. A. Lawrence's[10] notorious *Guy Livingstone* (1857). Although the element of mystery was introduced very late in the novel and the detective worked entirely off stage, there was a portentous insistence upon the exceptional and exciting which certainly resembled the sensation novel of the 1860s. But it was impossible to conceal the commonplace nature of the story of a man of legendary muscle, drawn one way by the physical seduction of Flora and another by the gentle virtues of Constance. Lavish with historical and literary analogues for his hero, the author succeeded only in creating a sustained, stagey charade of extremes of riot and devotion, heroic violence and preternatural self-control. The background of 'high society' in Paris and by the Mediterranean was far away from the setting of common life characteristic of the sensation novel proper.

The sub-genre first developed fully in the work of Wilkie Collins.[11] In the fifties this consisted of magazine stories, stage melodramas (*The Lighthouse* 1855, *The Frozen Deep* 1857, *The Red Vial* 1858), and three novels. The first of the novels, *Basil* (1852), a Gothic experiment, attracted attention by the very contrast between its 'violent action' and its 'smooth commonplace environments' (to quote the review in *Bentley's Miscellany*). Details from

[10] George Alfred Lawrence, 1827–76, another barrister who preferred literature as a profession, was born at Braxted rectory, Essex, and educated at Rugby and New Inn Hall, Oxford. He attempted to serve as a volunteer on the confederate side in the American Civil War.

[11] William Wilkie Collins, 1824–89, was born in London. Son of a painter, he exhibited at the Royal Academy himself, 1849, but was called to the bar, 1851, became a friend of Charles Dickens, and began to write for *Household Words*, 1855. Collins suffered much from ill-health and from the drugs he took to deal with it. Both of the women he had lived with were acknowledged in his will, together with his three children by one of them and the daughter of the other, whom he had adopted.

common life—a suburb of unfinished streets, a girl first seen in an omnibus, a villain who drinks tea—were forced to yield an uncommon accumulation of dread. The opening references to an 'enemy ... ever lurking' already showed Collins's characteristic device of a first-person narrative written while the catastrophe was still unknown. The trouble was that the facts given hardly accounted for the fears raised—in the hero's amorous dreams, or in the apparition of a 'warning figure of dumb sorrow' presiding over his conversations with his wife. The facts revealed only the fictionally threadbare situation of a man of ancient lineage obsessed by a shopkeeper's daughter and so falling into the power of a family enemy. The centre of it all, the behaviour of the daughter, was left half-comprehended in easy phrases—'her own sensual interests, her vulgar ambition, her reckless vanity'—while all that was exhibited was her petulance. The villain, his face illuminated by lightning, to give it 'a hideously livid hue ... a spectral look of ghastliness and distortion', would have been quite at home in a Bulwer novel of the forties if he had had more to do, and that less indistinctly. Fear, still under heavy authorial prompting after the revisions of 1862, was apparently all the reader could expect.

Hide and Seek (1854), the second of these novels, showed the marks of his friendship with Dickens and a resolve, parallel to Bulwer's after the horrors of *Lucretia*, to cultivate 'the art ... of creating agreeable emotions'. *Hide and Seek* followed the success, from the fifteenth number of *Bleak House*, of Dickens's detective Bucket, and it was in Collins's detective, Marksman, that Dickens saw his own influence, as he said, 'most reflected'. The detection, however, was singularly tepid, helped on, in a way that Collins learned to avoid, by the ease with which Marksman became acquainted in London with the two people he had come from America to seek. A characteristic Collins motif appeared when the search ended in the guilt of the conventionally pious.

The Dead Secret (1857) paraded the common ingredient of sensation novels, a secret, without giving the reader sufficient reason to be interested in it. Even the most selfconsciously charged 'atmosphere' failed to arouse curiosity—another defect from which Collins learned. The mystery, however, arose from the device which gave him his first big success and to which he repeatedly returned, the assumption by one person of the identity of another. The resulting situation, which perhaps had its origin in Charlotte Brontë's *Shirley*, anticipated the one from which Mrs Henry Wood extracted all the pathos in *East Lynne*, that of a mother who, while concealing her own identity, has the care of her own offspring committed to her.

Neither devices of this kind, however, nor the experiments in crime, terror, and mystery conducted in the magazine stories—from the simple particulars of 'A Terribly Strange Bed' (1852) and the prosaic detection in 'Anne Rodway' (1856) to the exotic Radcliffean horrors of 'Mad Monkton' (1855) and the mocking first-person narration of 'A Rogue's Life' (1856)—prepared readers for *The Woman in White* (1859–60). Critics might still complain, justly, about 'the general tendency of the book to sacrifice everything to intensity of excitement'. But now it was excitement carefully controlled to penetrate secrets with exact circumstantial detail. Moreover, it was excitement about the nature and relationships of strongly marked personalities: the comradely vigour of an unconventional heroine, Marian Halcombe, ranged with a conventional hero and heroine against a pair of villains, the one, Fosco, zestful and engaging like the narrator of 'A Rogue's Life', the other, Glyde, conventional and cowardly. At least as much fear was evoked by questions about the villains' ability to commit crimes as by any they did commit. Was Glyde a person to 'hesitate at nothing to save himself'? What was to be made of Fosco's half-mocking bravado: 'Vast perspectives of success unroll themselves before my eyes. I accomplish my destiny with a calmness which is terrible to myself'? It was an enormous advance upon the attempt in *Basil* to evoke unaccountable fear in the language of the Gothic novel, though the old rhetoric lingered in Marian Halcombe's dream of 'the unknown Retribution and the inevitable End', or in Hartright's readiness to invoke 'the Hand that leads men on the dark road to the future'.

The narrative technique itself, as in *Basil*, assisted the evocation of fear; for threats to peace and safety were recounted while still present, in interlocking first-person accounts, a method derived no doubt from the eighteenth-century novel in letters, though it was supported by the more recent example of *Wuthering Heights*. It was also a method of diffusing through each section the evidence of a distinctive personality, each contrasting with the others in relatively simple ways, while each presented different and changing views of the main villain, Count Fosco. These techniques of characterization would have attracted more notice if they had not been overshadowed by Dickens's powers in this line.

Collins had been one of Dickens's young men, taken on to the staff of *Household Words* because, said Dickens, he was 'very suggestive, and exceedingly quick to take my notions'. He was now becoming a rival (and the success of this novel enabled him to resign from Dickens's staff, after serving for five years). He even seemed to have surpassed Dickens in the ingenuity of his plot—at least until Dickens went one better in *Great Expectations*, which followed *The Woman in White* in *All the Year Round*. Collins managed three plots, one inside the other, each of them depending, as had many of Dickens's, upon a matter of contemporary interest—in this case, one upon the activities of Italian émigrés, one upon the treatment of the insane, and one upon illegitimacy. The Italian plot, important at the beginning and the end, tapped a vein of interest arising from the first phase of the unification of Italy in 1859–60—a phase begun by a member of a secret society who failed in his attempt, in 1858, to assassinate Napoleon III, but brought down Palmerston's first government. Public attention for the treatment of lunatics and their property came to a head in the inquiry by a Select Committee in 1859, leading up to the legislation of 1862. Illegitimacy was no doubt an old stock device, but its legal aspect had also been recently much discussed at the time of the Legitimacy Declaration Act of 1858.

The three plots grew one out of the other, in a symmetrical construction which returned to its starting-point: the Italian professor of the opening became necessary for the concluding climax which secured Fosco's confession and confirmed Laura's disputed identity; the second plot, the mysteries connected with the Woman in White, was the first to be resolved and led to the third, Glyde's Secret, after discovery of which the action moved back to the second plot in pursuit of the evidence which would restore Laura's identity. It was a triumph of control over the details divulged: the resemblance between the manner of their narration and the giving of evidence in court may have suggested a newspaper report. Reviewers, however, as in *The Times*, observed the 'astonishing precision' with which 'the several parts dovetail into each other'. By 1862 they might observe the contrast with a mere heterogeneous accumulation of evidence in the anonymous *The Notting Hill Mystery*, serialized in *Once a Week*, where the reader was left to construct the story for himself. Collins provided not actual documentation but the fictional pretence of it, sometimes amusingly extended to make particular acts of writing into important events, as with Miss Halcombe's illness, brought on by sitting up when wet through to write her diary, or Fosco's final act of bravado in his written self-justification.

The construction of the book accentuated the pleasure of skilfully realized immediate impressions, by the persistence of anxiety about their outcome. The combination was as effective in sequences of simplest activity, like the pursuing of the hero when about to discover the Secret, as it was when description of the setting aroused expectation, when the oppressive London summer gave way to the freshness of the north and 'the distant coast of Scotland . . . with its lines of melting blue'. The direct dramatization of the resemblance between two women on which the second plot depended took place as one of them walked in the moonlight outside, while her drawing-master, by lamenting that most people require 'the beauties of the earth' to have a 'human interest' before appreciation of them is possible, drew attention to the instance the two women furnished. Debate about the matter was carried on in conversation beside the lake at Blackwater, one villain taking the sentimental English view, the other the practical Italian view, and both giving the reader the pleasure of the hopes and fears which may be stimulated by 'the beauties of the earth'.

The relations between the parts of the whole construction gave rise to unusual overtones. The courage and comradeship of Marian Halcombe, who gained nothing for herself, quite outshone the virtues of Laura, the conventional object of idealizing devotion; the chivalry and probity of a hero with the punning name of Hartright seemed pale beside Fosco's vigorous, insulting admiration for Miss Halcombe, his ingenious plotting in the belief that 'crime is . . . a good friend to a man and to those about him as often as it is an enemy', and his mockery of public attitudes to his fellow-villain's activities—forgery, the crime of Chatterton, 'the English poet who has won the most universal sympathy', and loveless marriage, which, even though 'for gold', secures unquestioned respectability. Hartright could beat such a man only by entering what looked like Count Fosco's own world, and undertaking an apparently subversive 'struggle against Rank and Power . . . and fortified success' which took him outside the law: 'The law would never have made Pesca the means of forcing a confession from the Count.' It was melodrama no doubt, but melodrama which stirred the beginnings of reflection. In suggestiveness of this kind Collins approached Dickens more closely than ever again.

No Name (1862) presented the subject of the limitations of the law more directly. This time it was the law preventing illegitimate offspring from inheriting even though their parents had later married. Again the search for a remedy meant going outside the law. A typically resolute young woman, Magdalen Vanstone, attempted to do so with the help of a self-confident down-at-heel rogue called Wragge; but she failed, while her sister, who patiently endured the wrong, found by accident the evidence which restored the inheritance (after multiple accidents had overtaken the other heirs). The large part played by accident, on top of the anomalous circumstances in which the parents had at first been unable to marry, meant that the action had little weight as a contribution to controversy, despite its mocking view of what marriage can and cannot accomplish. Collins was, characteristically, fashioning his plot from the materials of current argument rather than advancing arguments himself. Sensation was the object, secured on the one hand by comically detailed plot and counterplot and on the other by the sufferings of a heroine degraded to the point of suicide by her own plan of a marriage of convenience. On the one hand the particulars of wills, dates, movements from place to place demanded the kind of attention which might be given to a careful newspaper narrative. On the other hand there was a selfconscious attempt to make the reader share the heroine's sufferings by distilling them into the description of their setting—the dull airless evening in which Magdalen faced marriage with a despairing sense of loss: 'What is it? Heart? Conscience? . . . I feel as if I was forty':

> Now and then, the cry of a sea-bird rose from the region of the marsh; and, at intervals, from farm-houses far in the inland waste, the faint winding of horns to call the cattle home, travelled mournfully through the evening calm.

Nothing could mask the glaring incongruity between this kind of sensibility and sensational farce. But at least the result was better than Collins achieved in the novels of his last twenty years, when sensational particulars almost entirely displaced feeling.

Two major successes separated him from that decline, the one his most baroque and overblown attempt at tragedy, *Armadale*, the other his most lucidly contrived comedy, *The Moonstone.* In *Armadale* (1864–6) the essential element of fear came from the possibility that dreams might foretell the future and that human freedom might be an illusion. It was curious that a narrative so patently controlled by its author should be able to arouse this fear. He contrived that the son of a murderer and the son of his victim should bear the same name and that efforts to keep the sons apart should end only in their closer association, until they found themselves on the very ship where the murder had been committed. The interest, however, came not from such a monstrosity of coincidence or from the later fulfilment of the dreams which afflicted one of the young men in this place, but from the way in which the one who was more inclined to superstition actually endangered the other by trying to keep out of his way. The action seemed both to call into question these fears of some fatality attending their association and to endorse them by the fulfilment of premonitory dreams, with the result that the reader was led to expect some even more extraordinary resolution, not only of the plot but even of the metaphysical question itself. The latter expectation, like the threats of violence in the two previous novels, remained unfulfilled; but expectation was enough. A cleric's letter asserting eternal providence removed, if temporarily, the young man's 'paralysis of fatalism'. The same letter moved the novel's striking villainess, Miss Gwilt, to give up, if only temporarily, her resolute plans for enrichment by murder. The denouement turned upon a final exertion of free will by the fatalistic young man and a triumph of love even in the villainess, who took her own life when the alternative would have been to take his. It was a strange confection, stuffed with the material for a dozen ordinary novels, clumsy in its devices of documentation, especially Miss Gwilt's long, incriminating diary, and yet cumulatively stirring fears more than sufficient to draw the reader on, and even moving him with glimpses of Miss Gwilt's love and self-hatred. It may be another example, too, of the symbiotic relationship at this time in the sixties between Collins's work and Dickens's: Dickens's long preoccupation with the sense of fate and the achievement of freedom came to a head in *Our Mutual Friend*, launched, after long preparation, in May of the year in which *Armadale* began to appear in the *Cornhill* for November.

The Moonstone (1868) was conceived of as comedy, everything depending on, and in turn affecting, the course of true love. The comic idiosyncrasies of the first two narrators, who were responsible for over half of the story, established a tone of holiday quite different from the previous three novels. Certainly the prologue, giving the Indian history of the moonstone, supplied a background of fear but it was a romantic fear, no doubt assisted by the Indian romances of Meadows Taylor which had appeared in the mid-sixties, inferior though they were to his spellbinding *Confessions of a Thug* (1839). The menacing reference, in the preface, to the Koh-i-Noor diamond, which had been surrendered to the British Crown in 1850, was an afterthought tenuously connecting the story with contemporary fact. In the novel, contemporary matters were mentioned, only to point to the contrast, tongue in cheek, between a conspiracy connected with 'a devilish Indian Diamond' and 'an age of progress ... in a country which rejoices in the blessings of the British constitution'. Although Collins's justification of the tale as tracing 'the influence of character on circumstances' (since the 'conduct pursued, under a sudden emergency, by a young girl supplies the foundation') was reasonable, it only served to underscore the primary motive of entertainment. For the 'circumstances' began with the keeping of a secret which, if told, would have put an immediate end to the game before it began. Nevertheless the heroine's eventual reasons, arising from her regard for the culprit, touched the story with genuine emotion and gave colour to the author's claim for the importance in the whole work of her character and conduct.

The same might be said for that of the other young woman attached to the colourless hero, Rosanna Spearman. Although, as a means of diverting suspicion, she did not entirely break the holiday mood, the part of the plot centred upon her was full of Collins's characteristic sympathy for servants and for women. There was a moving short story in the understated account of a thief of obviously unusual education and ability who, when reformed, found honest people a reproach to her, spent her spare time reading, and made a friend only of a crippled girl with whom she had pathetic plans for an independent life. It was a pity it had all to be subordinated to the lucid giving of evidence, for the calm and self-possession of the letter in which she gave it were hardly appropriate to despair and, moreover, showed no sign of her anguish at the prospect of separation from her friend, Limping Lucy. Collins had hold of a subject, but it belonged to a different kind of novel in which all would not be subordinated to the investigation of a single crime.

The Moonstone is a notable early example of a long fiction entirely given over to such an investigation. Collins had already written shorter pieces with this limitation of interest, such as 'A Stolen Letter' and 'The Biter Bit', the models for which were no doubt stories like De Quincey's 'The Avenger' in 1836, or those of Edgar Allan Poe in the 1840s, or the series in *Chambers's Journal* from 1849 to 1852 of 'Recollections of a Police-Officer', by one 'Waters', which were collected in 1853 and supplemented by another volume of them in 1859. The whole latter series purported to be the work of one of the plain-clothes detective police (as distinct from the uniformed preventive) recently established at the time the series began.

Longer fictions with some degree of detective interest, though without a professional detective, are found, of course, well before Collins. The crime, the criminal proceedings, the wrong person accused on circumstantial evidence, appear in Irish novels like the Banims' *Crohoore of the Bill-Hook* (1825) and Carleton's *The Black Prophet* (1846) and (with clues gradually followed up, but no trial and only minimal mystery) Marmion Savage's *My Uncle the Curate* (1849). The investigator who seeks out the evidence to prove a convicted man innocent is there in the Edgar Adelon of G. P. R. James's *The Convict* in 1847 and the parson of Warren's *Now and Then* in 1848. Early in the sixties, Miss Braddon, prompted by Dickens and Collins, saw how the miscellaneous sensationalism of her first fiction could be clarified by limiting a novel to the careful investigation of one crime by her amateur detective, a friend of the victim, in *Lady Audley's Secret* (1861–2). The professional detective had entered longer fiction in the previous decade with Dickens's Inspector Bucket in the novel on which she drew for her main action.

In *The Moonstone*, then, Collins was not founding a new sort of novel, but selecting from existing materials, and playing, with mischievous ingenuity, upon the expectations which readers might already have. This ingenuity led him to many of the features of later detective fiction. The principle, for instance, later enunciated by his detective in *My Lady's Money*, 'suspect . . . the very last person on whom suspicion could possibly fall' was carried to such a length in *The Moonstone* as to prove guilty the person who had turned amateur detective when the professional, Sergeant Cuff, had failed. For a time Sergeant Cuff, by claiming '*Nobody has stolen the Diamond*', seemed to be giving the story something of the interest of *The Woman in White*, most of which was concerned with whether crimes had been committed and what they might have been. In *The Moonstone*, however, this was a mere diversion, and the acceptance by the other characters that there had been a crime limited the action to the search for the culprit. As has been suggested in the case of Rosanna Spearman, so determinate a subject inhibits the cross-relations and comparisons by which, in the novel proper, the reader is led to discover for himself what constitutes its action.

Of Collins's subsequent fictions only *The Law and the Lady* (1875) is concerned entirely with one piece of detection and even this takes the older form of an amateur detective's search for evidence to clear a clouded reputation. The greater detail of the search comes from Collins's own habit of construction out of particulars exactly identified, as in a newspaper. There is however none of the skill of *The Moonstone* in evoking and sustaining the reader's interest in the search, while the quantity of evidence furnished by one grotesque witness, an unreliable and histrionic cripple, makes much of the investigation a crude horror story. In the novella *My Lady's Money* the eccentric professional detective plays a quite subordinate role. Collins's other detective stories are of this mixed kind; he does not succumb to the simple interest of the mere conundrum. The detection in *I Say No* (1884), for instance, which appears to point to a crime, reveals something more interesting, a series of efforts to conceal a father's suicide from his daughter.

It was the sensational crime story that Collins developed rather than simple detection. He was ready to use recorded crimes (as he had done in taking the identity trick in *The Woman in White* from an eighteenth-century French case); the extraordinary events in the novella *The Dead Alive* (1874) were factual, he told his readers in a Note in Conclusion: 'Anything that "looks like truth" is in nine cases out of ten the invention of the author.' He was ready to revamp early work, rehandling in *Jezebel's Daughter* (1880) his stage melodrama of 1858, *The Red Vial*. By 1880 such stories were being syndicated in the press, reaching vast audiences in America and the antipodes as well as in the north of England. *Jezebel's Daughter* was the first of Collins's books to become a *feuilleton* like this, followed by *The Evil Genius* (1885–6) and *The Legacy of Cain* (1888–9). M. E. Braddon had been publishing in this way for the same newspaper proprietor, W. F. Tillotson of Bolton, since 1873. Two of the most successful sensation novelists thus played their part in the rapid development of the provincial press after the abolition of the paper duty in 1861.

A similar lucrative fate would have been appropriate for other novels of Collins's last twenty years. Their unusual plots turn upon the materials of the sensational press—physical or mental disabilities like those of the blind heroine of *Poor Miss Finch* (1871–2) and her fiancé, for whose illness nitrate of silver is prescribed, giving him a blue-black face, or of Mrs Gallilee in *Heart and Science* (1882–3), eventually sent mad from having 'deliberately starved her imagination and emptied her heart of any tenderness of feeling'; doctors and their treatments are naturally prominent, like the callous scientific experimenter of *Heart and Science*, Dr Benjulia. Collins assumes a reader interested in contemporary discussions about marriage and divorce in *Man and Wife* (1869–70) and *The Evil Genius* (1885–6), prostitution in *The New Magdalen* (1872–3) and *The Fallen Leaves* (1878–9), and 'the hereditary transmission of moral qualities' in *The Legacy of Cain* (1889). He appears to have believed himself to be contributing to the discussion of such matters, but again the situations he deals with are so unusual as to make generalization from them impossible. At least he leaves his events to speak for themselves, with only a little rhetorical or sentimental urging in *The Two Destinies* (1876), or an occasional Dickensian sarcasm, like 'Done in the name of morality', as the heroine of *Man and Wife* departs with her now-hated husband. In prefaces and footnotes, however, Collins is ready with the facts or corroborating authorities, to suggest not only that things recounted in the fiction could have occurred, but that they may therefore be taken as evidence for or against hypotheses about the real world. His hold over curiosity remains, but it is curiosity about strange events—some of them variations upon those of his main successes, like the confusion of identity which recurs in the extraordinary circumstances of the twins who are both in love with Miss Finch, as well as in the crimes of 'The Haunted Hotel' (1878) and *Blind Love* (1889–90). It is rare for the reader to be moved, as he is, in *Man and Wife*, by Anne Silvester's ability to suffer with weary courage. The author's ability to feel and to make others feel becomes increasingly attenuated, so that Walter Besant can take over Collins's plans for the last quarter of the unfinished *Blind Love* quite efficiently, without any need to preserve a distinctive emotional texture.

On the other hand, feeling, of a somewhat effusive kind, is the distinguishing mark of Collins's immediate follower, M. E. Braddon.[12] 'Wilkie Collins is assuredly my literary father,' she said in an interview of 1887. But earlier in her career she believed she should aim higher, writing to Bulwer in the first decade of her long fame: 'In you—and Dickens—the art of the novelist has reached its highest perfection—and you alone will I choose for my Gods.' Dickens's methods of over-emphasis fostered some of her own worst habits, but it was to Bulwer that she wrote, 'I shall always consider myself in a manner your pupil.' It was an appropriate sponsorship, even though some of the reasons for its acceptance on both sides will not stand examination: flattering him as 'the greatest writer of the age', Miss Braddon believed his example could lead her to peaks of '*poetry* or *truth*'. Actual similarities between them arise rather from the elaborate attention they give to crime and from their habit of rhetorically heightening emotions which already tend towards the extravagant through their lack of subtlety. Her rhetoric is not, however, quite of Bulwer's kind, for Miss Braddon has too much common sense to believe the sublime to be within her reach and she has the genuine impetus of her own astonishment to protect her from writing which is merely meretricious: it is as if, although these bizarre forms of life have obviously been recorded and arranged to entice the reader, their attendant emotions force her to cry out. Yellowplush justly reproaches Bulwer for 'sham morality', and 'sham poatry' but Miss Braddon's

[12] Mary Elizabeth Maxwell, née Braddon, 1835–1915, daughter of a failed London solicitor from whom her mother separated in 1839, went on the stage in the provinces, 1857–60, and began to write verse and fiction. From 1860 she cared for, and added to, the children of her publisher John Maxwell, whom she married on the death of his invalid first wife, 1874.

writing, however fustian its breathless adjectives and rhetorical questions, is not mere fake. She believes what she says, with an enthusiasm the force of which her reader cannot easily resist, even when laughing at it.

The whole performance, in book after book, is of a piece, the clear, unsubtle events matching her decisive moral distinctions and her exclamatory division of the life of her characters into fateful phases. Without this honest abandon the events would be, as in much of the later Collins, merely ingenious. The smaller proportion of her characteristic rhetoric in *The Trail of the Serpent* (1861), the revised form of her first novel *Three Times Dead*, makes it indeed much like inferior Collins. And although she felt, when writing to Bulwer, that without 'a strong coarse painting in blacks and whites, I seem quite lost and at sea', her handling of commentary, action, or speech often shows her endearingly aware of the pitfalls of a high-flown manner.

Such habits, however they set her apart from novelists who depend entirely upon extraordinary incidents, cannot obscure the fact that her power over the reader is founded upon an appeal to curiosity about the events rather than about their causes in the motives and potentialities of the characters. In *Lady Audley's Secret* (1861–2) she selected from the heterogeneous variety of *Three Times Dead* the few events she needed—bigamy, a husband's survival and unseen departure after his wife's attempt to murder him, and a deception about the identity of the dead. These things were fitted into a framework derived from *Bleak House* —a mansion, a baronet doting on a wife of unknown antecedents, the wife's 'exhaustion' when anything reminded her of that earlier history, and the grave warning she received from the lawyer who had investigated it.

But Lady Audley's predicament had none of the overtones of *Bleak House*. Although the social reasons for her resentments were eventually revealed—poverty, and the limited range of occupations open to her, when left by her husband to support their child—the book was not so constructed as to develop such matters. The reluctant detective might ruminate in chapter 24 about the energy of women, the 'self-assertive sex. They want freedom of opinion, variety of occupation, do they? Let them have it. Let them be lawyers, doctors, preachers, teachers, soldiers, legislators . . .' But none of this was imaginatively developed in the fiction, while Lady Audley's own energy in a crisis was accounted for by hereditary madness. However much her provocations might suggest something wrong with society, they hardly made her 'representative' when all the emphasis fell not upon social cause but upon criminal result and her hereditary madness broke the connection between the two. The reader knew far too little about the lady to do more than accept this 'taint' on trust; all he had was her judgement upon herself, 'I suppose I am heartless', and the picturesque contrast between her crimes and her golden hair. By the same token he did not know enough to take her as sane either. The arrangement of these materials so as to appeal primarily to the reader's curiosity made quite superficial any resemblance to Dickens's use of suggestion and implication.

For *Aurora Floyd* (1862–3) a character was devised who should be as arresting, but for the opposite reasons; instead of the unspecified inner demons within a golden head, the new heroine, who gave a nickname to the author at the time, Miss Aurora, was endowed with 'ebon hair . . . shining like snakes', and a dubious familiarity with hostlers, only to mask quite conventional feelings, like the 'reverent admiration' and 'romantic fancy' of her first mature attachment and her even stronger affection for her father. More subversive passions appeared in her notorious horsewhipping of a stable-hand—her tiny jewelled weapon 'stinging like a rod of flexible steel', her tangled hair falling to her waist. Nevertheless, 'this simple drama of domestic life' allowed the heroine to survive both bigamy and the suspicion of murder and dismissed her bending over the cradle, her second marriage legitimized. She was already the 'fast' girl with whom Rhoda Broughton would soon provoke her public. Because it was only appearances that were against her she was easier to present with clarity than the more complicated Lady Audley, who was finally left as no more than the 'kind of half-illusion' of Henry James's sympathetic account. Sensual complications in Aurora's 'sovereign vitality' were hinted at in images—the satisfaction in the whip 'stinging like . . . steel', the threat and the attraction of the groom the heroine married, 'a vulgar, everyday sword' in a 'beautiful scabbard'.

A parallel simplicity of conception in Olivia, the villainess of *John Marchmont's Legacy* (1862–3), failed because of the complicated actions demanded of her by the plot—the 'killing' of her passion for the hero before she contracted a loveless marriage, and the jealousy which subsequently moved her to take revenge upon him for marrying once and then to reunite him with his kidnapped first wife when he was about to take a second. The large terms of the insistence upon unrequited passion, exacting joyless duties from a woman of 'genius, indomitable courage, and iron will, . . . her powerful mind wasted and shrivelled for want of worthy employment' were mere bombast when they could not explain her actual effect upon the course of events. Pastiche of *Great Expectations* in her final ruined state of mind ('around her . . . ruin and decay, thickening dust and gathering cobwebs') only invited adverse comparison with even so limited a Dickensian conception as Miss Havisham. In the male villain, too, an urbane failed painter, the author was valiantly attempting, as she wrote in a letter, to make 'the story . . . subordinate to the characters'. But although he was unusual for the role, he was hardly adequate for the prolonged inner view required by a climax of suicide. Her problem was not in making the story subordinate to the characters but in imagining them fully enough for its demands.

In fact Miss Braddon worked better when less ambitious, developing her ability to sharpen the reader's appetite for what came next by engaging his sympathies. In *Henry Dunbar* (1864), which she depreciated to Bulwer after he had admitted enjoying it, the solution to the mystery was plain to the reader from an early point and the interest centred upon the manner of its discovery by the characters, especially the criminal's daughter. There may well be a debt to Caroline Clive's *Paul Ferroll* for this motif, but the working out of it, the girl in heroic collusion with her father against a detective who was her lover, showed the yoking of curiosity to strong, simple emotions which gave the author her power over readers common and uncommon.

Her most extraordinary novel, *The Doctor's Wife* (1864), tried to make what amounted to an answer to critics who were by now protesting about the materials of 'sensation' fiction and their supposed moral effect. Here were methods of self-mockery more like Thackeray than Bulwer, and a framework which came from Flaubert's *Madame Bovary*, out only seven years and not at all a common quarry for popular writers. The dreams of her heroine, Isabel, were nourished by 'everything that was romantic, and different from her own life', especially by Byron, Shelley, Dickens, and writers on the French Revolution. Even the poetic lounger about whom these dreams crystallized had 'delicious' passages of sentiment 'worthy of Ernest Maltravers, or Eugene Aram himself'. So it was 'the works of great men' (the author's phrase when writing to Bulwer about the novel) which were most dangerous to this immature reader, because it was all too easy for such a reader to simplify their relationship to her own life. The stories which her friend Sigismund Smith wrote, on the other hand, about characters such as Count Montefiasco, the Demon of the Galleys, kept fiction to a realm of its own, with the result that Smith and, it was implied, his readers remained uncorrupted by it. Eventually, however, even Isabel's desire to find in life the things that attracted her in literature was strictly limited: her poet's suggestion that they fly to Italy together left her grieving that on his side 'there had been no Platonism, no poet-worship' such as would have allowed her to give him 'the poetry of her soul' while giving to her husband her duty and obedience. Impudently, the third volume led the poet to his death through the action of a woman whose love for him was less anomalous. While it could not be said that the language established an attitude of any subtlety toward the heroine or her plight, it did evoke for her a unique mixture of sympathy and amusement.

A second novel to forgo reliance on crime for its main interest, *The Lady's Mile* (1865–6), was greatly inferior. Miss Braddon's habitual hyperboles hardly made for penetrating social comment. She could point to the contrast between the Hyde Park parade, 'the mighty tide of fashion's wonderful sea', and 'poverty and crime prowling side by side in their rags'; her popular novelist, now become Sigismund Smythe, could liken 'the lives of the women of this age' to 'this drive they call the Lady's Mile. They go as far as they can, and then go back again.' But the fiction itself gave only the most obvious account of the monotony of these lives or of the 'women who lose themselves in some region beyond'. The author's enthusiastic manner, which assured the reader he was being introduced to the most striking of scenes and people, was not the one for satire or even for the distinctions of tone and emphasis required of a 'light' social novel (her word for it). Lavish description of dress and décor in the homes of the newly rich could be considered as disapprobation, but need not be. Occasionally her familiar excess of adjectives might archly define a character's state of mind or her own attitude towards it, as with the fair-haired girl who 'would have preferred to be a heavy-browed person of the masculine order, with blue-black hair and aquiline nose, instead of that dear little insolent *retroussé* which seemed perpetually asking questions of all humanity'. But the commonplace tale of mercenary marriage and financial disaster which followed left both the state of mind and the author's attitude unexplored. There was a moral of a sort in a neglected wife's settling down to a life of devotion to the husband she had nearly deserted; but the wife remained neglected and the author, by obviously approving, showed she had simply dropped the charge of monotony implicit in her title. The artist, Crawford, whose activities went uncriticized, remained quite irrelevant to the social theme. The conclusion seems unavoidable that a 'social novel' was beyond her powers.

With *Sir Jasper's Tenant* (1865) she had already returned to what she called, in writing to Bulwer, 'the old sort of thing, mystery, murder, and so on'. Sensation was her métier and by now she knew it. Criticism of this kind of writing reached a climax with a long article by Mrs Oliphant in *Blackwood's Magazine* for September 1867. Beneath the moral outrage here could be seen the valid recognition that the genre, because it dealt in a greater range of human experience, had brought to English fiction some of the freedom which characterized French. There was admittedly only a slight movement in that direction, as is clear when *The Doctor's Wife* is compared with *Madame Bovary*, which suggested it. Indeed, what most offended Miss Braddon in the moral denunciation of her work was that she believed herself to be in agreement with her critics in matters of morality. But the range of what might be presented in fiction was undeniably extended. For the tasteless horrors of *Birds of Prey* (1866–7) and its sequel *Charlotte's Inheritance* (1868), in which the villain slowly poisons his friend and tries to do the same to his friend's daughter, now his own stepdaughter, the exemplar appeared to have been Balzac, whom the murderer himself praised, saying 'Give me a book that is something like life.' But any further connection with Balzac was hard to find in a story where two of the birds of prey repented effusively, and the murderer, a dentist turned stock-jobber, appeared to have only the most limited appetite for life. Long before her career finished with the posthumously published *Mary* in 1916, she had seen the scope for melodrama afforded by the subject of Zola's *Les Rougon Macquart* novels, 'Histoire naturelle et sociale d'une famille'. But the new ingredients, for instance, the varieties of Parisian life and political opinion in *Ishmael* (1884), did not radically alter her superficial and exclamatory habits.

The same must be said of late work of hers in an entirely native tradition, for which large claims have been made. The bizarre relationships, for instance, developed in *Joshua Haggard's Daughter* (1875–6) from the marriage of a Methodist local preacher to a girl less than half his age who then attracted his daughter's fiancé were not qualitatively different from those in *John Marchmont's Legacy*. The daughter's devotion to her father, though more insisted upon, was much like Aurora Floyd's. Curiosity as to the outcome kept the reader engaged, but he was not prompted to any more than commonplace understanding of the relationships. Nevertheless, Braddon never quite lost 'the rush, the flow, the animation' which the *Spectator* still found 'very remarkable' in 1884. There was equally no slackening in her habits of unmistakable emphasis and contrast, the 'strong coarse painting in blacks and whites' which made her sort of work the goal for Henry James's would-be popular novelist in 'The Next Time' (1895), as he lamented, 'I haven't been obvious—I must *be* obvious.'

Obviousness is equally the mark of Mrs Henry Wood,[13] but there is no animation. The coarse blacks and whites in character and incident seem to be there only for the author slowly to draw out their emotional consequences. In her most massive commercial success, *East Lynne* (1859–61), the events were striking and strained enough—the heroine's adultery with a man eventually convicted of murder, her divorce, her disfigurement in an accident, and employment to tend her own children, unrecognized, in the household of her former husband. The distinctive thing was the relentless elaboration of her sufferings to the accompaniment of formal reprehension.

'Reprehensible!' groans a moralist. Very. Everyone knows that, as Afy would say. But her heart, you see, had *not* done with human passions: and they work ill and error, and precisely everything they should not.

[13] Ellen Wood, née Price, 1814–87, was born in Worcester and lived abroad for twenty years after her marriage in 1836, writing for London magazines. On her return to England she began a thirty years' career as popular novelist and magazine editor.

The heroine's death became the climax both of the torments and of the morality.

There was a topical element in these torments in so far as several turns of the screw depended on questions of divorce such as had been debated at the time of the 1857 legislation and the establishment in 1858 of a Court for Divorce and Matrimonial Causes. The villain, concealing the fact that the heroine had gained a divorce, refused to marry her in time to legitimize their child; in consequence, remarrying, though legally possible, meant nothing to her. Neither the author's warnings to women who were unhappily married against 'the alternative, . . . far worse than death', nor her championing of the husband who, although divorced, would not remarry until he believed his first wife dead, could conceal the horrified excitement she secured from the adultery of the one and the notional bigamy of the other.

After this, the success in 1862 of two quieter novels, *Mrs Halliburton's Troubles* and the *The Channings*, was even more extraordinary. It was not secured only by the notoriety of *East Lynne*, for the other two remained second and third in the size of their sale (if figures in the publisher's advertisements may be credited) until well into the twentieth century, followed by *Roland Yorke* (1868–9), the sequel to *The Channings*. Each of these depended on mystery unresolved until near the end, but the murder in *Roland Yorke* was apparently never as popular as the mystery in *The Channings* concerning an inked surplice and the theft of twenty pounds. While it was more usual for Mrs Wood to rely on the morbid emotion generated by presenting mortality in the bizarre forms which its causes and its consequences might assume—in *Verner's Pride* (1863) the mistaking of suicide for murder, and of one man who had survived his apparent demise for another who had not; in *St Martin's Eve* (1866) the 'decline' of one personage after another, hereditary insanity, and customs like the reception held by the dead person, in this instance attended by her lover in ignorance both of her fate and of the custom—the ability to stir emotion and curiosity apparently remained undiminished when the means were less outlandish than these. Curiosity about the varieties of character and behaviour, simply conceived, against a background of the customs and the church of England, was enough. The entire absence of humour or of aesthetic appeal in the story or the style may even have helped. The whole phenomenon, continued until the posthumous publications of the nineties, is mysterious. Something must be credited to the very lack of verve. Where Miss Braddon was impetuously eager to tell all, Mrs Henry Wood narrated with admonition and regret.

Of the writers who followed the fashion once sensation had become notorious, one, R. D. Blackmore, could claim also to have been among its initiators, for his first novel, *Clara Vaughan* (1864), a murder mystery, had apparently been 'in manuscript' as early as 1853. It belongs to an earlier mode in so far as the reader is not seriously expected to attend to the facts and reasons in order to solve the problem. Clues are merely loose ends until a confession, the result of a reconciliation with the chief suspect, makes all plain. Although surprise and a climax of suspense, arising out of the attempt to forestall another murder, are clearly Blackmore's aim, his eye for sensuous detail, which attracted Gerard Manley Hopkins, curiously blunts the edge of the events. The heroine, trying to burn through the door of her prison, watches the fire 'licking deeper and deeper (with ductile wreathing tongues and jets like a pushing crocus) the channels prepared to tempt it'. And Blackmore has ambitions befitting an admirer of Bulwer. Fear and surprise must not be pleasurable ends in themselves, but evidence of divine warning to a heroine who has 'groped and groped, with red revenge my leading star, no breath of love or mercy cheering the abrupt steps of a fatalist'. As the ill-judged language here suggests, she has, in fact, been too little of a credible presence, too much a creature of melodrama in her vehemence, for the quality of her motives to count.

In *Cradock Nowell* (1865–6) he aspired to tragedy, working hard to evoke sympathy for a guilty man while detection (exactly narrated) closed in upon him. But having begun with exaggerations like M. E. Braddon's—almost amounting to pastiche in the culprit's 'cubic mass of ... forehead, the span between the enormous eyes, and the depth of the thick-set jowl, which rolled with the volume of a bulldog's ... In the towering of his wrath how grand a sight ...'—he could neither maintain the force nor modulate into anything more subtle, while the meanderings in the title-character's story, its main idea based on Henry Kingsley's *Ravenshoe*, made the whole construction less pointed. The simpler action of *Lorna Doone*, which followed, suggested that Blackmore had learned how to clarify a crime story by keeping it in touch with country pursuits, but when he tried to repeat the experiment in *Cripps the Carrier* (1876) there was an indistinctness of tone and even of event, in the absence of the sharpened antagonisms of history and legend. Without these he seemed unable to generate a subject: reworking a plot from Le Fanu's *Uncle Silas*, the detention of an heiress in order to subject her to an unwelcome proposal of marriage, adding suspected and attempted murder and carefully detailed suicide, and writing it all up with his accustomed care, he still left the reader without any idea why this sequence should have been displayed. Cripps himself was too slight a creation for his point of view to matter much and he was in any case ignorant, though not comically so, of what was going on. The implication of the epigraph from Aristophanes, that this little fable would be wiser or more cunningly devised than commonplace comedies, was mere pretension.

Similarly the lines from Sophocles prefixed to *Erema* (1876–7) to explain the Greek title (meaning lone, destitute) promised more than this crime thriller could offer. Repeating the culprit's illegitimacy from *Cradock Nowell* gave a double application to the subtitle, *My Father's Sin*, and explained why the heroine should be urged to leave the Secret undiscovered; but in her attempts to discover it, she was not 'lone' but befriended at every turn, as she was equally in the American adventures which turned out to be largely irrelevant to that search. The sensation form, of course, required only that relevance should be expected, in a story which need not stand a second reading. It seems to have been the kind Blackmore enjoyed reading himself, to judge by the preference he expressed in a letter of 1876 for characters like M. E. Braddon's 'golden-haired homicidals' over the heroine of *Daniel Deronda*. Sensation was indeed something of a trap to him, taking his attention away from the country life which was the only subject he really cared about. The parts of his first novel which had hauled in an irrelevant Devonshire farmer and made him talk had been worth more than the rest.

Christowell (1881–2) and *Perlycross* (1893–4) brought such characters into the foreground as Blackmore re-created the countryside of his youth. It was seen as nostalgically free from the anxieties he expressed in *The Remarkable History of Sir Thomas Upmore* (1884). The vision was slighter and rosier than Hardy's produced under the same shadow of British agricultural decline. Wheat from the huge mechanized prairie farms of America, brought in in increasing quantities by shipping which progressively reduced costs, and at a time when world depression was reducing meat and dairy prices, had made rural life before 1846 seem a haven of stability. Here Blackmore sought the legendary enchantment of *Lorna Doone*, going back to the same century for the account of a disaster on which he could base the climax of *Christowell.* He developed it with a fine sense of country superstition, imagining the Evil One as a lean, dark figure in whose throat the ale 'hissed ... like the quenching of iron'. Country festivals, the gardener's 'zodiac of clustering tasks', the shooting season, with 'puffs of smoke in the breezy distance, and far away sounds of feeble pops, such as a little boy makes with a fox glove', were affectionately described to mark the time-scheme of the melodrama. In *Perlycross* country descriptions were fewer and less mannered, and country customs less nostalgically introduced:

> A five-barred gate was flung mightily open, half across the lane, with a fierce creak of iron, and a shivering of wood; and out poured a motley crowd of all sorts and sizes, rattling tea-kettles, and beating frying-pans, blowing old cow's horns, and flourishing a blown dozen of Bob Jakes's bladders, with nuts inside them. . . . it happened to be the Northern party of the parish, beating bounds towards the back of Beacon Hill, and eager to win a bet . . .

The action, based on a blacksmith's half-comprehended evidence, tinged with superstition, was pursued, however, mainly among the author's own class, leaving the rustics to the periphery—except for the picturesque ones who were smugglers or were 'of very ancient race, said to be Norman'. It was romance again; little of the actual business of the district was important to the action once suspicions had been started as a farmer dug his potatoes. Nevertheless, it was perhaps the nearest Blackmore came to a fiction the subject of which grew out of what he could do best.

Le Fanu seized the opportunities offered by the vogue of 'sensation' by writing *Wylder's Hand* (1863–4) and working up some of his shorter fictions to fit the full-length genre, 'Passage in the Secret History of an Irish Countess' (1838) via 'The Murdered Cousin' (1851) into *Uncle Silas* (1864), and 'Richard Marston' (1848) via 'The Evil Guest' (1851) into *A Lost Name* (1867–8). It was his distinction to be obsessed with fear as Reade was with pain. In his short stories it took the form of self-persecuting phantoms, most often with their origin in submerged torments of conscience. In the best examples in the late collection *In a Glass Darkly* (1872) this form allowed him to make the phenomena clear with casual realism, but not to account for them with the same clarity. The *frisson* was enough, especially when so economically conveyed. The longer fictions forced him to adopt a different mode in which the world would be seen more in the round and not merely through the fixed ideas of his victims. Attempting to conform to the conventions of the sensation novel meant dealing in facts and reasons of a more obvious kind than in the short stories and arranging them in a more obvious way, first defining the mystery and then making the reader wish to learn the facts which might resolve it. Such procedures he appears to have found much less congenial. Indeed, his most successful novel, *The House by the Churchyard* (1861–3), used them only intermittently, among a number of vigorous and incongruous interests which made the result difficult to classify. Although set in the mid-eighteenth century it was so much informed with the sense of an unchanging Ireland as to be barely a historical novel. Although its supernatural element produced terror by Le Fanu's quiet, matter-of-fact methods, there was no attempt to develop the passing trepidation into apprehension for the future. Convivial good spirits and sympathy for the sheer impudence of ill-doers complicated the human feeling stirred by both raw pathos and pastiche of Sterne—in a father's thanksgiving 'for blessings not real but imagined' seen as 'expressions of thy faith recorded in Heaven, and counted—oh! marvellous love and compassion!—to thee for righteousness'. Conceivably this was the kind of novel Le Fanu would have gone on writing if left to himself, and conceivably its headstrong verbal impetus and emotional diversity might have been mastered without loss. But when, after the republication of the book in London, Le Fanu negotiated with Bentley to publish another, the publisher, perhaps with an eye to the reigning fashion, included among his stipulations an 'English subject' and 'modern times'.

The result, *Wylder's Hand* (1863–4), was a relatively uncomplicated sensation novel. The Gothic horror which had been taken to guarantee interest in the shorter fictions was not enough for a longer form which required him to control curiosity and engineer surprise. He created mystery simply by leading up to but not telling what took place before Wylder disappeared, but he was not yet adept at making the reader wish to fill the gap. No use was made of the limited viewpoint of the first-person narrator; indeed, no distinction was made between episodes at which he was present and those at which he was not, and for most of the time he could be forgotten. To leave a gap, while dwelling on inner and outer events surrounding it, was an arbitrary procedure when some of these events needed a knowledge of the solution to be intelligible, that is, to be of interest, within the convention—as with the encounter with Wylder's double when the preservation of the mystery could not allow the resemblance to be seen by the reader. A secluded setting and its 'ghost', obviously described to generate apprehension, appeared stagey in the absence of adequate reason why one of the novel's two heroines should live in such a place. But then the author was clearly more interested in the phenomenon of fear in his characters than in what caused it.

The contrast of the apparent heroine, Dorcas, dark and unsmiling, her inner nature concealed, and the golden-haired actual heroine, Rachel, with her suppressed animation and elaborate, unspecified terrors, formed more of a subject than Wylder's behaviour. The effort expended on the inner states of the villain, brother of Rachel and subsequently husband of Dorcas, left him inferior in interest to the women because it failed to illuminate or develop the initial vividness of the outer view of 'his odd yellowish eyes, a little like those of the sea-eagle, and the ghost of his smile that flickered on his singularly pale face, with a stern and insidious look'. The secret being a simple one which must all come out at once, the author had the problem of what to do till it emerged, apart from increasing the bombast about its frightfulness; the financial sub-plot hauled in for the purpose was insufficient, despite the exaggeration of the lawyer managing it into 'a priest or a magician', confronting something 'supernatural and talismanic' in Wylder's letters, 'like Cornelius Agrippa's "holy book"—a thing to conjure with . . .'. The facts and reasons revealed could be adequate to such preparation only by transcending the mundane conventions of the sub-genre.

In *Uncle Silas* (1864) Le Fanu attempted such a transcendence by working up the terrors of an earlier story, the tautness of which it was still possible to prefer to the repetitions of the novel. Vivid caricature for the title role and for the narrator's governess (who appeared only near the end, as Silas's 'creature', in the short version), apparently employed by a loving but inscrutable father, centred the mystification upon the opacity of the character, especially to a girl of 17. The characterization of the subsidiary roles was filled out with great virtuosity but nothing more, since the relations between them had to remain those of the original plot. Upon it were thrust religious overtones which certainly added something to the mystery before being lost in the factual snap of its resolution. The author claimed, in his Preliminary Word, to 'observe the same limitations of incident, and the same moral aims', as Scott, and so to belong to the 'legitimate school of tragic English romance'. But moral aims were well hidden in this story of a fallen gentleman turned religious hypochondriac, exercising his 'silvery smile of suffering' upon an inexperienced heiress. The emotional upholstery of her morbid terrors imposed heavier restrictions than Scott's melodrama, which had been at once simpler and more wide-ranging. Some possibility of tragic treatment, in retrospect, for both Silas and his brother often seemed about to take the novel beyond the reach of the 'degrading term' sensation, against which Le Fanu protested, but the possibility was lost when the secrets of the past revealed only the commonplace relation of deceiver and deceived. The distinction in the writing which had added these overtones generated the impression that the whole thing ought to mean more than it appeared to, without making clear what this was. So the book lapsed into a morbid melodrama much beneath the kind of sensibility implied by the language. At the same time, the human content of the situation was not substantial enough for the continually altering judgement by the reader which, in its longer form, it appeared to require.

It is a big drop from here to the other writers in this mode. James Payn[14] found in 'sensation' a formula to attract readers, but he had no strong vein of his own to cultivate. He at first exercised a slight humorous gift upon echoes of Dickens and the Brontës. As *Lady Audley's Secret* had taken its leading motif from *Bleak House*, so Payn imitated in *Lost Sir Massingberd* (1864) the fears evoked in

[14] James Payn, 1830–98, was born at Cheltenham and educated at Eton and Trinity College, Cambridge. He edited *Chambers's Journal*, Edinburgh, 1859–74, and the *Cornhill*, 1883–96.

The Old Curiosity Shop that Quilp might make a sudden appearance in the country, and the fears from *Jane Eyre* that persons associated with the forceful central character might be attacked to the accompaniment of strange laughter. The result, however, was not deft or ingenious enough for a parody of these authors, still less of Wilkie Collins. For the construction was feeble. The maniac wife appeared too soon, a Bow Street officer's investigation was quite without point, and the villain's disappearance had no connection with either. A spoof that led nowhere was no doubt a possibility—though parody should seem more than a feeble attempt at the real thing—but hardly a strong possibility when the love story which held everything together was clearly meant to be taken seriously. Successive novels revealed this to be Payn's own formula, love romance with second-hand sensational trimmings—in *Mirk Abbey* (1866) the mother shadowed by scandal, who returns to her family, like the heroine of *East Lynne*, with a different appearance and another name; in *A Woman's Vengeance* (1872) a confusion about a dead person's identity; in *By Proxy* (1877–8) the calamities attending the theft of a jewel from a temple in the Far East. These were ambitious narratives, but the stiff propriety of style, in dialogue, description of scene, and analysis of motive, meant that their power over the reader came only from the bizarre events. In *By Proxy* there was a routine quality about the terrifying states of mind of the villain and the man he wronged, no less than about the pattern sentiments of the young lovers. Payn seemed to do better the less his plot required him to account for extremes of conduct—in the action of *Carlyon's Year* (1867–8) or of *Less Black than We're Painted* (1878), for instance, which depended on the simpler kinds of villainy or on behaviour explicable by reference to common matters of religion, from Carlyon's scepticism to the charity of an Anglican Sister of Mercy. Compared with the main sensation novels, Payn's had simply not enough of a distinctive outlook for a rank above transitory entertainment, though of course people might, obviously, be entertained by worse confections.

Mortimer Collins[15] had a little more personal quality. At least he gave the impression of writing to please himself as well as his

[15] Edward James Mortimer Collins, 1827–76, was a mathematician and a Bohemian Tory journalist, friend of R. D. Blackmore and R. H. Horne.

readers. 'Miss Braddon and Mr Trollope', he said impudently in *Who Is the Heir?* (1864–5), 'are our most admired novelists because they never stop writing. But there are some men born to be amateurs, like Lord Rochester among the poets and Mr Congreve among the writers of comedy ...' While hardly justifying such comparisons, this was a novel which had some pretensions to amateur parody of the professionals. Not only was its perky style obviously mocking, but comparisons of procedure were openly encouraged. 'I cannot, like Mr Wilkie Collins, ... perplex my readers with a mystery which shall be undiscoverable till I choose to reveal it.' But the person he went on to expose at this point in *Who Is the Heir?* concealed another mystery, the identity of her husband, which was held over, of course, till the third volume; there, moreover, she was revealed to be insane and eventually succeeded in murdering him, as she had tried to do before. Needless to say, of the other principal characters one was recalled from abroad by hearing his beloved, who remained in England, call his name, and the other had an illegitimate half-brother for whom he could be mistaken. Here the game was not in doubt and was played for all it was worth. His later vein of comedy as in *Marquis and Merchant* (1871) was of a different kind. Its events were startling enough—the gypsies reappeared as a romantic alternative society where mysteries might be created and resolved—but mockery seemed not to be the aim. Collins even attempted, as the confessed admirer of Disraeli, to give the fiction some slight standing as a criticism of life by means of epigram and paradox. He had a certain success, not so much from what was said as from its urbanity.

Joseph Hatton,[16] like Mortimer Collins a journalist cultivating fiction as a sideline, produced quite a different effect. *Clytie: A Novel of Modern Life* (1872–4) stood out for the evidence it offered of the inattention which could apparently be expected of readers in pursuit of sensation; for the criminal trial in the second volume was filled out with two sets of verbatim repetition, each several pages long. Mystery and crime were laced with heavy remarks written for the most casual reading—on Fate, or on a society in which 'blood' is out of fashion: 'The new franchise has turned

[16] Joseph Hatton, 1841–1907, became a popular dramatist and editor of newspapers (including the *Sunday Times*) and of the *Gentleman's Magazine*.

England topsy-turvy'; 'Mammon has laid his hand on us all.' Yet the obligatory secret of sensational fiction was made eventually to yield ingenious plotting and counterplotting, not only between the hero and the villain but between the villain and his lawyer, and the latter's impudent survival and prosperity constituted a cynical comment, in Wilkie Collins's manner, worth all the earlier portentousness.

The tendency of 'sensation' to encourage ambitions beyond the writer's capabilities was equally clear in the case of William Gorman Wills.[17] As early as the preface to *Life's Foreshadowings* (1859) he was aware that 'stirring incident' might hamper his 'faithful studies of the human heart'. Efforts to 'select my types of character from the minds which I have myself come into contact with, and felt I understood, to the exclusion of letter-press hero and heroine' merely exposed him to the native melodrama of contemporary Irish life, and like Carleton he succumbed. He even used the story of *Life's Foreshadowings* over again in *The Love that Kills* (1867), to the extent of making most of the second volume of the two novels identical down to the very solecisms. The closest he came to the new sensation genre was *The Wife's Evidence* (1864), which resembled Wilkie Collins's development of 'stirring incident' to illustrate a thesis. The story kept no Secret from the reader, but held him by the claustrophobic family situation of a husband who incriminated himself in attempting to conceal his mother's crime and a wife who was debarred from giving evidence at his trial. The question of the wife's legal position there, in contrast to the case of bankruptcy in which she had been compelled to give evidence, clearly had precedence over the 'study of the human heart', and melodrama over both.

With Edmund Yates,[18] 'sensation' did not so much stimulate false ambitions as inhibit the fulfilment of those he might reasonably have held. He benefited, at first, from the widened range of behaviour which the new convention encouraged novelists to present, but the mere stimulation of curiosity became so lucrative

[17] William Gorman Wills, 1829–91, was born in Kilmurry and educated at Trinity College, Dublin. He came to London in 1862 and quickly made a name as portrait-painter, novelist, and dramatist.

[18] Edmund Yates, 1831–94, combined a career in the Post Office with journalism and novel-writing. He retired from the Post Office in 1872 and became the successful proprietor and editor-in-chief of the *World* in 1874.

that he failed to develop the little of his own that he had to say. His first novel, *Broken to Harness* (1864), 'a commonplace story of every-day life' as *Aurora Floyd* had been 'this simple drama of domestic life', was centred upon another 'fast' heroine, this time 'equivocal' in reputation because by profession a teacher of riding and breaker of horses. Like Aurora's, this reputation concealed a fundamental moral conformity. Although the plot consisted of a variety of marital complications, these were conventionally resolved. The flavour, however, was of scandal. The knowing, assured, rattling manner and the gossiping allusions made the novel almost continuous with actuality—not only with what the *Athenaeum* called 'Bohemia, club-rooms and masculine absurdities', but with city office life and the loadstone activities of women. Like Wills he promised more 'discourse' of the 'thoughts' and 'motives' of the characters than he supplied, but there was individuality along with the raw sentiment in his presentation of an unhappy marriage from the woman's side and in the combination of directness and deviousness in his heroine.

Running the Gauntlet (1865) developed the slight element of mystery already present in the first novel into a matter of crime, blackmail, and a pale, 'noticeable and fearsome' woman with dark green eyes. Stronger sensationalism followed, bigamy in *Land at Last* (1865–6) and murder in the novel which was for long named as his most striking, *Black Sheep* (1866–7). Progressively there was less scope for the slight originality evidenced in the strangely disinterested motives of his apparent villainess in *Running the Gauntlet* or the quarrels with herself of the bigamous heroine, bored with respectability, in *Land at Last*.

IV

Charles Reade[19] made his name by applying some of the techniques of sensation fiction to the social novel, but his real successes took a different form. He described himself, at the end of his life, when writing his own epitaph, as 'Dramatist, Novelist, Journalist'. He tried to be a dramatist first, and never entirely gave up that

[19] Charles Reade, 1814–84, son of the squire of Ipsden, Oxfordshire, went to Magdalen College, Oxford, of which he was a Fellow from 1835, though living and working mainly in London. He wrote successfully for the stage from 1851, both alone and in collaboration with Tom Taylor.

ambition. In fiction he worked at what he called the 'matter-of-fact romance', 'a fiction built on truths'. These were contemporary truths, gained from written reports, especially in newspapers, of which he came to keep extensive extracts and cuttings, with some unsuccessful attempt at systematic arrangement. 'I propose', he wrote in 1853, 'never to guess where I can know'—overlooking that to write fiction is to invent, and that for much of the time this means guessing, approaching the truth by means of hypothesis. Even his collections of facts involved guessing at what was worth picking out and at the principle of arrangement which would show why (a problem which his crude headings never solved).

He began by turning his first play into a novel. *Christie Johnstone* (1853) championed 'a truth that was struggling for bare life, in the year of truth 1850', against 'a cant that was flourishing like a peony' (as he put it in his concluding 'Note'). The cant was that of Carlyle's *Past and Present* (1843) and the truth that of the Pre-Raphaelites and their periodical of 1850, *The Germ*. 'The world, after centuries of lies, will give nature and truth a trial', claimed the artist-hero, echoing the 'adherence to the simplicity of nature' and the 'attachment to truth' of *The Germ*, number 2. In the rather childish story, the hallmark of truth was the actual instance: as Gatty, the hero, aimed to paint not 'abstract trees' but 'various trees of nature, as they appear under positive accidents', so he contrasted his mother's 'general rules' about matrimony with his own individual needs, and the beauty expected of ladies with the beauty of a fishergirl, 'one of Nature's duchesses'. In the story of Gatty's patron, Lord Ipsden, the medieval ideal of courage held by his Lady Barbara stood in contrast to an actual instance of the lack of courage in a rival who claimed to hold the ideal, and the justice on which Lady Barbara prided herself to the actual injustice of which she was guilty. These afforded opportunities for arch and obvious humour and also for something better in the comic duel of the fifteenth chapter. Some of the best of the individual instances, however, were not comic—though they were also not necessary to the story—like the reluctance of a fishing community to tell a wife that her husband and son were drowned, or another widow's unenthusiastic reception of sympathy and help. The naïveté and freshness here, as of things observed for the first time, never quite left him. Nor did the jerkiness of the translated play and the brusque selection of surface particulars to convey thoughts and feelings. The short paragraphs and tabulatory methods in the non-dramatic parts also became steady mannerisms, the marks of a man impatient with continuous discourse in the literary form which to him was second best.

Another novel, *Peg Woffington* (1853), was a much smoother rehandling of a play of 1852, in which the established dramatist Tom Taylor had collaborated with him. Both play and novel developed the notion that this eighteenth-century actress and demi-rep had been striving all along to become a worthy mid-Victorian. The novel added further instances of her goodness, her pious end, and some delightful by-play to show her professional virtuosity. It was an agreeable confection, technically more adroit than the drama in so far as it concealed from the reader, until more than halfway through, the fact that the lover whom Peg touchingly accepted was already married. The dedication spoke of her as 'Falsely "summed up" until today', overlooking the possibility that even though the items selected might be factual, their effect as a summing-up might nevertheless be false.

In his diary for 17 June 1853, he accepted the charge that there was 'too much criticism' in *Christie Johnstone*, and 'too much dialogue' (presumably straight from the play, which has not survived). 'I lack', he wrote, 'the true oil of Fiction.' It was in the next entry, for 20 June, that he proposed 'never to guess where I can know'. The collocation was ominous, for by this time he was again rehandling a play of his, *Gold*, and the first-hand information with which he filled it out made the success of the novel *It Is Never too Late to Mend* (1856). Stimulated by the furore over *Uncle Tom's Cabin* —not to have read it, said Reade, 'was like not to have read *The Times* for a month'—he found material for a similar work in the malpractices at a Birmingham prison. Following *The Times* reports back to their source in the inquiry of a Royal Commission gave him a villain. Gaols were visited, criminal courts attended, blue books read. The result, in its shrill insistence upon physically painful details, many of them exaggerated, as Fitzjames Stephen showed in the *Edinburgh Review* in 1857, seemed barely sane. It suggested paradoxically why Reade was unlikely to become a sensation novelist. Sensation in the newspaper sense was as deadly serious to him as to any reporter or editor. Far from being a possible source of fictional pleasure for the reader, it was an obsession to the writer, a source of strong emotion which it was difficult for him to master. His obsession undeniably gave a crude force to the accumulation of torments in his notorious eighteenth chapter, but to give it shape he had only the melodramatic contrast of the sadist with a saintly preacher, and the iteration of the vague word 'heart', applied to the reaction now of the writer, now of the reader, now of the observers in the story. Intemperance damaged the standing of the result as denunciation and destroyed it as art. A contrasting simplicity of implausible logic, in the prisoner's thoughts before his suicide, moved, if not to pity, at any rate to astonishment: 'God lives up there! . . . I know he can't be so cruel as Hawes; that is my only chance, and I'm going to take it.' The exposure of Hawes, the prison governor, brought about by a warder's 'boiling heart' and the representations of the chaplain, made a sequence of febrile excitement which Reade then had to keep up in the Australian half of the novel. If this was little more than a series of yarns, they were under banner headlines: 'the sudden return of a society . . . to elements more primitive than Homer had to deal with; . . . a desert peopled and cities thinned by the magic of cupidity; . . . a huge army collected in ten thousand tents, not as heretofore by one man's constraining will, but each human spurred into the crowd by his own heart'. With a journalist's enthusiasm for what he called 'THIS GIGANTIC AGE', he believed it a theme for 'great epic'. What he offered, however, was the crime story of his play *Gold*, with descriptions lifted from William Howitt's recent *Land, Labour and Gold, or Two Years in Victoria* (1855) and whole episodes from earlier sources like Peter Cunningham's *Two Years in New South Wales* (1827). His rapidity and decisiveness in narration kept attention awake, but could not supply any principle of coherence.

The interest in female character which had impelled his first two fictions was absent for most of *It Is Never too Late*, although it had first been called *Susan Merton*. Something of that early plan was visible in the prominence, at the end, of her refusal to give reasons when giving her trust, her pride 'that would not let her pine away', and her alarm at the power of her pity. In the smaller novels that followed, except for the *Autobiography of a Thief* (1858), an overspill from *It Is Never too Late*, female character was dominant. He made his version of Maquet's play *Le Château Grantier* into a serial, *White Lies* (1857), not only to boil the pot but because its foolish story focused upon female suffering and devotion. In *Love Me Little, Love Me Long* (1859) the arch comedy of the love of Lucy Fountain was subtler in intention, though not in presentation, so that the author had to explain that no male would understand her 'who does not take "instinct" and "self-deception" into the account'. With its discussions of speculation and inflation, its footnotes quoting *The Times*, answering critics, or defending the author's 'Baconian method', it was a characteristic Reade production, lightened by its brevity and the freedom of its main action from thesis-riding.

In *The Cloister and the Hearth* (1861), however, another huge subject, Europe on the verge of the Reformation, beguiled him away from the understanding of female character as a novelist, by offering scope for his interest, as a man, in celibacy (the condition on which he held a college fellowship). Here his obsessions drove him to accumulate rather than to explore. The story of the separation of the solemnly betrothed parents of Erasmus was expanded, with the help of an antiquarian colleague and reputedly well-paid undergraduates, into four volumes of miscellaneous fact and adventure. All the time the domestic hearth was assumed to be the summit of satisfaction, in contrast to the hero's experience on a journey to Rome and the cloister. The details, arising from Reade's obsession with physical pain, seemed often irrelevant, like the torture of the 'iron maiden', or the terribly strange bed taken over from Wilkie Collins's story. But the relevance of the climaxes of intensity became all too clear in the horrible varieties of capital punishment which the hero, Gerard, had to witness as a test on becoming a friar, in the first chapter headed 'The Cloister', and his self-inflicted torments, near the end, in order to escape from dreams of his betrothed and then from her presence—flinging himself into icy water, putting on a penitential garment of bristles, and fulfilling the most exacting religious duties, while crying 'Mea culpa! mea culpa!' The woman at the centre had not quite the same reality as the sufferings she occasioned, but Reade's insight took every opportunity—from the logic of her pity at Gerard's misbehaviour to her irritability at his cheerful success as a priest. The 'self-denial' of these two people was set in contrast with 'the well-greased morality' of the sixteenth century when vows would be taken as 'sacred in proportion as they are reasonable'. In 1861 the combination of all these things was apparently irresistible.

Hard Cash (1863) Reade thought his best example of 'the story that cuts deep into realities of the day'; but there was much in common between his historical novel and this treatment of a topical issue, the control of admisssions to private lunatic asylums. Like the previous hero, young Hardie was forcibly separated from his bride on his wedding-day and then subjected to tortures which exquisitely increased his frustration. The author's efforts on behalf of an actual victim in 1858–9 had set him accumulating facts from published and unpublished sources about the commital of the insane. The impulse to turn it into fiction, however, came from the contemplation of interrupted domesticity. Sane, Hardie was chained amongst the mad much as Gerard had preserved his devotion to Margaret amongst the professionally celibate. As Gerard, 'wrapped up in his Margaret', had suffered the wooing of a Roman princess, so Hardie had to contend with nurses in love with him. The 'regal beauty', 'mellow voice', and caressing 'white hand' in Rome corresponded to the 'beautiful back, with magnificent head and shoulders, and a skin like satin', the 'massive but long and shapely white arm', the 'soft, moving voice' and 'tender convulsive hand' in the asylum. This degree of elaboration of the earlier hero's predicament went with a greater vivacity in the presentation of the women's feelings. Even those of whom the author approved were seen with a sharper eye. The Lucy of *Love Me Little* reappeared, now twenty years older and married: exemplary in bringing up her family, she was yet capable of unreasoning enmity, while the diary of her daughter revealed fears not usual in a heroine:

> Marriage ! what a word to put down ! it makes me tingle; it thrills me; it frightens me deliciously: no, not deliciously; anything but: for suppose, being both of us fiery, and they all say one of them ought to be cold blooded for a pair to be happy . . .

Because they gave the novel such logic as it possessed, these matters were more important than the narrative excitements in which Reade was too prodigal, for invention without constraint merely drew attention to his freedom as puppet-master. Even the materials of which the preface spoke, 'gathered by long, severe, systematic labour', imposed no constraints. The Muse of history came to him, as he said in a Notebook, 'with her apron full of melodramatic truths'. They did not generate a subject for the novel, but merely opened up further possibilities of arbitrary adventure.

His best work, in *Griffith Gaunt* (1865–6), because it was done, as Henry James saw, by 'guessing', was more coherent. The less considerable part of the plot involved bigamy and a murder mystery, stock materials, by now, of the sensation novel. The whole even shared a common source with Wilkie Collins's 'A Plot in private Life' (collected in *The Queen of Hearts*), in an actual seventeenth-century case of which an account had appeared in *Household Words* in December 1856. To lead up to the sensational events, Reade developed his own story of the first wife's marriage, with a consecutive grasp which had been inhibited hitherto by his concern with documentary 'truth'. Kate's turns and changes became the signals of an inner uncertainty which gave freshness to the fictional commonplaces of courtship before marrying and of attraction to another man afterwards. She was shown, with her character still unformed, dreamy, but hungering after fixity and firmness, distrustful of passion, unable to yield to Gaunt's suit until to do so seemed a means of endowing him as a queen might. With the minimum of analysis, the reader was admitted to a sympathy with her fear of mere feelings and her contrasting inclination to be convinced by objective things, like the bullet in his arm and the inheritance which he had expected, had lost, and would not accept if leaving her alone were the condition. The contrast in appearance between her scarlet habit and the white of her horse or the snow of the background, references to 'chastity of snow', and the melodramatic violence of the scene in which she was restored to life by a transfusion of 'bright red blood, smoking hot', from Gaunt's arm, 'corded like a blacksmith's, and white as a duchess's', these were the devices of sensationalism, no doubt. They appeared in simpler events like the early disarming of duellists by this 'scarlet Amazon', who then 'sank slowly forward like a broken lily, and in another moment . . . lay fainting on the snow'. Their obvious excitement suggested her power girt round with weakness and made pictorial a decisiveness in outward action and appearance at variance with her limited self-knowledge and self-command, present by implication and waiting to be developed.

The development came as the novel moved closer to the obsession of *The Cloister and the Hearth*, love restrained by antecedent vows. But this time it was love incompletely acknowledged, upon which the artist could work. The priest idealizing Kate, as he saw her sitting under the sound of his oratory and stimulating his ascent, did not, like Gerard in the same position, simply believe he was seeing the ghost of a dead woman, though he did imagine he was seeing an 'angel' and not 'a woman of the world'. From the woman's side, instead of recognizing, like the betrothed in *The Cloister and the Hearth*, that she was 'wading in deep waters', Kate was in a rapture with the preaching which concealed from her the possibility that the whole current of 'her dreamy, meditative nature' might be directing itself upon the preacher, officer as he was of the church which satisfied her desires for stability. Despite her alarm at learning he 'perhaps admired her more than was safe or prudent', she took pleasure in the implied praise of herself when her presence animated the sermon and in the communication thus established with the preacher 'in public, yet, as it were, clandestinely'. Once she could be seen as self-deceived, her explanations of the crucial scene in which she became suddenly healed of an illness to get up and talk with the priest seemed one more subtle example of the knife edge she walked, especially after Reade had insisted that 'were I to analyse the heart . . . I should be apt to sacrifice truth to precision; I must stick to my old plan, and tell you what she did'.

But the author's best insights did not furnish the organizing principle for the whole of his plot. The second half of the novel became, rather, a justification of his heroine in events much more like those of an ordinary sensation novel, with Kate on trial for murder. No secret was made, however, of the more important matter of the bigamy which had been the object of detection in Wilkie Collins's version of the same plot. Nevertheless, as with similarly strange events in Hardy, the reactions of the characters repeatedly arrested a different kind of attention from that which the turns of the plot would arouse in Collins or Mrs Henry Wood (in whose magazine, the *Argosy,* the novel was serialized). The stress on jealousy, as Gaunt's 'ruling passion', was mere tautology compared with the interest of seeing him 'rather flattered' to hear Kate might kill him if he approached her. The reader, too much incited to admire Mrs Gaunt's fidelity, was nevertheless moved by her humility when justly praised: 'She could not understand justice praising her: it must be love', or by the response to her refusal to forgive: 'Who asked you to forgive him? . . . Your own heart.' Amid much that was embarrassing from the nature of the plot, from the idealization of the two wives, and from Reade's confidential, buttonholing manner of address, it was, at its best, a novel of unusual reach and grasp.

After this, his best work behind him, Reade relapsed, like the later Wilkie Collins, into topical variations upon earlier themes which failed to engage all his talents. The vehemence of *Put Yourself in his Place* (1869–70) against trade union violence did not prevent it from being another story of love thwarted until the last chapter. This time the hero's frustration was increased by the villain's success in marrying the heroine, though, as it turned out, in an invalid ceremony. Again the heroine believed her hero false and dead; again his letters were intercepted, and so on. Despite the epigraph from Horace stressing 'the power of sequence and connection', these were the very things the novel lacked, strong as it was in restless physical action. Reade's case cohered only as an unrelieved indictment of all trade unions, though the warrant he might have for each individual instance did not, as he apparently thought, extend to an accumulation of them when all were unfavourable. The repeated formula of the title, which might recommend sympathy, if only for practical ends, also implied a threat such as had appeared at the end of *Hard Cash*, chapter 33: 'Pray think of it for yourselves, men and women, if you have not *sworn* never to think over a novel. . . . Alfred's turn today, it may be yours tomorrow.' In either case the formula was feeble in effect compared with the adventures of a hero of unlimited energy and resource. The narrative was more professionally conducted than Mrs Gaskell's in *Mary Barton*, but it had nothing to compare with her compassionate attempt to understand the problems from more than one side.

Foul Play (1868), written in collaboration with a popular dramatist, Boucicault, was more like a strict sensation novel, complete with detective. It was on the old pattern, however, of an investigation to clear a tarnished name, while events in the present time of the story were not kept as a mystery from the reader at all. The insurance swindle on which the plot depended (scuttling a ship and then claiming indemnity for its cargo of lead as if it were gold) was appropriate to the year in which Samuel Plimsoll took into parliament his campaign against 'coffin' ships. But the matter was not in the novel to provoke thought; it merely took its place among other seafaring excitements.

Reade's decline into a writer of diverting magazine tales had become more marked by 1873 with *A Simpleton* and *The Wandering Heir*. Into the first, the story of a husband lost and found, were introduced 'causes' like the dangers of tight-lacing and the importance of testing mother's milk. Topicality in the second was a question of resemblances between an unknown person's claim to the Tichborne estates in 1871 and a vaguely similar eighteenth-century case. Like the Tichborne claimant, the novel and the play founded upon it attracted great public attention for a very short time.

Reade's whole career shows a war between the journalist's concern for immediate sensational fact and the artist's concern with what may be *made* of it. Only rarely was the artist dominant over the journalist, and even then most often in sequences of rapid and direct narrative—no mean gift. The humour to be found in his first two fictions disappeared once the facts on which he chose to base his fiction had become matters which engaged his indignation—the tormenting of prisoners, clerical celibacy, trade union violence. Where he justified his indignation by presenting extreme physical pain, the pain took precedence, endowing the narrative with a crude force for which it is not easy to find parallels outside Dickens. But where in Dickens such force was, as Sidney would say, 'the groundwork of a profitable invention', in Reade its expression became an end in itself. It was equally far away from fear and suffering as used by the sensation novelists, to stimulate curiosity concerning the outcome. Reade's control of the reader's curiosity was a matter of short spasms and repeated tricks, not extended and sedulous mystification in order to accentuate surprise. The instances which he chose as the basis of 'matter-of-fact romance' were not such as to reinvigorate his whole vision of life. They seemed more often to keep before him his position as a man, suffering and celibate, than to stimulate him as an artist to transcend his limited self. He approached such self-transcendence in the presentation of women. Only in *Griffith Gaunt* did his power to imagine female character—showing rather than explaining it—join with his obsession with pain to produce a work which, even if only in parts, was of anything like major status. The degree to which Reade's best work here, though small in quantity in proportion to the total amount he wrote, stands up to the strong competition of his contemporaries means that such critical attention as he has received has tended to undervalue him.

V

In the best of the later religious novels, following the example of Harriett Mozley, polemics become secondary to life itself. The tendency shows clearly in the development of Elizabeth Sewell's[20] fiction. An early novel like *Margaret Percival* (1847) simplifies the main action to the choice between two forms of Christianity. In later work, such as *Katharine Ashton* (1854) and *Ivors* (1856), the action arises from individual character and is more directly concerned with results in individual character. *The Experience of Life* (1853), though falling short of the full dramatic interest of *Katharine Ashton*, effects the transition.

Much of the material in *Margaret Percival* is of a cheaper kind, more common in fiction than is usual with this author—a Countess from Italy, even if not Italian, a mysterious great house, a Roman priest scheming to convert the heroine, an Italian girl jealous of her. There is discretion in the treatment—the priest, for instance, is a dignified character, quietly presented, despite the 'editing' of the book by Miss Sewell's brother William—but the plot is resolved at the level of polemic, with the return of a clerical uncle to put the Anglican case. The domestic life which is the author's proper sphere is of importance in so far as the successful mentor is the one who has the strongest links with the heroine's home and upbringing. But the effect is merely to give her decision overtones of an opposition between romance and reality which the author soon learns to transcend.

The title of the first of her more distinguished trio of novels is symptomatic, *The Experience of Life*, though it seems rather large, with its definite article underlining what is representative in the adversities of the first-person narrator. In fact the narrator's position in her family is extraordinary, as its most discerning and

[20] Elizabeth Missing Sewell, 1815–1906, daughter of a solicitor on the Isle of Wight, dealt with the family's financial difficulties by the proceeds of her writing and interested herself in female education, especially in the school she founded, 1866, St Boniface Diocesan School, Ventnor.

practical member though neither the eldest nor the most robust. In forgoing controversy, except about dissent, which, 'as an active progressive power', is relevant to the heroine's practical interests, the author leaves her inner life indistinct. Her 'impious doubts' are unspecified, so that the 'racking misery of a mind striving to satisfy itself by its own reasoning' remains vague. The emphasis is all on her practical duties; even the reasons for rejecting matrimony are outward and practical. It is fortunate for the novel that the aunt whose advice she takes, to pray rather than think, should be its most vivid character, whose advice and actions are capable of surprising the reader and whose home is most strongly present to him. That the narrator has not the same understanding of the characters who deceive and despoil makes a virtue of a deficiency of the author's, for it helps the narrative suspense and gives importance to the few actions which decide between the guilt of the character and the possible prejudice of the observer.

Katharine Ashton, on the other hand, is notable for its pellucid clarity, even in presenting the heroine's inner experience. There is not the same melodrama about this experience as in the previous novel, no vaguely terrifying doubts, or obscure self-questioning with little effect on the action—'Were they dreaming, or was I? . . . What was this existence . . . ?' Katharine Ashton is puzzled, but about things which are distinct to the reader, such as people's view of the plight of the poor. Larger questions of the 'unity' of society are made specific by the comedy of a Union Ball: it fails of its pretentious object, but allows the heroine to see a duchess and her party, 'all rather wonderful to Katharine, because they were so like everyone else'; it is their exclusiveness which destroys the 'unity' which the ball was meant to encourage.

The fullness with which Katharine's experience is realized and the admirable control of a small number of naturally unfolding events stimulate the reader's curiosity about the two marriages which Katharine closely observes and about the effect of her disinterestedness upon the possibility of her marrying herself. Although the author takes the opportunities for preaching, this does not blunt the keenness of her insight or break the restraint of her irony—for instance, when Mr Forbes goes for sympathy to his wife, who is seriously ill, though he believes (wrongly) her illness to be merely the result of her own 'imprudence':

Then he went to Jane, and she pitied him, and made him sit down by her sofa, and put eau de Cologne to his forehead; and as she was looking better by that time he did not feel as much irritated as he had done, and had therefore the pleasant consciousness of acting a magnanimous part in forgiving her.

It is not necessary to share all the presuppositions of the preacher in order to understand her claim that 'as we work upon the outer surface of the world we see, we are at the same moment indenting ineffaceable lines upon that true and spiritual world which lies beneath it.' The tone of such a comment resembles that of George Eliot, with its quiet urgency and its confidential 'we'. But in the working out of the notion there are decided limits to what Elizabeth Sewell can contemplate as an artist. She sees, for instance, what is admirable as well as what is petty in Jane's husband, just as she sees that Jane has 'spoiled' him by being over-submissive. Despite the acute anatomy of his selfishness, however, it is exposed as a moral predicament rather than understood as the life of a person, so that his reformation appears simplified and sentimental. There is more of individuality in the vulgar malice of Crewe, the butler, who does not have to be reformed. It is only with people like Katharine, resolutely trying to be good under great strain, that Miss Sewell allows herself the freedom to round out individual character. Even the idealization of Jane in her religious life, or of Katharine in her cheerful service as Jane's maid, do not conceal the insidious quietism of the first or the skill of the second in vindicating herself.

With Helen, one of the 'two cousins' of the subtitle to *Ivors* (1856), Miss Sewell comes nearer to contemplating fully a form of life of which she does not approve. A refined sort of selfishness is understood in its pleasures and its miseries, and not merely in its right and its wrong, so that a character emerges and not merely a case. The more subdued experience of the contrasting cousin, Susan, is delicately touched in, her less active part made lively by the subtleties of her disinterestedness: the defeated hopes from which she eventually suffers would, after all, not have been fed if she had been less willing to act as an intermediary. But although Susan's 'vivid sense of existence' at the climax of her hopes is moving, as are Helen's initiating jealousy and the intricacies of her half-willing deception, it is rather beneath this novelist to limit a plot to the fictionally commonplace question of marriage to a young Sir Charles Grandison, unexposed to criticism. She has other modes, approaching the sensation novel in *Cleve Hall* (1855), but *Katharine Ashton* remains her most considerable work, with its extraordinary vitalization of the mundane matters of small-town life as seen from the vantage point of a girl of questing energy, anxious to learn 'what use one is in the world'.

Charlotte M. Yonge, whose fiction forwards the same kind of religious views, is more prolific in the invention of lively characters and dialogue to embody them, but less able in selection and arrangement to make these things really tell. Whole progenies are imagined, capable of surprising the reader with both the distinctness of individuals and their resemblances within families, but the imagining of an action for a whole novel is less successful. The author recognized this in the preface, written later, for an early work called simply *Scenes and Characters, or, Eighteen Months at Beechcroft* (1847): plot, 'a matter of arrangement', was the more difficult in proportion to the ordinariness of the events; hence Jane Austen's pre-eminence, where 'another can do nothing with half a dozen murders and an explosion'; 'of arranging my materials, so as to build up a story, I was quite incapable. It is still my great deficiency; but in those days I did not even understand that such an attempt was desirable.'

Miss Yonge was perhaps too hard on herself, for her earliest attempt to meet the difficulty showed in the subtitle to her first novel, *Abbeychurch, or Self Control and Self Conceit* (1844). The moral abstractions were well enough supported by a cautionary tale. Characteristically, however, the interest came more from the conversations about the action, continuing right to the finish where the moral itself and its exemplification were discussed with some humour—for instance, at the expense of the character who was to be praised for self-control:

'She may have every virtue upon earth for aught I know,' said Rupert; 'I can only testify that she has *un grand talent pour le silence*.'

Even the drawing of morals was dramatized:

'Nonsense,' said Rupert; 'do you think that if anyone read its history, they would learn any such lesson unless you told them beforehand?'
'Perhaps not,' said Sir Edward, 'as you have not learnt it from your whole life.'

'No,' said Lady Merton; 'that lesson is not to be learnt by anyone who is not on the watch for it.'

It is possible to think of her talent as taking her in two directions, the one in novels written for adults, like *Hopes and Fears* (1860), the other in a wide range of work for readers not yet adult. The former were the more selfconsciously shaped by a lesson which those on the watch for it might learn, while the shape of the latter work, at its best, emerged rather from the 'whole life' of the youthful characters themselves.

Selfconscious shaping at first gave her her most resounding success, *The Heir of Redclyffe* (1853), which might well have had the same subtitle as her first novel. The self-control was that of the young inheritor of Redclyffe, Sir Guy Morville, quelling 'high animal spirits' with a serious ascetic temper, the self-conceit that of the next heir, his cousin Captain Philip Morville, whose Grandisonian rectitude and confidence had the effect of hastening Guy's marriage and then his death. Although the author had accepted that Guy's self-discipline resembled that of a noted tractarian, J. A. Froude's brother Hurrell, the moral was made picturesque in terms readily acceptable in the third quarter of the century, during which the book was reprinted again and again. Guy's Byronic impetuosity, his admiration of Malory (especially of Sir Galahad), and his background of an apparently hereditary curse made him a figure of romance. His conquest of 'the besetting fiend of his family—the spirit of defiance and resentment'—was insistently connected with Fouqué's *Sintram* (1814). So the melodramatic self-communing which dates him for later readers was accepted not only as dramatization of an active conscience but as the appropriate mode for a romantic hero.

There was a ready appeal therefore to readers who might not share the author's religion. Moreover, religious significances which appeared in picturesque suggestion, the brightness of the Easter moon, for example, after Guy had 'suffered and conquered suffering', appeared also in conduct more mundanely observed, 'the soberer manner in which he spoke of his faults . . . without the vehemence which he used to expend in raving at himself'. There was even some attempt to make his religious life, however highly coloured its extremes, credible in its psychological processes, using mundane as well as religious terms: 'Guy had what some would call a vivid imagination, some a lively faith.' All this is far from sufficient to recommend the romantic hero to most readers over a century later. The parallels with Fouqué have lost their efficacy and the religious doggerel which mercilessly accompanies the hero's death and its aftermath raises a formidable barrier, not to mention the tendency to bombastic misapplication of ancient formulas, like the *Nunc dimittis* at the christening of Guy's daughter.

It is quite otherwise with Philip, the Grandisonian prig. It is not only that the author understands him more clearly as a form of life, not a mere figurehead, but that her attitude towards him is proportionately more varied. It includes even touches of an irony approaching Jane Austen's: for instance, Philip, who had been discovered applying a different standard of openness to Guy than to himself, was glad to be forgiven by a fool whose reason was 'I have been young and in love myself':

> That Captain Morville should live to be thankful for being forgiven in consideration of Mr Edmonstone's having been young!

The irony could be called structural in so far as it was Edmonstone who, as guardian, had yielded too readily to Philip's advice to delay Guy's engagement and then, with exaggerated self-assertion, when circumstantial evidence proved Philip wrong, had hurried on not only an engagement but marriage. But the book was finally too didactic to secure all the advantage that it might from such material. The plot could not stress the ironies of Edmondstone's position when his wife and family were allotted so much more space and so much more positive a role, especially his daughter Amy, Guy's widow. The aim was not the civilized pleasure arising from the selection of just what would sharpen the ironical point, but the heavier, more didactic irony that Philip, by inheriting the very estate he had once thought desirable, should suffer so much.

In *Heartsease* (1854) the author's religious interests were subordinate, from the reader's point of view, to the remarkable presentation of the wilfulness of the chief character, Theodora. Her pride was understood in its good manifestations as well as its bad, even though religion might have depreciated the former, and her reformation was stimulated, as it might have been in George Eliot, primarily by the influence of another person, the 'brothers' wife' of the subtitle. The title, referring to the effect of the wife upon all surviving members of the family into which she married, meant that hers was the role of heroine—a role which left her vulnerable to idealization once the delicacy, and even comedy, in the treatment of her early hesitations and anxieties, after marriage at 16, gave way to the more improving spectacle of her moral strength. Nevertheless, the novel is a strong candidate for revival.

The same could not be said for *Dynevor Terrace* (1857), where a contrast, as in *The Heir of Redclyffe*, between two young men, the one 'fiery and sensitive', the other 'cool and impassive', yielded not a single plot but two stories, neither of any great force. The restraining of religious and high romantic sentiment in order to present an action much concerned with money, in a period just before and just after 1848, seemed to have left the author with less that she wished to say.

In *Hopes and Fears* (1860) the story was made from the difference between two such periods. At the conclusion the middle-aged heroine looked back to the 'mediaevalism, . . . chivalry, symbolism, whatever you may call it', of her own generation, who were 'fed on Scott, Wordsworth, and Fouqué' and 'moulded 'moulded their opinions and practice on the past'. The younger characters, adopted by her 'widowed heart', 'were essentially of the new generation, that of Kingsley, Tennyson, Ruskin, and the *Saturday Review*. Chivalry had given way to commonsense, romance to realism, . . . the past to the future.' It was a fruitful subject when the heroine tended to turn into 'idolatry' her love, first for a man who fell short of its demands and then for his son. The faltering of the son's faith, his marriage as a result of 'the theory of fusion of classes', and the development of his sister into a 'fast' girl of 1860 further identified the new generation. But the sister's suitor brought with him his whole family who then swamped this subject with quite another novel. It had some contemporary interest—a young woman attaining 'liberty of action and independence of judgement', her brother refusing to enter the family distillery and devoting himself to the redemption of the patrons of gin palaces—but it left the heroine with little to do apart from approving those actions which enabled her to regard three of the family as each in turn 'a child of her own'. Her 'fluffiness . . . figurativeness and dreamy sentiment' were amusingly criticized by the fast girl: 'I cannot like mutton with the wool on!' But for most of the time the heroine's viewpoint was simply too close to the author's own to have any prominence in the multifarious action. This ambitious book suffered most from the principal feature of her work: it was all foreground. Her very prodigality in providing immediate 'scenes' left her with a correspondingly smaller power of attention to organizing principles and interpretive ideas.

The early chapters of *The Clever Woman of the Family* (1864–5) gave promise of the ability to develop the point of the whole novel from the heroine Rachel's opinionated self-reliance and 'sententious dogmatism', active in dialogue and in scenes which showed how easily eagerness could prevent her from being observant. If the terrible events which disciplined her had grown more specifically out of this dramatization of her defects the result might again have challenged comparison with Jane Austen. The materials were there, but their potentialities for comedy were forfeited in the interests of narrative excitement and a didactic completeness of reformation. Comedy lurked when the most important revelations were brought about by the practical energy and good sense of a young widow on whom Rachel had looked down, determined 'to put Fanny upon a definite system . . . counteracting the follies and nonsenses that her situation naturally exposes her to'. Attempting to be more serious morally, Charlotte Yonge became artistically less so in her crime story and in the suddenly accumulating illness, despised love, death, and birth, in another runaway family chronicle.

In the fiction where she had younger readers principally in mind, their interests placed limits, which she found congenial, on the kind of organizing ideas which might bind the whole together. The foreground was the chief thing and, as long as it was sufficiently lively, a subject could be left to grow out of it. The 'family chronicle' in *The Daisy Chain* (1853–6) and its sequel *The Trial* (1862–4) is the prime example. The first of these began as a series of 'conversational sketches'; after Part I had appeared in monthly instalments, the author, writing ahead, had as much material again in hand and this had shown her her subject. In the complete novel she thereupon published, everything came together, centred upon a widower, Dr May, and his third daughter Ethel, who, having chosen to live single in his interests and those of the family, 'had begun to understand that the unmarried woman must not seek undivided return of affection'. The subject was not unlike the one the author would develop in *Hopes and Fears*, but it consisted to a much greater degree in a pattern that was not recognized until after the main events had occurred. So they could be left free to happen and the habits which prevented a more pointed development of her subject in *The Heir of Redclyffe* and *The Clever Woman* could reign:

> I have taken a sheet of paper and turned my *dramatis personae* loose upon it to see how they will behave.

Naturally, the standards of an urgent, self-examining morality continued to apply; but they did not inhibit humour and they left plenty of scope for surprise, especially in the detail of the reactions which distinguished the two chief personages. The range of their feelings, their aims—church-building, schools, missions—indeed the very existence of large families among the professional classes, may now seem as remote as jousting; a less intense kind of sensibility than Elizabeth Sewell's may be at work, concerned with less painful affection and self-scrutiny; the reader's participation may be limited to observing what is immediately there to be noticed, without being invited to make inferences from it; but the ability of that omnipresent foreground to startle and enchain, once these limitations are granted, is as little diminished in late work like *The Two Sides of the Shield* (1884–5) as it is for the reader a century later.

Religion is so quietly ubiquitous in Charlotte Yonge's novels that one may overlook it in one's interest in the people for whom it is unquestioned. With an eccentric like Laurence Oliphant,[21] on the other hand, it attracts direct attention from both its incongruity in surroundings of satire, and its oddity in itself. The management of *Piccadilly* as a novel is equally unorthodox. 'In the first place,' Oliphant half-seriously informed his publisher, 'it is a caricature, and not intended to be quite natural or possible. I look upon it as the highest form of art to supplant the natural with the imaginative; but of course it runs the risk of failing by reason of its extravagance.' He felt driven to this by opinions, which, he said, 'could be called extravagant in many points'; they were

[21] Laurence Oliphant, 1829–88, son of a Scottish Attorney-General at the Cape of Good Hope, travelled extensively in Europe, Russia, and the Far East, where he undertook missions for the Foreign Office. From 1867 he came under the spell of an American pseudo-prophet.

'based, nevertheless, upon truth and rectitude, which two principles are so extremely dry and distasteful that nobody would care about a novel conveying such an old-fashioned moral unless it were put in some new-fangled form'. His method was to avoid 'doing it too seriously'. So that the result would not be 'too much burlesque', however, he 'determined on making Vanecourt [the narrator] more or less mad'. Things like his loud attack upon 'Churchianity', his vision of a fiery cross, or his discourteous denunciation of fashionable religion at a *conversazione*, could, then, be received according to the reader's predilections, while the unreliable narrator whose depressions, fantasies, and racing thoughts were described from introspection and whose difficulties with the narrative were frankly discussed with the reader, gave an impression either of being engaged in living the story as the instalments appeared in *Blackwood's Magazine* in 1865 or of making it all up without much sense of direction. The gaiety and satirical wit at the expense of people like Lady Broadhem, forcing her daughter into a mercenary marriage, or Spiffy Goldtip the Stock-Exchange tipster, the strange sensibility of the observer (female flounces in a coach were seen 'surging up at the windows as if they were made of some delicious creamy substance and were going to overflow into the streets'), all thrown into the same fiction as the mysterious, urgent injunction to 'LIVE THE LIFE', made an arresting, slightly mad novel. It contrasted extremely with the lugubrious *Masollam* (1886) about the deceptive power of the prophet who appeared at the end of *Piccadilly*. *Altiora Peto* (1883), which had as great a success as *Piccadilly*, however, showed Oliphant still capable of the earlier vein, as did his spirited autobiographical *Episodes in a Life of Adventure* (1886–7).

The final example for which there is space, George MacDonald,[22] is a strange one. It is hard for the reader of novels to have patience with his work. He has something of Scott's ear for memorable speech and even something of his vivid awareness of individuality in the speaker and, as with Scott, these vernacular speakers can quicken the most tepid narrative; but MacDonald is always looking less at them than beyond them, even in the books

[22] George Macdonald, 1824–1905, born in Aberdeenshire and a graduate (in chemistry and physics) of the University of Aberdeen, became a preacher without a pulpit when forced to resign from the Congregational ministry on account of his unorthodoxy. He became Professor of English at Bedford College, London, 1859.

like *Alec Forbes of Howglen* (1865) or *Robert Falconer* (1866–8) where their enlivening effect is strongest. The exact nature of objects, people, and events, in the relationships of actuality, mattered much less to him than certain recurrent moral activities and aspirations. He preferred not to imagine these in a world which looked like actuality but to reveal them more directly under less restricted conditions. He began his long writing career with this kind of direct revelation. Even where *Phantastes: A Faerie Romance* (1858) drew upon actuality for images, it related them to each other according to what they signified for him. Throughout his career, in works like *At the Back of the North Wind* (1871), the Princess books of 1872 and 1883, or *Lilith* (1895), he returned to this mode, aiming at the direct creation of religious myth. They have been appreciated, then and since, by readers predisposed to receive their revelations—H. G. Wells for the latter work, C. S. Lewis for the earlier ones.

But MacDonald returned also, again and again, to realistic fiction, at first not merely because there was a wider audience for the three-deckers which circulating libraries might buy in quantity, but because, apparently, he needed to reaffirm the connection between mundane actuality and what he called in *Robert Falconer* 'the heights where all things are visible'. For this purpose the English and German Romantic poets who, he believed, had helped him in the creation of myth proved something of an impediment. Tutored by them, he knew too easily where he was going; the exact circumstances which are the usual business of fiction were of less importance.

At first Wordsworth's *Prelude* supplied a plot mould, the development of a dedicated individual. Although such a plot too easily admitted almost anything, including the stock materials of circulating library fiction, it also turned his attention to childhood. The result was episodes marked by a simple authenticity and the narrative point of short stories, those for example in which Forbes and his friends tried to deal with a biting dog, or Falconer, practising the violin in a derelict factory, was mistaken for the proprietor's ghost by a blind woman who had formerly worked there. The trouble was that he could not use what was distinctive in such episodes to advance the action, while the more conventional matters which did carry the story forward—the debts of

Forbes's mother or Falconer's search for his father—tended to be resolved quite perfunctorily. It made an uncomfortable mixture.

The materials of routine fiction could hardly be said to hamper the central theme, when it was expounded rather than seriously explored. It tended to appear in versions of the central Wordsworthian experiences on which, without any further development by the novelist, theological conclusions were imposed. Falconer's outdoor pleasures as a boy, before he had 'looked nature in the face, or begun to love herself', made him 'dimly conscious of a life within these things' and even fearful 'lest he should fall off the face of the round earth into the abyss'; they were immediately taken as 'the first dull and faint movement of . . . the need of the God-Man'. Alec Forbes first became aware when sailing his boat 'of a certain stillness pervading the universe like a law, a stillness ever being broken by the cries of eager men'; nothing more was done with this awareness until the huddled conclusion where, to Forbes facing death in the Arctic ice, 'the silence seemed to be God himself a' aboot me', and he was led to safety. In the same novel an interesting alcoholic, within three pages of his decisive reform, became subject to 'the menstruum of the earth-spirit', as 'a prodigal returned at least into *the vestibule* of his Father's house'. Macdonald's (usually motherless) heroines, beginning with the daughter of *David Elginbrod* (1863), were even more defenceless against the imposition of his own opinions. It was only in the early childhood of characters like the Annie of *Alec Forbes* that specific Scots detail filled the outlines of the author's sentimental approval. There was more of genuine fictional exploration in the case of characters like Falconer's grandmother or Forbes's friend Crann, of whose Calvinism MacDonald disapproved but whose human peculiarities contrasted with their theology. The attempt in later work to show how the central truths of his own more humane religion might be discovered in ordinary life led to the moralizing pathos of stories like *Paul Faber, Surgeon* (1879) and *Salted with Fire* (1897). Their exclamatory kind of melodrama approved, without illuminating, the process by which a sceptical surgeon might achieve some sort of faith through forgiving his wife, or a minister might find his ministry renewed as a result of marrying the vilified mistress of his youth. At the same time, in the later novels, the writing of fairy stories seemed to have led him to reduce the saving realism of his presentation of childhood; from their earliest life both the heroine and the hero of *The Elect Lady* (1888), for instance, were ludicrously idealized. But even such a hero could be touched with vitality when speaking Scots: 'Ye h'avenly idiot! . . . Will ye be my wife, or will ye no?'

7. Elizabeth Gaskell and the Kingsleys

Elizabeth Gaskell and Charles Kingsley first gained attention in the late eighteen-forties because of their concern for the condition of the people. Of the earlier novelists who shared this concern, some, like Mrs Trollope and Mrs Tonna, had had difficulty in making viable fiction out of their social subject, others, like Harriet Martineau, in integrating it into a novel about something else. Elizabeth Gaskell and the Kingsleys had difficulties of the same kind. For Charles Kingsley, born to be a novelist like Surtees, with a strong feeling for the open air and the sociabilities of sport, the problem was to reconcile such interests with his didactic aims as a clergyman and, for a while, a 'Christian socialist'. Elizabeth Gaskell had more success in weaving the strains and tensions of a changing society into her best work. Henry Kingsley began writing after these two had published the most notorious of their 'social novels'; he fitted some of their concerns into his rather feckless fictions, while seeming to take both the fiction and the concern much less seriously than they did.

Charles Kingsley[1] had an unusual command over the sort of language which could convince a reader that certain things were happening before his eyes in a particular natural setting, so that the clarity of the action and the frequent felicity and exactitude of the description stand out among the minor novelists. He was too impatient, however, in pursuit of his aims as parson and educationist, to face the formal problems of realistic fiction, and readily took the short cuts of melodrama and allegory. He first attempted, in the series of episodes in *Fraser's Magazine* (July–December, 1848) which became *Yeast*, to interest the members of his own class in a sample of the questions 'fermenting in the minds of the young'. He acknowledged 'the fragmentary and uncon-

[1] Charles Kingsley, 1819–75, was born in Devon, son of the curate at Holne, went to King's College, London, and Magdalene College, Cambridge, and became rector of Eversley, Hampshire, 1844, a royal chaplain, 1859, Regius Professor of Modern History at Cambridge, 1860, a canon of Chester, 1870, and of Westminster, 1873. As a follower of F. D. Maurice, Kingsley interested himself in social problems and, as a keen naturalist, in popular education.

nected form of the book', but justified it as 'an integral feature of the subject itself, and therefore the very form the book should take', for the subject was the 'Yeasty state of mind' of the central character, Lancelot Smith. Nevertheless, even in its six instalments in *Fraser's*, the work was given the form of routine fiction by a climax which resolved all Lancelot's doubts. Some of his states of mind were well realized—when, for instance, after he had been forced to notice 'oppression . . . want . . . filth . . . grinding anxiety from rent-day to rent-day' in the countryside, he turned to his fishing:

> All his thoughts, all his sympathies, were drowned in the rush and whirl of the water. . . . how it spread, and writhed, and whirled into transparent fans, hissing and twining snakes, polished glass-wreaths, huge crystal bells, which boiled up from the bottom, and dived again beneath long threads of creamy foam, and swung round posts, and roots, and rushed blackening under dark weed-fringed boughs, and gnawed at marly banks and shook the ever-restless bulrushes, till it was swept away and down over the white pebbles and olive weeds, in one broad rippling sheet of molten silver, towards the distant sea.

The intimate sense of the attractions of his habitual way of life in such a passage contrasted well with the harsh conditions in the classes below him. For the most part, however, despite the occasional satire at Smith's expense, his reactions came to the reader in the clichés of mediocre fiction and counted for much less than the brisk and often surprising narrative of the matters that caused him to react—the ragged poacher hauled out of a stream to deride the notion that anyone might give him work at his time of life, the open sewer in a village without water, the miserable village 'fair'.

The hero's personal history, which held the whole together, was hinted at in trite regrets for wild oats sown in the past. There even seemed some intention of a contrast with a certain Colonel Bracebridge, who thrust such regrets from him until his mistress murdered their child. But here the melodrama was too sudden and rapid, compared with the longueurs of Lancelot Smith's devotion to the heroine, Argemone. Something could have been made of the contrast between the apparent 'completeness' of the colonel and the stumblings of the ludicrous learner who might yet be the better man, but this remained only one line of interest among many.

In an attempt to link the personal and public constituents of his 'problem' (the subtitle of the volume edition), Kingsley killed off

Argemone with a fever contracted while working in that waterless village. The interpretation of her fate as part of an ancient curse upon her family drew attention to the inevitable spread of infection from fetid country places. This was as near as Kingsley could get to making fiction out of the contrast between social practices which can be changed and the 'laws of the universe' which cannot. In episodes concerned with other people who had power—landlords, bankers, churchmen—he widened the social attack while dissipating the attention that fiction should have concentrated. In trying to concentrate at the end upon his hero he only succeeded in getting further away from the specific details which were his main strength: Lancelot's insistence upon working to pay his creditors, though refusing a commercial situation as 'pay without work' and, instead, writing pamphlets and resolving to become a painter, was the merest moonshine. The attempt to resolve not the immediate issues of the book but all the issues of the world was little better: Lancelot was led back to the church 'like a child' by a mysterious stranger whose preaching might just have been accepted by the reader if the presentation of the hero's inner life had been steadily successful. As it was, this 'very mythical and mysterious dénouement' was a characteristic leap into allegory. To justify it on the specious grounds that 'sooner or later, "omnia exeunt in mysterium"' was a glib evasion for the novelist, whatever it was for the preacher. To invoke Bunyan at the end was only to accentuate an inability to follow him in wedding the general to the particular.

Anxious in his next novel, *Alton Locke* (1850), to manage more coherently his preoccupation with art, religion, and reform, Kingsley not only finished writing it before publication, but revised it after taking advice. Some of the advice, we know from his letters, concerned 'the interpenetration of doctrine and action'. He knew his weakness. All the same, he had little success in controlling it, excited as he was by what he called, in the third paragraph of the novel, 'the terrible questionings, the terrible strugglings of this great, awful, blessed time'. It had been a time of European-wide revolution, while in England the failure of the third and final attempt to have parliament accept a petition for the People's Charter was followed, in the summer of 1849, by the worst-ever outbreak of cholera. The press was full of the latest of the investigations into public health for which the eighteen-forties were famous, the reports of Dr John Simon, the Medical Officer appointed under an Act of 1848 by the City Commission of Sewers, supervening upon the reports of the Royal Commissions of 1844–7 and of Chadwick in 1842. In 1849, too, the *Morning Chronicle* began to print the articles by Henry Mayhew subsequently collected into a volume as *London Labour and the London Poor*. With characteristic energy Kingsley declared himself a Chartist, preached sanitary reform from the pulpit, and interested himself directly in the formation of the first Co-operative Association of Tailors and in the supply of fresh water to a notorious London slum.

His journalism as 'Parson Lot' in the columns of *Politics for the People* showed, in addition, a passionate and somewhat half-baked interest in the social mission of the arts. At first he planned a sequel to *Yeast* (which had not yet been revised for issue as a volume): 'looking at the Art of a people as at once the very truest symbol of its faith, and a vast means for its further education, I think it a good path in which to form the mind of my hero, the man of the coming age.' Acquaintance with a Chartist poet, Thomas Cooper, gave a more definite outline to his ideas and *Alton Locke, Tailor and Poet* began to be written, in scattered episodes—a letter shows Kingsley writing his hero's 'Conversion at the end' just before 'his becoming a poet towards the beginning'. There was an original subject in the poet's temptation to desert his own class, but it was in view for less than a quarter of the novel, and he was in any case not enough of a character for the reader to care much about this, any more than about his love for a girl above him in station.

Nor did the conditions of work from which Kingsley started afford a steady subject. They were realized memorably in the tailors' attic, smelling of new cloth, gin, and sweat, its windows shut and streaming with condensed breath, but these conditions were the source of the main action only in so far as it was from them that the hero was freed, by the help of benefactors from the classes above. His imagination was nurtured by a bookseller, Sandy Mackaye, with the accent and ideas of Carlyle, the nearest to a memorable character in the whole of Kingsley. 'His very dialect', said Carlyle himself on reading the manuscript, 'is as if a native had done it.' Certainly this was an achievement, but talk was all the character was given to do. In a violent attempt to develop an action from the specific social conditions of chapters 2–10, Kingsley returned to didactic melodrama, as in *Yeast*: Downes, the man who had earlier betrayed the tailors' attempt to combine against their employer, was discovered to be working at home in conditions so bad that his family were all dead from typhus; he himself, out of his mind, eventually drowned in the open sewer whence they had drawn their water, while the garment they had been working on carried the infection to classes above them. Downes's death-scene, for all its realistic detail, was more the work of the topical journalist than the novelist: the excited repulsion, for instance, in the repeated 'foul'—'a spark of humanity after years of foul life . . . quenched at last in that foul death'—contained only the minimum of reference back which might have reinforced the crude local effect, since Downes had made only one incidental appearance in the intervening twenty-four chapters.

The closeness of the journalist to the man of letters, however, showed in a fascination with 'last words', especially when they took the form of disordered monologue, after the example of Jacobean drama which he had already followed in verse in *The Saint's Tragedy* (1848). Downes, hurrying back to his tenement 'quite demented', was present to the reader as never before:

> 'The rats!—the rats! don't you see 'em coming out of the gullyholes, atween the area railings—dozens and dozens? . . . I hear their tails whisking; there's their shiny hats a glistening, and every one on 'em with peelers staves! Quick! quick! or they'll have me to the station-house.'

Sandy Mackaye's final speeches grotesquely developed suggestions from Shakespeare, although with a more tendentious aim ('Canst thou administer to a mind diseased? . . . D'ye ken a medicatum that'll put brains into workmen—?') There had been three examples of the same thing in *Yeast*: the Colonel, before his suicide, hearing again his child 'under my pillow, shrieking—stifling—two little squeaks, like a caught hare', and Argemone and the dying gamekeeper recalling the legendary curse upon the Lavingtons. Obviously Kingsley reconciled, by such means, a certain imaginative freedom with the realism to which his social subject committed him. In *Alton Locke*, chapter 36, the long sequence of Locke's delirium in his nearly fatal illness fulfilled a similar function, this time to allow the narrative to become allegory while remaining realistic. Locke in his fever believed he had been 'turned again to my dust', and then had begun a progress from 'the lowest point of created life . . . through ages to its top'. Once he developed stirrings of a 'higher consciousness', as an ape, he was threatened with the danger that 'the animal faculties in me were swallowing up the intellectual'. Rescued from this fate by a woman, he grew to be a man and joined with others in a society of equals, with 'one work and one hope and one All-Father'. When this unity in turn fell under threat, the prophecies of the same woman revealed to them that, though men had 'left the likeness of your father for the likeness of the beasts', it was their destiny to return to a Paradise of 'liberty, equality, and brotherhood'. The groundwork here came from the 'development theory' popularized by Robert Chambers in his *Vestiges of the Natural History of Creation* (1844) (which Kingsley mentions by name in *Two Years Ago*). The vivid speeding up of evolutionary and other changes gave, by its very incongruities, a new interest to the personal situation of the hero—when, for instance, as one of the 'jests of nature', a soft crab just emerged, 'naked and pitiable', from its 'armour', he had to receive the mocking attack of his successful rival. His love, which had seemed mere fictional routine when presented by the light of day, evoked a more complicated response when Locke in his dream crawled to the foot of a tree 'miles in height' at the top of which appeared 'a vast feathery crown of dark green velvet' and 'great sea-green lilies', from which girls' 'white bosoms and shoulders gleamed rosy-white'. As the character capable of such visions as these, he came nearer to fictional identity than in his adventures by day. But the interest was dissipated with the return of 'actuality': his prophetess turned into the aristocratic benefactor who had for long been watching over him and who now sent him, as a Christian, redeemed from making an 'idol' of the People's Charter, to 'train to be a poet of the people' among 'the primeval forms of beauty' in Texas. He did not survive his distant glimpse of 'the great young free new world', and the reader had cause to be grateful, for Kingsley had not found the means to make his hero's sensibility of interest, except when it almost coincided with his own in a chapter of allegory masquerading as delirium.

Kingsley's third wide-ranging social novel, *Two Years Ago* (1857), preached patriotism, public health, opposition to slavery, and much else, in what George Meredith, when reviewing it, called 'a hearty sermon with illustrations'. Kingsley was content with character and motive of the most obvious kind, compensating with surprises of external action—a physician foot-loose throughout the world, a female freed slave, a well-to-do American whom she converted to the cause of anti-slavery, a painter who had become a photographer, and a poet out of touch with his earlier self and with ordinary life. The energetic illustrations to the sermon did nothing to mitigate what Meredith called the characters' 'hopeless subjection to purpose' (the palm must surely go to the attempted sexual assault upon the heroine, called Grace, by a drunken Tory opponent of sanitation called Trebooze). Kingsley had too little interest in the intimate texture of human experience—except where it concerned our experience of Nature in the open air—and too much in the general tendencies of life in his time.

He was much more at home, in fact, with a framework that was not realistic but allegorical, as in the one book of his to survive for general readership, *The Water Babies* (1862–3). It was an original kind of allegory, drawing upon natural history to evoke wonder and amusement at the strange things that were true. At the same time, it started from a matter of profound personal interest to the author, the need to bathe and be clean, which enabled him to give importance as fiction even to the 'cause' he took up. When he wrote, the Children's Employment Commission was investigating the chimney-sweepers' climbing-boys. Their cause differed from the abuses in Kingsley's earlier fiction in that it had a specific relation to the subsequent allegory. Tom's escape, unlike Alton Locke's, developed something that was present in the original situation. The open country and 'a real North-country spring', in clear and simple language, made cleanliness attractive even to children. Kingsley was now writing for an audience he knew thoroughly, children of a certain kind of upbringing and education. He could express their more limited range of pleasure and hope in the form of actual things—like the ambition to 'be a man, and a master-sweep, and . . . wear Velveteens and ankle-jacks, and keep a white bull dog with one grey ear, and carry her puppies in his pocket . . .'. He was no doubt over-anxious to point out the proper way of becoming a man; in the first half of the book, however, he could do this by suggestion, in a development of Alton Locke's dream. Tom was 'like the beasts which perish; and from the beasts which perish he must learn' in a sequence of underwater episodes at the climax of which, having freed a lobster from a pot, he became at last able to find playfellows like himself. The wonder of the actual world, such as Kingsley had conveyed in *Glaucus*, his handbook for sea-shore naturalists, allowed him to make play with our relative ignorance about these marvels, and to turn our ignorance into the ground for accepting other marvels more properly the domain of the writer of fiction, including the marvel of a change in Tom's ability to see.

In the second half, the allegory came more under the domination of explanatory labels and didactic formulas. The Irishwoman who had befriended Tom changed into Mrs Bedonebyasyoudid, setting him on to become one of those who 'do what they do not like, and help somebody they do not like' and may even 'learn to like' this. But the sequence of fantasies which followed showed no progression toward this ideal. Tom was instantly converted to pursue it when shown, in the history of the Doasyoulikes, that evolution might work in reverse, and from there any sense that he was doing anything became lost in the variety of his encounters. The subtler aim of 'learning to like' it was coarsened into the mere determination to 'see the world, like an Englishman', much in the way that the inner development of Kingsley's earlier heroes had been submerged in the varied spectacle before them.

The humour which had played over the natural history of the first half of the book lightened much of the didacticism in the second. It was mere high spirits in the boisterous satire of the Pantheon of the Great Unsuccessful or the Isle of Tomtoddies, but there were places where it suggested a greater range of attitude towards his subject than Kingsley had shown elsewhere. It was here that his appeal to adults as well as to clever children was strongest. At first sight there seemed, for instance, to be a simple inconsistency between the pattern scientists of the first half, 'wise men ... afraid to say that there is anything contrary to nature, except what is contrary to mathematical truth', and those of the second, 'the men of science who get good lasting work done in the world' whom Tom eventually joined—for he could then not only 'plan railroads, and steam-engines, and electric telegraphs' but also know 'everything about everything'. Yet there were humorous exceptions to Tom's omniscience and they were accompanied by the same formula, 'till the coming of the Cocquigrues', as had earlier withheld full approval even for the author's own pet notion of evolution operating in reverse. The scientists of whom he seemed to be approving at the finish, the children of Epimetheus (who was 'always looking behind him to see what had happened, till he really learnt to know now and then what would happen next'), were as humorously satirized, in the figure of a naturalist, as were the children of Prometheus ('who kept on looking before him'), the fanatics and the theorists, the bigots and the bores. Even the point for which Kingsley had been working from the start, Tom's fitness to marry, was lightly handled: 'Don't you know that no one ever marries in a fairy tale, under the rank of a prince or a princess?' For all its didacticism, the book was in fact Kingsley's nearest approach to a work of the imagination, without the hectoring rhetoric which in most of his work left little for the reader to do but flinch or submit.

In most of his fiction Kingsley tended to confuse artistic intensity with didactic vehemence. Yet again and again the illustrations to his argument contained more than he had time to pursue; he did not see that the examples constituted his primary material, and that as a novelist his main attention should have gone to developing its possibilities. Except in *The Water Babies* he did not really try to understand what his vigorous imagination was putting before him.

His younger brother Henry,[2] with similar gifts, seemed more of an artist in so far as he was concerned primarily to interest the reader in the situations and settings he had created. The situations were not as a rule of the kind that would develop from the momentum of the characters themselves; the romancer was always ready with his conjuring tricks to secure the development he wished. Yet within that development he was capable of episodes which did unfold as if in a novel, strictly from the nature of the participants—the making and breaking of the engagement between the heroine and an honest, asinine lord in *Leighton Court* (1866), for instance, which was freshly discovered, even though set in a framework of the inventions of romance. These inventions were usually those of *Ravenshoe* (1861–2), the story which an unpublished letter shows to have been the first he attempted to

[2] Henry Kingsley, 1830–76, left Worcester College, Oxford, without taking a degree, to spend 1853–8 gold-digging and wandering in Australia. On his return, besides novel-writing, he edited the *Edinburgh Daily Review* for eighteen months, in the course of which he also went as war-correspondent for the paper to France, 1870.

write, involving disputed parentage and the hero's long, uncommunicative sojourn, like the author's, in a place unknown to relatives and friends. He was equally ready to use stock devices from the social novel, as with the hero's single-handed rehabilitation of an island community in *Austin Elliot* (1863). Yet even here, as in most of his other works, there was a relative freedom from open didacticism, 'divesting', as Henry James said in a review, 'the name of Kingsley of its terror'.

It was not that interests like those of his brother were absent, but that he did not preach about them. Characteristically, in *Leighton Court* he derided both those who 'wander sentimentally . . . bemoaning the state of a world we have never raised a finger to mend' and those who put a 'shoulder honestly to the wheel. Where they are going to shove us to is a question which has all the pleasures of profound uncertainty.' He might call a chapter of *Reginald Hetherege* (1874) 'Old Friends with a new Face', reversing the tendentious terms of his brother's subtitle of *Hypatia*, but it was a chapter of roguery. Where a worthy cause like the condition of the people appeared, its importance was shown by the course of the story itself, though he would sometimes—for he would not be a Kingsley if he did not want to improve even the improvers—draw attention to the absence of preaching, or mock its faint presence. In *Ravenshoe* John Marston and an Australian went 'down into the dirt and profligacy of Southwark, to do together a work the reward of which comes after death', and the author was curt: 'We have no more to do with it than to record the fact, that these two were at it heart and hand.' But when many were seeking Charles Ravenshoe, it was Marston who found him, ill and on the verge of suicide. Similarly when Austin Elliot, in gaol after a duel, had a plan, patently like those of Parson Lot, of elevating 'the moral tone of the convicts around him' by talking 'about Shakespeare and the musical glasses' ('fashionable topics' in *The Vicar of Wakefield*), although the only immediate result was a quarrel, it led again to the crucial encounter of the plot.

The 'cause' which Henry knew most about, from his own five years in Australia, that of colonization, already taken up by Carlyle, Dickens, and Bulwer, was interesting to him as the chance for a tale rather than the text of a sermon. Certainly the vigorous, grim melodrama in England in the first volume of *Geoffry Hamlyn* (1859) threw into relief what was idyllic in the two Australian volumes; yet the whole ended with an impudent accumulation of prosperity, most of which had nothing to do with colonizing. There was a different attitude towards colonization in his second Australian story, *The Hillyars and the Burtons* (1863–5), in so far as it contained important characters for whom Australia was or became their home. And the contrast between the new country and an 'old' Chelsea, crowded with antiquities, underlined the opportunities of status and recognition which a blacksmith and his family gained by going out. The opportunities were recounted with relish, and sometimes casually, *pour épater*—the immense copper-find, the hunchback brother's success in parliament, the pleasure of his sister in going to the opera. All the same, the sombre ending included disaster in a region of great beauty but 'not ready for human habitation', 'still in its cruel, pitiless phase'. In fact, 'causes' like colonization were subservient to the author's enthusiasm for life itself, admittedly most often for its sudden possibilities of good fortune, but also for the spectacle of what people have it in them to be and to do. The examples were scattered improvidently without their effect being focused upon the main concerns of the story—the vigorous female farmer in *Stretton* (1868–9); or Gerty, the dazzling, witless Australian who became Lady Hillyar; or the marriage of the Welters in *Ravenshoe*, modelled on Becky Sharp's, but slowly developing as hers did not. In so far as the wishes of the romancer were apparent in the encouraging possibilities he selected, it was the territory that Bulwer had tried to enter in the fifties, impeded by the elaborate corsetry of an artificial style. Kingsley's enthusiasm for the whole spectacle appeared, on the other hand, to be guaranteed by his flow of slightly raffish colloquial comment, less protean and more simply personal than Thackeray's, but owing much to his example. *Geoffry Hamlyn* began:

> Near the end of February, 1857, I think about the 20th or so, though it don't much matter; I only know it was near the latter end of summer, burning hot, with the bushfires raging like volcanoes on the ranges, and the river reduced to a slender stream of water, almost lost upon the broad white flats of quartz shingle.

It was an old trick by 1859 to draw attention to a February summer. (G. P. R. James was already doing so in one instalment of *The Convict* in 1847.) Kingsley made it seem new by the off-hand colloquial grammar, the breezy exaggeration in 'raging like volcanoes', and the discreet use of local words ('ranges' and 'flats'), the whole pegged down by a final phrase which was technically specific and rhythmically skilful. It seemed artless, too, to begin at a time of drought and fire—at least if, as was to be expected in the decade after *The Caxtons*, Australia was being recommended (though drought appeared in the story not as the steady enemy it was in fact but as a mere occasional visitant). The favourable possibilities of human nature (except among the indigenous people, towards whom the harsh conventional attitudes went unquestioned) were proclaimed with equal breeziness. It was characteristic that even the villain, the convict George Hawker, should preserve some regard for his wife, and that the fears of violence which had been caused, reasonably enough, when he broke into her house should eventually be shown to be groundless.

In *Ravenshoe*, where the Machiavellian resident priest, Mackworth, was not eventually repudiated, any more than the hero's self-confessed 'evil genius', Lord Welter, wickedness tended to resemble the supernatural in Mrs Radcliffe, an appearance to be eventually explained away, enjoyable for the pleasure of subsequent relief. This impression was the result of his very skill in the arts of romance. He knew that the threat of wickedness was quite sufficient, schooled as he was by Scott and the Romantic poets in the power to suggest it by the activity of plants and animals and the state of the light and the weather. So in the thirteenth chapter of *Ravenshoe*, what turned out to be by no means extraordinary events seemed menacing from the suggestions of witchcraft conveyed by the presence of a black hare and by the change from a bright morning, when 'the deer were feeding among the yellow fern brakes, and the rabbits were basking and hopping in the narrow patches of slanting sunlight', to a midday 'dark and chilly' with 'a low moaning wind . . . through the bare boughs, rendering still more dismal the prospect of the long-drawn vistas of damp grass and rotting leaves'.

In the most characteristic effects of this kind, the menace in the setting was conveyed by an observer well accustomed to estimate it. The surroundings of Leighton Court were seen through the eyes of a huntsman at the edge of a 'vast melancholy bog' high up on Dartmoor, those of Oakshott Castle by fishermen on a wild night, and those of Ravenshoe by 'a lazy yachtsman, sliding on a summer's day, before a gentle easterly breeze, over the long swell from the Atlantic', and grasping well the difference between the 'shoreless headlands of black slate' and the amphitheatre of wood and lawn where the house and its village lay. From these surroundings, and especially from the sea, the story in *Ravenshoe* gained an emotional dimension which the main characters were hardly substantial enough to sustain. When, in chapter 14, the heroine walked with Lord Saltire, while 'the misty headlands seemed to float on the quiet grey sea, which broke in sighs at their feet', the language distinguished a moment of more promise for the story than it fulfilled; as so often, Kingsley was tantalizing the reader once he had brought forward an unusual variety of people, with an enthusiasm which seemed to portend some equally unusual event to involve them all. Again in this instance he presented a distinctive observer, Lord Saltire, commenting in his old age so as to draw attention to the author's own technique:

> 'The new school of men,' said Lord Saltire at last, looking out to sea, 'have perhaps done wisely in thinking more of scenery and the mere externals of nature than we did.'

The crucial event for the story at the end of the second volume, Cuthbert Ravenshoe's drowning, gained its emotional emphasis from three weeks of genial weather: 'All before them the summer sea heaved between the capes and along the sand, and broke in short crisp surf at their feet, gently moving the seaweed, the sand, and the shells.' Such emphasis could obviously become too selfconscious, thinning out the feeling for the character into general sentiment; nevertheless, it underlined the author's skill in interweaving these slight characters with 'scenery and the mere externals of nature'.

The removal of the middle part of the novel to London supplied the simple contrast of setting on which nearly all Kingsley's romances depended for their imaginative coherence. The result was to bring into relation with the first volume the Crimean chapters at the beginning of the third, with their return to 'the free wind of the sea' and 'the state of mind called romantic'. This state of mind was brusquely shaken in an economical narrative concerned with cholera among the troops, the relative pointlessness of the presence of the cavalry in which Ravenshoe served, and his fortunate delirium in hospital in Scutari. It all formed a strong contrast with Charles Kingsley's thumping of the patriotic tub in *Two Years Ago* and *Westward Ho!* The return to London accentuated Ravenshoe's outcast state. Throughout, the events gained status in the memory, and the whole its shape, from contrasts of hope and fear matched with contrasts of setting; but Charles Ravenshoe was not known subtly enough for his obstinate attempt to stand alone to seem at all inevitable.

Kingsley continued to make his plots out of variations upon the main constituents of this story. *Austin Elliot* (1863) gave a jejune version, with striking seascapes which remained set-pieces, quite separate from the intrigue. The version in *Silcote of Silcotes* (1866–7) obscured with restless, superadded inventions the long disappearance of the squire's son, and managed to make nothing at all of the grandson's feelings at the discovery first of his mother and then of his father. This absence was all the more marked when descriptions of the setting had made the grandson the centre of a potentially imaginative inner biography: at first there was an extravagant enchantment, appropriate to his first adventure, as he followed a party seeking out some poachers.

> The beech forest was blazing in the glory of the August moon. The ground, golden all the year round, by daylight, with fallen leaves, was now a carpet of black purple velvet, with an irregular pattern of gleaming white satin, wherever the moonbeams fell through to the earth.

Later, working as a shepherd, before getting the chance of a formal education, he watched the Thames become 'a chain of crimson pools' as the sun set behind 'dim mysterious wolds' beyond which, he knew, lay Oxford. When eventually he was sent to a city school which was suddenly moved into the country, the new place held new hopes, though it was on a promontory thrusting into one of the 'desolate Hampshire lakes', beside which the 'bittern startles some solitary cow in its flapping and noisy flight, and the snipe bleats in the place of the lamb'. In such descriptions, all bearing upon the same character and upon an important social concern like education, it seemed as if a novelist were trying to emerge from the routines of the romance-writer.

Kingsley could deride his preoccupation with scene—for instance, in *Leighton Court*: 'the place which we, with our fine imagination, have hitherto called "the glen of a hundred voices," but which is marked on the Ordnance map as "Crab's Gut"'—

but it was the main source of his hold upon the imagination, stirring habits of sentiment and memory such as it required a major talent like George Eliot's to direct towards subtler human sympathies. Although description might sometimes be merely portentous, it often seemed about to reveal something worth knowing about life. But, writing over a dozen fictions of some length in only sixteen years, Kingsley had not enough time to think about it. What is remarkable is that only a minority of his books were as disjointed and spiritless as *Oakshott Castle* (1873) or as pretentious as *Mademoiselle Mathilde* (written to order, in 1867–8, the first fiction to be admitted to the *Gentleman's Magazine*). At least he understood, in most of them, that success depended more upon what he imagined than upon anything he might wish to teach. And even when what he imagined was flimsy or foolish, his flow of talk about it could arouse the sort of interest which a familiar letter may have though the persons and motives mentioned are less than fully known.

With Elizabeth Gaskell[3] character and motive were the essentials, even for the advancing of a social thesis. While it was unusual for her to write without some concern for the effects of social change—with reference now to industrial relations, now to manners, now to education—this concern seldom appeared, as with Charles Kingsley or Disraeli, in detachable 'thoughts and opinions' (her phrase in a letter). If she had difficulty in dissolving her opinions and recommendations into the fiction, it was less because they were stronger than the fiction than because the fiction was too forceful for them or even conveyed something different.

Her concern in *Mary Barton* (1848) was that the sympathy which could easily be shown to be evident between the members of one class, the poor, should become equally evident between two classes, the poor and the rich—even though the gap between them appeared to her, as to Disraeli, to be greater or at least more apparent in the economic difficulties of the early 1840s, than it had been before. The importance of such sympathy was to emerge from the nature of the character she called 'my hero', John Barton. 'I had so long felt,' she wrote in a letter soon after the book was published, 'that the bewildered life of an ignorant thoughtful man

[3] Elizabeth Cleghorn Gaskell, née Stevenson, 1810–65, was born in Chelsea and brought up by an aunt at Knutsford, Cheshire, which is recognizable as the scene of many of her fictions. In 1832 she married William Gaskell, Unitarian minister in Manchester, her home for the rest of her life, though she travelled extensively in England and on the Continent.

of strong power of sympathy, dwelling in a town so full of contrasts as this, was a tragic poem . . .' To a considerable degree the claim was borne out by the book, which gave startling evidence of an advance in the power of the novel in the 1840s to deal with tragic issues, in this case a conflict within the hero which could be resolved only by his death. As with greater tragic heroes, the forces opposed within him were mutually interdependent, his 'power of sympathy' and his bitterness. Superimposing the hard times of 1839–40 upon common human vulnerability to bereavement, Elizabeth Gaskell found the origin of his bitterness in the very strength of his affections—for his son (the loss of whom he believed to have been caused by his own unemployment at a time when the boy was convalescent and in need of help his father could not afford), for his wife, after whose death he became 'harsh and silent', and for his fellows, whose distresses were refused a hearing by the reformed parliament. His sympathies nevertheless remained active, and his tragic mistake was made when under the influence of compassion—compassion, not only for the children on whose behalf his trade union had struck against starvation wages, but for a particular country worker who had been attacked when coming in to take the work offered at that figure. Starving and overwrought, Barton was too easily moved, by means of an argument about what is and what is not cowardly, to suggest violence against the supposed cause, 'the masters'. Eventually he became (by lot) the one to commit a murder.

His development up to that point was clear in the contrast of silence with vehement speech, forced out of him and going to the heart of his position:

> '. . . I've thought we han all on us been more like cowards in attacking the poor like ourselves; them as has none to help, but mun choose between vitriol and starvation . . .'

Spoken by a man 'slightly made', with 'almost a stunted look about him', though sharing the 'acuteness and intelligence of countenance which has often been noticed in a manufacturing population', speech like this took tragedy in fiction beyond the reach of the rhetorical stilts of Bulwer. But the programme Bulwer had set out in 'On Art in Fiction' in 1838 was certainly fulfilled, especially his emphasis upon 'stormy and conflicting feelings'. Their climax was finely conceived, the hero 'crushed by the knowledge' of having 'forfeited all right' to offer sympathy to a father who was like Barton bereft of a son, but now by Barton's own fault.

The climax, however, was marred by exclamatory fervour and, more seriously, by the wish to make moral recommendations and in their interest to interpret the action, even through dialogue, in ways at variance with what had been shown. Barton had been movingly observed, for the first time after the murder, sitting by the grey ashes of his empty grate with his hands crossed and fingers interlaced, 'usually a position implying some degree of resolution or strength', but now 'so faintly maintained, that it appeared more the result of chance'; his face, 'sunk and worn,—like a skull', echoed poignantly the features that had established him in chapter 1, his pinched appearance, 'resolute either for good or evil'. The author's inexperienced language for the probing of inner processes even had an original oddity: 'all energy . . . seemed to have retreated inwards to some of the great citadels of life, there to do battle against the Destroyer, Conscience.' His agony, in being 'haunted . . . with the recollection of my sin', he summed up with typical terseness: 'As for hanging, that's just nought at all.' But in the interests of a vague plea for the education of working people, Barton was made to elaborate this with 'I've kept thinking and thinking if I were but in that world where they say God is, He would, may be, teach me right from wrong, even if it were with many stripes. I've been sore puzzled here.' The words related to the claim of the commentary in chapter 15, at the time of Barton's desperation after parliament had rejected the Charter, that 'the actions of the uneducated' are like those of 'Frankenstein, that monster of many human qualities, ungifted with a soul, a knowledge of the difference between good and evil'. What had been shown, however, was not a man 'puzzled' about 'the difference between good and evil' and needing to be educated in it, but a man, and his associates, conscious of that difference and terrified by it. At the drawing of the lots the stress fell upon their consciousness of guilt, 'the suffering their minds were voluntarily undergoing in the contemplation of crime', and on the fact that the knowledge of who had 'drawn the lot of the assassin' could be possessed only by 'God and his own conscience'. This made it impossible to take Barton's unease before the murder—economically conveyed as he waited at home with his daughter, 'starting up and then sitting down, and meddling with her irons'—as anything but revulsion against the deed, mastered with difficulty. And such an impression was not to be obscured by explanations in the final chapters. There Mrs Gaskell was clearly working, like Charles Kingsley at the end of *Yeast* and *Alton Locke*, to draw from the case of her hero recommendations relevant to the immediate situation outside the novel. Barton was endowed in retrospect with a whole religious life of which the active part of the narrative knew nothing, so that his actions might exemplify his desertion of the 'Bible rules' which he saw he 'would fain have gone after'—if only he had found any one to satisfy his questions about them, and if only other people, masters and men, had cared about them. Direct tragic force was dissipated in polemical interpretation of what had never been realized as fiction at all. The importance of education to working people had been clearer in the character of Job Legh, with his scientific interests and point of view closer to that of the author.

In the mutual forgiveness between employer and employee (not even suggested in the extant draft plot of the novel), what was merely exemplary further encroached upon substantial fiction. No more than Charles Kingsley could Elizabeth Gaskell present her hero's problems without trying to solve them all too thoroughly. Moreover, the serial conclusion was amplified for the volume edition with further recommendations in two chapters which accentuated the tendency, repeatedly disturbing to the reader already, for the characters to speak from the needs of the thesis-maker rather than from their own—for instance, when recommending patience to working people in return for sympathy from the masters. In the best parts of the novel this tendency had been mitigated by the very strength of the impression of the characters' needs, the specific presentation of the outward conditions justifying their mutual sympathy and giving of advice.

The story of Barton's daughter, Mary, who eventually usurped the title of the book, was imagined as in almost every way the opposite of his—the girl crossing prematurely the barrier between the classes that he crossed only in his last moments, the father betrayed into crime in a helplessly distraught condition of mind, the daughter keeping her head, concealing the evidence of guilt and proving the alibi of the innocent. In Mary's case there was tragic potentiality, but it was not realized. Like the heroine of *The*

Heart of Midlothian, she held the essential clue to the incrimination of a member of her family, but, in addition, she was in love with the man falsely accused of the same crime. The author chose to turn her into the idealized heroine of romance, of unusual personal beauty, never coarsened by the conditions of her life, and lacking even the dialect of contemporaries, just as Oliver Twist had—and like him representing a principle of Good surviving through every adverse circumstance. There was considerable straining to keep her motives pure. The scene in which she destroyed the evidence of her father's guilt stressed not only her isolation and helplessness but also her 'prudence' and even her 'high, resolved purpose of right-doing'. However justified the comment might be by the heroine's attempt to clear the innocent, it hardly applied to her shielding of the guilty. Nor was such exaggeration of her moral purpose necessary to secure the reader's sympathy with her in the difficulties of her position. The patent wish to turn sympathy into moral approval by a short-cut which obscured those difficulties was the mirror image of the author's exaggeration of moral disapproval in denying a knowledge of the difference between good and evil to men who could be seen to suffer from their very recognition of it. In the tragedy the commentary was too hard on the hero; in the romance it let the heroine off too lightly. The artist had her triumph over the moralist, however, in the vigorous presentation of the 'lightly principled' Sally Leadbitter, plain and freckled, 'never likely, one would have thought, to become a heroine on her own account'.

The papers in *Household Words* which became *Cranford* (1851–3) developed subtler means of moral discrimination, because their purpose had been less carefully pre-arranged. The starting-point was the sense of social change apparent as the author looked back upon the time before she came to Manchester. She was prompted to do so, according to the reminiscences she communicated to an American periodical in 1849, by a regret that Southey had never fulfilled a plan 'to write a "history of English domestic life"', for 'even in small towns, scarcely removed from villages, the phases of society are rapidly changing'. Many of the same details as in *Cranford* appeared in this essay, but without much order or much play with attitudes towards them. The fiction developed both the order and the attitudes by forming out of the relation of past and present a simple action centred upon one character, Miss Matty, and the progressive gradations of amusement, sympathy, affection, and admiration which she might evoke. A narrator who had always 'vibrated' between Cranford and a modern city, Drumble, made possible exact shades of irony and tenderness towards manners and persons understood both from within their circle and from without.

As a self-consistent, cohesive little society of women, proud of themselves and of their single life, faced a series of invaders from outside, action in the present revealed more and more of Miss Matty's past, so that even those mores of Cranford which were a principal source of the comedy received a tincture of pathos. The strict code of gentility, for instance, observed by other Cranford ladies with such anxious deliberation, could be seen to have prevented her marriage. At the same time, the plot in the present brought her suitor, Holbrook, a yeoman who farmed 'four or five miles' outside Cranford, into relation with the other outsiders who broke into its closed circle. Most of them were modern and male, like Captain Brown who worked on the hated railway and preferred Dickens to Dr Johnson (so resembling Holbrook, who conspicuously appreciated Tennyson), or the conjurors who performed among the faded glories of the Assembly Room. For the 'Amazons' of Cranford, in the present time of the story, the eventual consequence of 'the invasion of their territory' might be matrimony—a comical threat in the present ('Two people that we know going to be married. It's coming very near!'), but a quite different matter in the past—or on the other hand the consequence might turn out in some degree to make up to Miss Matty for her unmarried state, as with the shop which gave her a regular contact with children, or the return to her of her brother, Peter. The narrative apparently so artless, even haphazard, all bore upon the one matter and, as with *The Rape of the Lock* (if greater things may be compared), small felicities flourished in its shade. Some of them had the selfconscious artifice of mock-heroic, like the term 'Amazons' for the inoffensive, unattached ladies, or incongruous juxtaposition, as in 'Hebrew verses sent me by my honoured husband. I thowt to have had a letter about killing the pig . . .' Some were a spontaneous overflow from the quaint economics typical of this becalmed enclave, like the cat that swallowed the lace. Others offered brief, dramatic development of character while effortlessly echoing the dominant note: the intricate little scene of matrimony hurried on by Miss Matty's servant, for instance, in which every touch told, the girl ready to risk being thought 'forward' in the interests of her mistress who would become their lodger, while her embarrassed young man, in 'no mind to be cumbered with strange folk in the house', echoed and so mocked the Amazons' belief that 'A man . . . is so in the way in the house.'

All the time, apparently trivial matters which stirred an affectionate smile revealed matters far from trivial—the affections denied full expression but expressing themselves nevertheless, the charity and contentment hard won, the goodwill that reigned 'to a considerable degree' despite 'strongly developed' individuality. And a discreet undercurrent of criticism of Drumble ways was suggested by their obvious contrast with those in the foreground—Miss Matty's climactic sacrifice of what looked like her last five sovereigns to redeem the note of a failed bank in which she was a shareholder, her imagining the directors of the bank would think 'poverty a lighter burden than self-reproach', or her reservations about depriving some one else of trade if she should set up as a seller of tea—a scruple which 'would put a stop to all competition directly', as the narrator's father, himself 'a capital man of business', pointed out. It is easy, moreover, to overlook the fact that the prime example of modernity breaking in, the failure of the bank, was typical of the period of the story, the years following 1836, when the banking system was under extraordinary strain, culminating in a crisis in which the Bank of England was only saved by the help of the Bank of France, and numbers of country banks stopped payment. The matter was still topical when *Cranford* began to appear, over ten years later, for Peel's Act of 1844 to control note issue had come under Committees of Inquiry of both Lords and Commons, following the renewed crisis in the late forties. It was highly characteristic of the author to make her plot turn upon such an important fact of contemporary history. The implied social criticism, however, was completely dissolved in the fiction, to form part of the demonstration of the goodness, innocence, and resilience which the chief character shared with Mr Pickwick and Dr Primrose.

The same, unfortunately, could not be said of *Ruth* (1853), which returned to overt moral recommendation. It made explicit what had been implied by the key position occupied in the careful plotting of *Mary Barton* by Esther, who became a prostitute:

anxiety about her caused the death of her sister, which left Mary motherless; John Barton's refusal to listen to Esther kept him ignorant of the danger to Mary from a manufacturer's philandering son, while Jem Wilson's willingness to listen led him to quarrel with that son and so to be suspected of murdering him; and it was Esther who found the clue to the actual culprit. The implication was plain: the woman whom most of the characters considered an outcast, cut off from all respectable relationships, still in fact had the status of a responsible human being and moreover of a member of a family (the point of Mr Peggotty's search for Emily and his virtual adoption of Martha in *David Copperfield*, Number XVI, for August 1850).

But to write in the belief, as Elizabeth Gaskell said later, that 'what was meant so earnestly *must* do some good' gave her only imperfect control over what she was capable of saying as a novelist. 'I could have put out much more power', she claimed, 'but that I wanted to keep it quiet in tone, lest by the slightest exaggeration, or over-strained sentiment I might weaken the force of what I had to say.' This completely underestimated the power of the quiet details at which she was so good. In her hands, the discomforts and humiliations of the kind of life approved by society for a girl who had been recommended to cultivate 'economy and self-reliance', when orphaned at the age of 15, were moving in themselves. They were so narrated, however, as to raise expectations that the factors predisposing her to her particular course of suffering might include the changes in that society. Not only had the town where she worked fallen from the favour of Tudor sovereigns and from the 'noble appearance' that went with it, but the very work-room used by the apprentices of a 'superior' dressmaker retained in its decoration evidences of aristocratic splendour like the Assembly Room at Cranford. But the story turned out to have no use for these details; it tended increasingly to make Ruth a heroine of romance like Mary Barton, with a 'noble head' and the face 'positively Greek' by which her seducer would recognize her years later, and with an additional ability, heroic in the circumstances, to stand back from her situation and articulate the author's own sentiments: 'I only ask to be one of His instruments, and not thrown aside as useless—or worse than useless.' George Eliot saw at once that the author was 'not contented with the subdued colouring ... of real life', in her 'love of sharp contrasts—of "dramatic effects"'. It was hardly the way to draw attention to a common predicament—only made even more uncommon by the author's efforts (understandable in 1853) to prevent the reader from refusing his attention altogether. As an unprotected child, without parent or even employer, Ruth must be as nearly as possible exempt from responsibility for being seduced and yet she must accept with 'her peculiar and exquisite sweetness of nature' the need to be 'purified ... by fire'. The humility elevated the character at the cost of exaggerating the stain. The elevation, completed when Ruth not only went nursing in a typhus epidemic but died as a result of tending the father of her child, was as much the work of the writer of romance as if she had offered the 'romantic incidents or exaggerated writing' which she wished to avoid.

The novelist proper could be seen at work questioning the degree of the stain, as the tale, though not the teller, invited comparison between Ruth's 'weakness' and the lie which her protector, the dissenting minister, Benson, told in order to pass her off as a widow and her child as legitimate. It was another instance of a temptation almost impossible to resist in the circumstances—and in the state of society, which the narrative stressed by recalling the sufferings of another illegitimate son. The reader could compare the effect of the 'unconscious tempter' upon a well-informed adult of disciplined conscience and upon an uninstructed girl, at first 'not conscious' even that the presence of a male friend 'had added any charm' to a country walk. Whether she knew what she was doing in living with him was left uncertain. Unwilling to resolve the matter realistically, the novelist offered instead images of innocence—the white camellia or the water lilies, accepted from her seducer, the white dimity of her room at the Bensons', 'the first snowdrops in the garden' put on her pillow as 'the baby lay on the opposite side'. In addition, Ruth's strong delight in natural forces, rain, river, sun, mountain, and sea, seemed to draw her into irresistible rhythms which spoke for her where the strict moralist would not.

It all failed to make Ruth of interest as an individual, despite the skill lavished on her states of mind. Her anxiety for her son's health, in conjunction with the profound disturbance of once again talking with his father, took the form of 'one sharp point of red light in the darkness which pierced her brain with agony, and which she would not see or recognise—and saw and recognised all the more for such mad determination ...'. Such experiences, for all their subtlety, remained disembodied, much as in Virginia Woolf, while the fictional interest was usurped by the indisputable individuality of the Bensons' servant, Sally, and of a daughter of the house where Ruth was accepted as governess, Jemima Bradshaw. Their championing of Ruth meant more than the tendentious sympathy of the author because it emerged from the presentation of their unique lives, their humour, their energy, their faults, their power to surprise. These lives, by contrasting with Ruth's, threw a light upon hers which seemed to be that of life itself and not merely the result of the author's design upon the reader: Sally's firm place in the family of the Bensons depended on her having been forgiven for an early fault. Her independence of the male sex was treated as comedy; nevertheless it made as strong a contrast with Ruth's vulnerability as did Jemima's plainness and her strong sense of her own individual worth. The novelist could even show what Sally had lost by her independence, in a delightful small scene with Ruth and her child out of doors.

It was these minor female characters who generated the liveliest things in the book. The male characters were merely adequate for the implications to arise—for instance, from the contrast between the excessive atonement believed appropriate for Ruth, and Benson's contracted pew-rents and occasional self-reproach. Bradshaw, the manufacturer who kept his justice and his charity separate but was forced by a speculating son to bring them into relation (like Dickens's Mr Gradgrind in the year following this novel's appearance), represented the class whose 'conscious righteousness' repudiated Ruth. A whole social novel was sketched, but only sketched, in his sharp commercial practices and uneasy sanctioning of electoral bribery. If the novelist had been prepared to allow imagination to deal with the book's central predicament in relation to the social actualities which were her strength, and not the romantic exaggerations which were her weakness, the materials were obviously there. But she would have had to take her own advice of 1859 to 'think' more 'eagerly' of her plot, and not use it 'simply ... as a medium'. In becoming only the means to a moral end, her plot betrayed the novel's own best insights. She devised, as Charles Kingsley might have done, the completest solution she could for its central moral problem, and so was taken beyond the stricter demands of realistic fiction into hagiography. Here the reader had to refrain from exercising the powers of judgement which the best parts of her plot had awakened. The summit of the author's ideal was that the heroine should remain unconscious of how well she fulfilled it, and specially conscious of how much she did not: nursing the sufferers from an epidemic of typhus, Ruth 'best knew how many of her good actions were incomplete, and marred with evil'. The matter was left like that, abstract, undramatized. The words, however, might have applied also to Ruth's final decision to risk infection again, after the local epidemic was over, by nursing her boy's father when she no longer loved him and had herself accepted sole responsibility for the boy's upbringing.

In *North and South* (1854–5), a moral impetus no less strong than in *Mary Barton* and *Ruth* yielded fiction which developed much more by its own logic. The energizing idea was again that of *Cranford*, an enclosed society facing the action of outsiders, the chief of them this time the heroine, Margaret Hale, more fully known than the previous two heroines, and less often merely admired. It was through her mind and sensibility, inflexible in self-judgement but charitable towards others, delicate in sensuous appreciation but strong in moral resolve, that the reader could take pleasure in discovering how someone from the superficially more civilized south might not only change her opinion of the manufacturing north, but even have her own effect upon its little-known 'Darkshire' habits.

The difference between such regions and those to the south of them which contained the centres of political power and official culture had been one of Disraeli's subjects in his novels of the mid-forties. Dickens approached it in the travels of Nell and again in *Hard Times*, which was appearing while Mrs Gaskell was at work. Unlike Dickens, she was writing of what she knew from long experience. Already in 1848 she had made ignorance in the south of conditions in the north a factor in the embitterment of John Barton. Differences of opinion between the two regions were highly topical during the debates about free trade in the seven years after the Corn Laws were repealed in 1846. Although Manchester (the Milton-Northern of the novel) did not speak with one voice in these matters, it was widely identified with the views of the radicals who had taken over its Chamber of Commerce, with their articulate opposition to the aristocratic and military centre of gravity in the south, their millennial hopes, and their confidence that industry, especially the cotton industry, was the pioneer of civilization.

Mrs Gaskell's skill in admitting the reader to share in the sensibility of her heroine exposed such views to a humane critique, translating the issues into the language of personal life. Arguments against 'the old worn grooves of what you call more aristocratic society down in the South' or in favour of the 'right' of 'the owners of capital ... to choose what we will do with it' became the expression of personality when advanced by a mill-owner, Thornton, who took the heroine seriously enough to enjoy arguing with her. Her critique was that of a woman without power (until the conclusion, when inherited wealth enabled her to assist Thornton), but with a capacity for feeling and for thought which made her the more able to reckon the human cost of industrial success.

From such a viewpoint—subtler, less partial, because it was feminine—Charlotte Brontë's chief characters in *Shirley* had watched the war between masters and men. Although Mrs Gaskell owned to having 'disliked a good deal in the plot of *Shirley*', it furnished her with the germ of her own, for *Shirley* had begun with Caroline Helstone's emergence from the 'marvellous fiction' of her early life into a 'school of Experience', the chief constituent of which was her relationship with a manufacturer. Margaret Hale, as a child, had promised herself 'to live as brave and noble a life as any heroine she ever read or heard of in romance'; her surroundings were those of 'a village in a poem—in one of Tennyson's poems'—beside which 'all other places in England that I have seen seem so hard and prosaic-looking ...'. The rewards of accepting what seemed hard and prosaic formed the substance of the book. It contrasted sharply with those earlier treatments of the new prose of nineteenth-century working conditions which had discovered ways to escape—in *Michael Armstrong* and *Helen Fleetwood* to the country, in *Mary Barton* to Canada, in *Alton Locke* into 'fresh conceptions of beauty'.

Reality in *North and South* involved less of the physical particularity of *Mary Barton*—those wet bricks of the Davenports' inhabited cellar through which 'the filthy moisture of the street oozed up'—and more of the subtleties of inward life. The poorest house in the book, that of Boucher, a workman who had incited to riot during a strike and, unemployed, had taken his own life, was not described at all; the energy of the scene went into the fine account of his widow's awakening to her plight. The physical particulars that did appear were there in order to define feelings which were central to the structure of the book. Country freshness and freedom contrasted with the confinement and ugliness of the town at the beginning of Margaret's acquaintance with working people: her offer to Bessy Higgins of a bunch of 'hedge and ditch flowers' led to the conflict of her father's northern independence with the manners Margaret had brought from the country and from the south. At the other end of the scale, when the house in which Thornton and his mother lived, next to the mill, was described, it was as the index of their pride in his career.

Mrs Gaskell's control of her descriptive powers quietly reinforced the thesis of the book that some change in the current habits of mind of both masters and men was demanded by the new facts of large-scale production. And the events, whether conveyed directly or by way of the narrative of participants, drew attention to the reciprocal relationship between outward facts and inner decisions. At the heart of the plot, the heroine told a lie in order to shield a brother vainly attempting to gain justice after he had resisted the tyranny of a captain over his crew—a tyranny much more familiar in the south than any arising from the new powers, in the north, of a small group of masters over large numbers of men. As with Benson's 'weakness' in *Ruth*, it was the state of society that made the lie difficult to avoid. In these circumstances even a virtue like filial love could drag the heroine down into mankind from her early romantic hopes of leading 'a life sans peur et sans reproche'. Similar resistance to 'oppression' and 'withstanding of injustice', and similar loyalties to family and fellows, led the cotton workers to their much more complicated decision to strike. 'It may be like war; along wi' it come crimes; but I think it were a greater crime to let it alone. Our only chance is binding men together in one common interest ...' was the claim of their spokesman, Higgins. This entanglement of justifiable aims with difficult circumstances made the connection between the private and the public subjects of the book. As a result, the trade union appeared in a vastly different light from that of *Hard Times*, or of Charles Reade's *Put Yourself in his Place* fifteen years later. Although union action might anger the humane observer ('Why!

what tyranny this is! . . . And you talk of the tyranny of the masters!'), her father could find its binding men together to be 'Christianity itself—if it were but for an end which affected the good of all, instead of that of merely one class as opposed to another'.

The conversation, at the beginning of the second volume, where these opinions appeared was characteristic of the novelist's anxiety to make public issues form part of the private action: Higgins had just lost his daughter and it was to comfort him for the lack of her sympathy with his cause that he was given the opportunity to explain it—even though this required the reader to believe him capable of putting his grief aside so soon and to believe Margaret Hale capable of relentlessly pressing the argument against him in these circumstances. Earlier discussions involving Margaret and Thornton had been more successfully absorbed into her story. There was a progression from her insistence on the 'religious' reasons which might override the merely 'human right' of the owners of capital to 'do what you like with your own', on to her anger when the trade cycle was explained to her 'as if commerce were everything and humanity nothing', and so to the culmination when she suggested Thornton should go out and speak to his rioting employees 'as if they were human beings' (before the soldiers arrived to drive them off) and then herself attempted to shield him from violence—as she said later, 'like a romantic fool'. Innocence tempered by experience formed the staple of the private action, its epigraph furnished by the heroine when she was herself misunderstood: 'Take care how you judge.' The application of this principle to her view of Thornton and his 'hands', and to their view of each other, was continuous with its application to her own family—to her view of her mother, whose declining health renewed their friendship, and to her view of her father, whose softness contrasted increasingly with her own steel.

Such an epigraph led naturally to an open-ended conclusion, far from the large semi-allegorical solutions of Charles Kingsley or from the parable of conversion and the reconciliation at the end of *Mary Barton*, and thus more in touch with the facts. If the author had lived another twenty years she would have seen prosperity bring some decrease in the intensity of the confrontation between masters and men while at any rate the larger employers attempted to promote better relations by methods developing from Thornton's at the end of the novel. Her prescience was plain in her choice of such issues for her plot—even the employment of Irish labour was of continuing importance, with its climax in the anti-Irish riots of 1861 in Oldham and elsewhere, and its longer-term effect upon the working-class vote, weakening the dominance of the Liberals in Lancashire.

As a novelist Elizabeth Gaskell showed her developing skill by the indirect methods she chose in order to make such matters important to the reader. Certainly George Eliot was right to be impressed with the direct presentation of working-class life in the early chapters of *Mary Barton*, but the strongest movement of the plot depended on romantic materials which did not grow out of the adversities of the opening. *North and South* on the other hand, by revealing the topical tensions largely through dialogue and allowing them to irrupt only occasionally into the immediate scene, drew the reader into an understanding of social changes through the changing experience of the central character. By showing her flawed when circumstances seemed overwhelming, though seared by conscience, able to forgive, to accept forgiveness, and to outface animosity, the author set the topical problems in a light which was more than topical, though it offered no simple solutions.

But the serious effort to handle such matters responsibly inhibited her talent for humour, except what appeared in an ironical exactitude of phrase—'I think I could support the honour of meeting the mayor of Milton.' The novella *My Lady Ludlow* (1858–9) returned to the pleasantries of *Cranford*, after three years occupied in presenting, in the *Life of Charlotte Brontë*, an isolated way of life only too little open to action from outside. Now Mrs Gaskell returned to the contemplation of a long-established enclave subjected to successful invasion, this time by a man who believed 'he was responsible for all the evil he did not strive to overcome'. The topical focus of the story, working-class education, was serious enough and had been the cause of some of the worst parts of *Mary Barton*. Carlyle had no doubt stirred her with his warnings in *Past and Present* in 1843: 'If the whole English People, during these "twenty years of respite", be not educated, with at least school master's educating, a tremendous responsibility, before God and men, will rest somewhere!' Mrs Gaskell surrounded the topic with a beguiling play of fictional banter, beginning with the lace ruffles with which the narrator's mother used to outface the rich democratic manufacturers, 'all for liberty and the French Revolution', amongst whom she lived. The innocent daughter of a gentlewoman fallen into such company was well chosen to tell of Lady Ludlow, to whom unpowdered hair was 'sans-culottism' and popular education 'was leading people to talk of their rights instead of thinking of their duties'. The story which proved my lady capable of changing her mind about education, through her gradual attachment to particular people, set a much greater distance between her and the reader than had been the case with the central character in *Cranford*, and the clash between the narrator's admiration and the effect of what she narrated was not quite marked enough to generate interest in itself. The issue of education was rather too limited to allow any more satisfying revelation. The echo of my lady's prejudices in the sharp, headstrong talk of Miss Galindo added little to them, despite the delightful free range afforded to absurdity—'It does not seem to me natural, nor according to Scripture, that iron and steel (whose brows can't sweat) should be made to do man's work.' A long inset story, well told, but not well enough to add to the characterization of Lady Ludlow, the supposed narrator, only bore out one of the few literary comments in the *Life of Charlotte Brontë*: 'it is a poor kind of interest that depends upon startling incidents rather than upon dramatic development of character'.

The two further novellas, *Lois the Witch* (1859) and *A Dark Night's Work* (1863), which occupied another long interval before the next novel, certainly depended upon startling incidents. But interest in the unfolding of character was never quite submerged. It appeared in the mental derangement of the youth who was obsessed with Lois and, when she was accused as a witch, was able to speak the anti-Calvinist truths which his compatriots in Salem in 1692 could not see. For the second piece the author's title had been 'A Night's Work', to stress that its subject was the effect of a single act of violence upon the whole life and character of the participants; Dickens, by inserting 'dark' into the title before publication in *All the Year Round*, deflected the emphasis on to the suspicion of murder. A derivative work, starting from a situation which owed something to that of Agnes Wickfield and her father in *David Copperfield*, and depending like *Silas Marner* upon the long-delayed discovery of a corpse, this second novella developed with some delicacy the sufferings of a heroine whose youth 'had gone in a single night'. There was melodrama, but it interrupted two sequences of stillness and quiet. That the second such interruption should result from the cutting of a railway was characteristic of an author whose 'startling incidents' tended to be representative of important contemporary economic changes affecting the fortunes of the individual.

Her greatest novel, *Sylvia's Lovers* (1863), dealt in the effect which social circumstances might have upon the individual's innermost choices and resolves. Here maturing talents worked on material which had some degree of continuity with *North and South* while also being indebted to Scott, to the Brontës, and to George Eliot. A historical novel, contrasting, with some irony, the habits of feeling of 'sixty years since' with those of the period of composition in 1859 and the years immediately following, it placed the action in a remote part of Yorkshire where 'the dramatic development of character' might occur without the 'selfconsciousness that more than anything else deprives characters of freshness and originality'. This emphasis is probably to be traced to her research into the background of the Brontës. It is noticeable in the second chapter of the *Life of Charlotte Brontë* that the emphasis on the 'irregularity and fierce lawlessness' of eighteenth-century life in the West Riding is much less relevant to Charlotte's work than to her sisters', an impression confirmed by the reference to 'the tales of positive violence and crime ... some of which were doubtless familiar to the authors of *Wuthering Heights* and *The Tenant of Wildfell Hall*'. The link between the *Life* and the major novel which followed it is clearest in the summarizing assertion made in the *Life* about people in remote parts of Yorkshire: 'fifty years ago, their religion did not work down into their lives'. The parallel claim in *Sylvia's Lovers* looked at the matter from the present time of writer and reader: 'our daily concerns ... are more tested by the real practical standard of our religion than they were in the days of our grandfathers'. Comment on the psychological differences which separated the reader from the characters backed up the claim, in order to stimulate understanding across that gap of time as the previous social novels had aimed to do across the dividing lines of geography and class. It is hard to believe that this culminating attempt was not encouraged by the example of *Wuthering Heights*, a novel selfconsciously constructed so as to view violent eighteenth-century action from a later standpoint.

Allusion to the 'wildness' of the people, landscape, and weather seemed distinctly in Emily Brontë's vein. And a similar kind of intensity to hers came from the limiting of the main action to two intertwined family groups to both of which the chief male participant, Hepburn, belonged; he had been brought up by his aunt and, upon her marriage, lodged with the second family, that of the widowed Alice Rose. The story of his unreturned love for Sylvia, the daughter of the first family, and the love for him of the daughter of the second, Hester—though he believed 'she and me is like brother and sister'—approached tragedy through the author's interest in social tension and change. The continuity with *North and South* was apparent: the lawlessness in *Sylvia's Lovers* arose from northern dislike of government from a distance. Thornton in the former novel had spoken of belonging 'to Teuton blood; it is little mingled in this part of England . . . we retain much of their language; we retain more of their spirit'. Separated like this, he said, 'We hate to have laws made for us at a distance.' The matter upon which the plot of *Sylvia's Lovers* turned, the use of force to recruit men into the navy—which was taken 'more submissively' by 'people living in closer neighbourhood to . . . the centre of politics and news' than by 'the wild north-eastern people'—was but a special case of Thornton's insistence.

The lawless reaction to 'all this tyranny', 'for,' said the author, 'I can use no other word', was analogous to the violence which had complicated industrial relations in her two earlier novels. All three novels, without abating a jot of the characters' moral responsibility, encouraged understanding of how they might be drawn into dubious forms of action by forces beyond their control. Already concerned with the social forces tempting her characters to deny or withhold the truth, Mrs Gaskell tried to clarify the way such forces might corrupt the individual's power to resist this temptation. The liar in this case, Hepburn, actually 'piqued himself on his truthfulness', but he had the habits of the region in resisting another set of the southern laws, those against smuggling. This gave rise to 'acted as well as spoken lies'. So his finding reasons to conceal the truth, when his rival, Kinraid, was pressed into the navy, was not merely to be shrugged off with 'All's fair in love and war'; it was part of the process by which, said the author with some special pleading, excise duties 'did more to demoralize the popular sense of rectitude and uprightness than heaps of sermons could undo'.

Sylvia too was presented as borne on the currents of conduct and feeling which arose from tensions within her society. At the opening of the French wars, a whaling port stood at the very point of tension between the country's need for oil, essential for industry, and for sailors, essential if the navy was to be quickly expanded. In preferring a skilled whaler, Kinraid, to a successful shop assistant, Hepburn, Sylvia was shown moved by the public indignation against the press-gangs and by the stories which made a hero out of Kinraid for resisting 'this tyranny' when, since Greenland whalers were exempt from empressment, local interests in fact had the law on their side. His own stories of whaling, which also played a part, were, it was carefully pointed out, like those by which her father had won his wife. War had only accentuated a tendency, natural enough, for women in this society to find whalers attractive. The report that Hester Rose herself was the offspring of a disastrous marriage with one of them, when her mother might have married a prosperous trader, even moved Hepburn 'to wonder if the lives of one generation were but a repetition of the lives of those who had gone before, with no variation but from the internal cause that some had greater capacity for suffering than others'.

To this long vista of successive lives the story opposed a subtler view of the constraints upon the individual's power to act—and ironically so when Hepburn was by his own action forwarding an opposite kind of calamity, Sylvia's marriage to himself, the 'safe' man, rather than the risky one. Public affairs appeared to co-operate with him in the 'blight . . . over the land and the people' caused by fear of the press-gangs and the spirit of vengeance against them which, taking 'a supernatural kind of possession' of Sylvia's father, led to his arrest and execution as a ringleader. Her mother's consequent mental enfeeblement made Sylvia increasingly dependent upon Hepburn despite the more congenial help movingly offered by her father's elderly hired man, to whom she was much closer in mind and habits.

The scales were in fact loaded as heavily against Sylvia as against Ruth before her liaison. But now the circumstances which brought this about were more steadily explored as a sequence of cause and effect. At the same time, the heroine's resulting states of mind formed a continuity unlike those of Ruth, which had been illuminated only intermittently. And what was left undisclosed about the relationship was unmistakably implied when the calamitous marriage was economically dramatized in the constraints of Sylvia's dark house and decorous life after the freedom of the farm and the open air, in the dreams of Kinraid haunting in different ways both herself and her husband, in her precarious serenity when walking alone with her child on the sands below her early home, and in her husband's irritation with this habit.

The return of Kinraid drove the novel back to a kind of didactic romance-writing characteristic of the author. The wife's oath never to forgive, like the resolve of the man wronged by John Barton, was made in order that the story might show it to be wicked. Even her moving reaction to the sight of the lover whom his rival, now her husband, had allowed her to believe dead was made exemplary: '. . . neither yo' nor him shall spoil my soul'. The husband's ambition to imitate Kinraid in war-service abroad led to events and a style of recounting them quite at variance with the penetrating realism for which the first two volumes of the novel had been impressive, especially in the presentation of the disharmony in Sylvia's marriage. To be sure, in so far as the events of volume iii were those of war, they could be said to stand within that system of social causes acting upon individual lives which the first two volumes (until just before Kinraid's return) had developed. But the whole point about the war had been in its remoteness from local concerns and in the local resistance to its demands; the consequent violence and bravado (finely presented in Sylvia's father) were imagined as parallel faults to the deceit encouraged by excise duties imposed from the south. To ask the reader's attention for the war at close quarters broke the consistency of the novel with turns of event which were simply more numerous than it needed for the realization of its tragic potentialities.

There were signs that the romance element was purposely introduced. Hepburn, for instance, was moved to return to live in hiding near his home by reading the old romance of Guy of Warwick's return from the Crusades—but there was none of the firm control of the marvellous within the structure of the whole book such as *Wuthering Heights* had shown, or *Villette*, both of which were very likely to have been in mind as examples. For the romantic excesses of atonement exacted from Hepburn, however, who must save not only his rival's life in battle but also his own child from drowning, the regrettable example of *Ruth* was enough.

The novel recovered with the return of mundane detail at the final exchange of forgiveness between husband and wife. Despite the forcing of the moral point, the scene approached tragedy: a genuine union between them became possible only in the moment in which the possibility was lost. There was some forcing of the emotion as well, for instance in the choice of physical setting, at dawn beside the 'ceaseless waves' whose action a repeated formula emphasized; but such a setting undeniably echoed the omnipresence of the sea in the best of the novel. The waves were now within 'the bar at the entrance to the river', a haven, that is, from the calamities of the shores to the west and the east of the river. Those calamities, coming from both within and without the two central characters, had stirred compassion for them and now stirred tragic awe at what was abnormal in their sufferings, compared with the less unusual adversities of Hester Rose and her mother.

The author had been concerned, it must be admitted, to reduce this abnormality and to stress, for instance, Sylvia's reformation through understanding the example of the other two women and through learning to read the Bible. So, like John Barton, she was made to serve the cause of education. Yet a less limited view of her situation was latent and survives for readers to whom the satisfaction of Sylvia's desire 'to penetrate the darkness of the unknown region from whence both blessing and cursing came' is a less simple matter.

Mrs Gaskell had not been quite able to keep control of the varied constituents of this full-length novel. She went on to produce her most nearly perfect novella, *Cousin Phillis* (1863–5). Success here depended on the story being allowed its own course from its own first premisses of character and situation. The events were not forced, nor was an interpretation forced upon them. Again outsiders broke in on a self-contained society, but while the story carried overtones concerning social change, they were audible only from the nature and strength of the fiction itself. Phillis, clearly and specifically present as an individual, was seen in an unusual light as the railway approached the idyllic rural society where her father was an educated pastor and an efficient farmer. An interest which, with him, she took in both literature and applied science, and hence in the inventions of the narrator's father, predisposed her to appreciate the qualities of the managing engineer Holdsworth. The outcome in Holdsworth's inability, with his cosmopolitan background and his enlarged prospects in Canada, to find his unspoken but returned love for Phillis more than an episode in a varied career, held personal and impersonal factors in a fine balance. Holdsworth, for all his gifts, seemed beneath her, just as the kind of life he represented was in obvious ways inferior to that in which Phillis had been brought up and which was described with moving economy. For Phillis's sufferings when Holdsworth failed to return, no blame was laid, except by the narrator upon himself for prematurely reassuring her. It was almost as if the author were reflecting on her own willingness, as she had looked into the future, to see a favourable outcome for the clash of the old ways with the new. Now the emphasis fell on what was natural in the whole sequence, on what, granted its primary situation, was hardly avoidable, any more than the attraction of the flame for the moth in the tiny dramatized cliché at the conclusion. The apparent simple naturalness, carried right through, set the piece apart from everything of the author's up to this point. The one touch of melodrama, Phillis's sudden desperate illness, did not seem too great a reaction to the discovery that her father could reproach her for being ready to leave home, as if their idyllic life were never to change. Elizabeth Gaskell's vision of that whole way of life had depended on the key position in it of a minister who was 'practical as well as reverend', but now his ability to accept change was called into question. At the same time, a glimpse of the stupidity of his fellow-pastors suggested how precarious was his own pre-eminence among them. Phillis's eventual resolve, on recovering, to 'go back to the peace of the old days' appeared impossible of fulfilment, and much was left open to doubt with it. But the author left that matter to the reverberations of the story.

On the other hand, the final novel, *Wives and Daughters* (1864–6), though unfinished at the author's death, was as uncomplicated in its overtones as *Cranford* itself. Again a self-contained society was stirred by the entry into it of people from outside.

Again the mode was that of comedy, though this time it was of a kind which could be reconciled with the most serious and consistent of Mrs Gaskell's preoccupations, education. In 'an everyday story' (its subtitle when first appearing in the *Cornhill*) the comedy depended less than in *Cranford* upon sharp contrasts and incongruities. The emphasis fell on the ordinary rather than on the oddities which were accepted as ordinary, on the degree to which those who broke into the charmed circle might become assimilated, and on the degree to which those within it who did not fulfil ordinary expectations, like Roger Hamley who became a scientist, might come to change them. The agent of change, as in *My Lady Ludlow*, was education, but its effects appeared, as in real life, stealthily, visible more clearly in retrospect than when occurring.

The conventional hero, Roger Hamley, was set apart by his scientific education, in contrast with the more usual 'literary' education of his elder brother. His father's objection to foreign words and ideas in an article of Roger's, answering one by a French writer, resembled the naïve comedy of the outsiders' reception in *Cranford*:

> 'I'd ha' let him alone!' said the Squire earnestly, 'We had to beat 'em and we did it at Waterloo; but I'd not demean myself by answering any of their lies, if I was you . . .'

But the squire transcended the stereotype of the Tory xenophobe in a moving central sequence concerned with his misunderstandings with the dilettante elder son, and the scientific work of the younger was more than matter for comedy. It led to travels which, in the author's plan for the novel, resembled those of Darwin. Changes in the attitude of other characters towards the man and his work showed the almost insensible changes in this stable society. At first regarded as something of an oddity, the scientist was by the end of the novel being eagerly sought as a guest by the local despot, Lady Cumnor, even though Lord Hollingford, her son, admitted that science was 'not a remunerative profession, if profession it can be called'. It was a pity that everything surrounding this nominal hero should be of greater interest than he was in person.

The original outsiders, a governess, Clare, and her second husband, Gibson, were more distinctly present as characters, though the action did not allow all their potentialities to be realized. They were skilled respectively in pseudo-education and in real research, neither of which, throughout the novel, was understood by any but a minority of those who welcomed them into Hollingford society. The research interests of Dr Gibson, a Scot who was said to have 'been in Paris', meant nothing to his patients, while only Clare's cleverest pupil could see that she was 'so useful and agreeable, . . . any one who wasn't particular about education would have been charmed to keep her as a governess'. Her incapacity for thought was dramatized with deft irony, though more perhaps for the delectation of the serial reader than that of the reader of the completed novel, for her talk failed to reveal new qualities in the speaker and it lacked the resources of absurd and imaginative detail which keep fresh the repeated *non sequiturs* of Mrs Nickleby. Even the disaster of the marriage of this pair, each widowed with one daughter, did not develop.

The most startling of the outsiders, Clare's daughter Cynthia, alone enabled the novel to move forward, so that the nominal heroine, Molly, was left relatively passive. Cynthia, shrewd, self-aware, following her own nature without any more restraint than came from her need to be admired, was sharply individualized, even though her inner existence, unlike Molly's, could only be inferred and much about her remained a mystery. Her contrast with Molly, the basic feature of the book's first plan, was a contrast of the unpredictable outsider with a girl securely brought up 'among these quiet people in calm monotony of life', thoroughly sound and thoroughly understood in her not unusual, though often subtly noted, inner life. The moral advantage might be allocated to the insider, but Cynthia was fully allowed her due. Such a character, though owing something to the Cecilia of Maria Edgworth's *Helen*, marked a decisive maturing of the power to contemplate as an artist, and to understand by directly exhibiting with the minimum of discursive generalization.

If this had been *Cranford*, Cynthia could simply have passed from Hollingford society, as she did, leaving no permanent trace upon it; but in this larger-scale treatment, there were matters connected with her which raised greater interest than all the strong and specific realization of that society's habitual texture. Her past life, glimpsed in fragments—her relation to her father ('he died when I was quite a little thing, and no one believes that I remember him'); the odd position of rivalry with her widowed mother, who would perhaps have been willing to marry the man who forced himself upon the daughter; the very strength of that man's attachment and the humiliations it caused him—suggested a more original action than the lover's preference for liveliness over solid worth repeated from *Mary Barton* and *Sylvia's Lovers*. The reader regrets it not merely out of curiosity but because, in the novel we have, the two best characters in Mrs Gaskell are not allowed action or interpretation commensurate with their potentialities—for Clare is a similar case: the action we have stops short of revealing the full danger to herself and others of her 'flimsy character', 'amiably callous . . . in most things', holding to 'her opinions so loosely, that she had no idea but that it was the same with other people'. The witty Lady Harriet, who has 'talked all the freshness off love' and values truth because 'Truth is generally amusing, if it's nothing else', suggests another missed opportunity. All these are things which the book itself supplies. It shows in its humour and ease not only how far the author had come in only seventeen years, but also what strange things her imagination could glimpse, given that freedom.

With Mrs Gaskell's relatively early death it seemed not that a career of secondary rank had been denied completion, but that a major one had been arrested as it was opening out.

8. George Eliot

'It sets a limit, we think, to the development of the old-fashioned English novel,' wrote Henry James of the work in which George Eliot's[1] powers were most fully expressed, *Middlemarch* (1871–2). This kind of novel has been the subject of the foregoing chapters. Its 'design and construction', which failed to gratify James, and its apparent lack of 'concentration', were appropriate to a form of art unusually close to life, unusually responsive to changes in society and in habits of thought and feeling outside of the art. George Eliot's work represents a culmination because she was exceptionally well informed about these matters and at the same time, from her work as a reviewer since 1852, about fiction. Her own fiction follows chronologically the main successes of the social novel in the decade before 1857. Looking back in 1859, after *Scenes of Clerical Life* and *Adam Bede* were out, George Eliot confessed to Elizabeth Gaskell that 'I was conscious, while the question of my power was still undecided for me, that my feeling towards Life and Art had some affinity with the feeling which had inspired "Cranford" and the earlier chapters of "Mary Barton". ' The grounds of the affinity may be seen in Mrs Gaskell's patent encouragement of the reader to feel sympathy and respect (even when appealing to his sense of the ridiculous) for a range of ordinary life of which his own position in society might have left him ignorant. Before this, George Eliot's article in the *Westminster Review* for July 1856,

[1] Mary Anne Evans, 1819–80, universally known by the name she adopted as an author, George Eliot, was the daughter of the agent in charge of the estates of the Newdigate family in Warwickshire, the setting for much of her fiction. Intensely pious and intensely attached to her father and her brother Isaac, Mary Anne had come to reject their orthodox religious views by 1842. Her wide-ranging education in European languages (largely through private tutors) enabled her to produce her first works, translations from the German of D. F. Strauss, *The Life of Jesus* (1846), and L. Feuerbach, *The Essence of Christianity*, in 1854 (unpublished). She wrote for, and virtually edited, the *Westminster Review*, 1851–3. From 1854 until his death in 1878, she was inseparable from George Henry Lewes and signed herself with his name, though the law of the time made it impossible for him to obtain a divorce so that they might marry. The degree of social ostracism which she suffered gradually lessened as her fame increased after the publication of *Adam Bede* (1859). Lewes was tireless in encouraging her talents, and was an acute business manager, while following his own career as a biographer, made famous by *The Life and Works of Goethe* (1855), as well as a journalist and scientific writer. Mary Anne was reconciled to her brother only when, in May 1880, seven months before her death, she married a long-standing devotee, J. W. Cross, who became her first biographer.

reviewing the work of a German social anthropologist, Riehl, had placed another writer of social novels, Charles Kingsley, and her two idols, Scott and Wordsworth, among those who have done 'more ... towards linking the higher classes with the lower' than 'hundreds of sermons'. Nevertheless, she was aware that Kingsley had 'two steeds, his Pegasus and his hobby' and even characterized his work up to 1854 as 'more imaginative than solid'. 'Solid' referred to the substantial presentation of what was true. For art was 'a mode of extending our contact with our fellow-men beyond the bounds of our personal lot'; any falsification when presenting 'the life of the People' meant 'sympathy ... perverted, and turned towards a false object instead of the true one'.

The true object included more than the 'external traits' of 'the People'. The work of Dickens 'would be the greatest contribution Art has ever made to the awakening of social sympathies' if only it rendered with 'truth' not just the 'idiom and manners' of some of the lower classes, but 'their psychological character—their conceptions of life, and their emotions'. While recognizing that 'external traits' might illuminate 'psychology' sufficiently to correct it where it was 'false', she clearly sought more direct explication of it than was furnished by Dickens's intuitive, non-realistic methods. Mrs Gaskell's more pedestrian analysis of the thoughts and feelings of such people obviously recommended her early work as furnishing the knowledge needed 'to guide our sympathies rightly, ... to check our theories, and direct us in their application'. One would give much to know how much of the tragedy of John Barton George Eliot included among the 'earlier chapters' of *Mary Barton*—some at least, one would expect, judging by the choice of the same surname for her own first unheroic hero, even though his tragedy, of irrevocably lost opportunities, was of a different kind.

The ambitions with which George Eliot came to the writing of fiction, then, were those of a social novelist whose 'solid realism' would not stop at the 'external traits' of 'the people as they are', but who would avoid 'hobby horses', what she later called 'the prescribing of special measures'. Charles Kingsley and Elizabeth Gaskell had begun to write during the revolutionary 1840s: George Eliot belonged to the apparently quieter succeeding decade and praised Riehl as a man who, after the 'failure of revolutionary attempts ... thinks it wise to pause a little from theorizing and see what is the material actually present for theory to work upon'. In particular, he understood the 'vital connection' of any present human society with its past. Here he touched an important aspect of her own sensibility as it had been formed by her reading of Wordsworth, who had memorably expressed in *The Prelude* similar 'social-political' aims—

> ... if Past and Future be the wings
> On whose support harmoniously conjoined
> Moves the great spirit of human knowledge ...

He had also made some of his most original poetry out of a similar connection of present with past in the life of the individual. Steady reading of Wordsworth, from the very beginning of her maturity, had reinforced her own introspective tendencies, at first as part of an intense evangelical piety. The painful inner history which followed that phase, the alienation from two individuals to whom she was devoted, her father and her brother, led her to set an even higher value upon feeling when its object was no longer supernatural. In her fiction she was to be satisfied only by the feeling which could be personally expressed as the experience of a narrator conscious of the relationship between present and past in himself and in what he saw.

The *Scenes of Clerical Life* (1857–8) were distinguished not by their slight stories but by the narrative voice with its gentle humour, its 'tenderness for old abuses', and its urging that 'the only true knowledge of our fellow-man is that which enables us to feel with him'. It was a *causerie* of the kind to which novel-readers had been accustomed by Thackeray, whom George Eliot thought 'on the whole the most powerful of living novelists'. It was a less playfully ironic voice than his, making a simpler emotional appeal: there could be no questioning of the feeling when, for instance, Janet's repentance was accompanied by 'delicious tears', and later, the insistent authorial tones were patently intended to bear down the reader's reluctance to sympathize with Janet in the trite dramatization of her further temptation and suffering. The voice was more successful in enlivening the presentation of the society in which her chosen clerics flourished. A full range of rural and small-town life appeared, from one-eyed Poll Lodge in the workhouse, who had produced a son 'in spite of nature's apparent safeguards', to Lady Assher with her small talk 'dribbling on like a leaky shower-bath'. The crude plots of the stories were much less impressive than the reactions of members of the society in which they took place. Despite the narrator's apology for 'being unable to invent thrilling incidents', there were all too many—a mother's early death, an attempted murder, the wife-beating of a husband 'fevered by sensuality'. But the reception of Barton's doctrinal formulas in the workhouse, the ability of parishioners to be more moved by his misfortunes than they had been by his sermons, the acceptance of the unpolemical Gilfil as belonging to the course of nature, and, in 'Janet's Repentance', the effect of Evangelicalism over an impressive social range in a small town—these weighed strongly against the tendency towards melodrama and uncontrolled sentiment. The setting in time was that of most of her best work, the thirties. They were seen in contrast either to a still earlier time, when church buildings were more picturesque than 'a well-regulated mind' would later approve, or to the 'enlightened' present time of the reader, when 'many of the younger ladies have carried their studies so far as to have forgotten a little German'. To make this kind of sport with the complacencies current at the time of writing became more and more characteristic of her treatment of the thirties until the culmination in *Middlemarch*.

As yet the author's dealings with the central characters did not match her awareness of her aim. Asked by her publisher, Blackwood, to modify their behaviour, she was 'unable to alter anything', because their actions 'grow out of my psychological conception': the example she gave concerned a heroine contemplating murder—'Those gleaming eyes, those bloodless lips, that swift, silent tread . . . Yes, there are sharp weapons in the gallery.' Clearly the doctrine, of great importance in itself, gave no guarantee of a good result. Indeed, her approach to the writing of fiction had at this time a solemnity that was ominous. 'Writing is part of my religion and I can write no word that is not prompted from within.' After the amusing criticism, written just before beginning the *Scenes*, of 'Silly Novels by Lady Novelists', it was disquieting to find that what was 'prompted from within' could be as aesthetically bad as Gilfil's love-story, and that it yet must stand in all its details.

The fourth of the *Scenes* became *Adam Bede* (1859), her first major success. Published as a book, not in instalments, its story was not revealed beforehand even to the publisher, 'on the ground that I would not have it judged apart from my *treatment*, which alone determines the moral quality of art'. Such a recognition, amounting to a criticism of much in the *Scenes*, would look like the fruit of the experience of writing them if one were sure what the author meant by 'treatment'. The novel itself suggests that 'treatment' now meant not merely the pleading with the reader, as in 'Janet's Repentance', to feel with the characters, but the more difficult process of making him do so by means of a pleasure which issued in understanding.

Country life was treated with an idealizing warmth which might seem at variance with her earlier insistence upon 'things as they have been or are', or even with her repetition in chapter 17 of much of the substance of the article on Riehl. Through the charm of the idyll, however, the reader was led to contemplate its foundations in the laws of cause and effect and the sombre result when these laws were seen active in human conduct. The golden light of a still afternoon—in 'just the sort of wood most haunted by the nymphs; you see their white, sunlit limbs gleaming athwart the boughs'—became 'a hazy, radiant veil' disguising the face of 'destiny'. The mutual attraction of a dairymaid and the future squire might be commonplace enough as a subject for fiction, but not when treated as an example of the operation of unalterable law. The laws of good husbandry which had brought the Loamshire countryside to its present beauty, and which Donnithorne, the future squire, was discussing with the parson at the time he first tried to confess his entanglement, were akin to the laws observed in his own behaviour. The same pitying, admonitory voice that dwelt on the charm of the idyll demonstrated also 'a terrible coercion in our deeds which may first turn the honest man into a deceiver, and then reconcile him to the change'. This coercion was all the stronger for the egoist's illusion of freedom while his actions were binding him down. It was to engage George Eliot's attention with increasing power until its climax in the Gwendolen Harleth of her last novel. The inward view in the case of Arthur Donnithorne was a matter of a new penetration into sequences of thought and feeling of which she might disapprove, but which were none the less contemplated with a compassionate lucidity.

In the case of the dairymaid, Hetty Sorrel, the pain of the self-made destiny which caught her was realized in the memorable form of the crying of her deserted child, from the sound of which she could not escape. As moralist, the author persistently looked down upon her, but the artist made her the centre of a Wordsworthian tragedy which was all the more moving for the girl's inability to understand it. 'Nature' had been spoken of as a 'great tragic dramatist' in relation to 'family likeness' within the Bede family. The same 'Nature' spoke a deceptive language in Hetty's 'sparkling self-engrossed loveliness': her looks of peach or apple and her elaborately described garden setting, when Adam approached her in chapter 14, falsely hinted at willing fruitfulness; her softness and childishness suggested, quite wrongly, affection for children. The point was in the complete change at the climax of her trouble, when she found herself unable to desert her child; its crying, still heard as she confessed, dramatized the inner pressure of facts which she had never expected. Her immobility when finding that, though 'I could hear it crying at every step', the child had been taken—

> My heart went like a stone: I couldn't wish or try for anything: it seemed like as if I should stay there for ever, and nothing 'ud ever change—

suggested an infinity of suffering, as Wordsworth had understood it—

> Suffering is permanent, obscure and dark,
> And shares the nature of infinity.

Of this part of the novel Dickens wrote, 'I know nothing so skilful, determined, and uncompromising.'

The treatment of the title-character depended on his role in relation to the tragedy. Adam Bede, who misinterpreted the language of Nature in the face of a girl, drew his force as a character from his association with the 'business' which maintained the stability and prosperity of country life. 'Business' required a man to be not ignorant of 'math'matics and the natur' o' things', and to work, to 'have a will and a resolution, and love something else better than his own ease'. He was the 'heroic artisan' all right, of the kind George Eliot herself had dismissed in the essay on Riehl as a 'misrepresentation' resulting from pursuit of 'what the artist considers a moral end'. But the heroic element in him arose from the foundation of the reader's pleasure, the novel's insistence upon laws which were in 'the natur' o' things' and were not to be evaded. In this way the artist's 'moral' might itself be justified as part of the 'truth' presented. And once this was recognized the character gained rather than lost in interest.

The apparently more spontaneous creations, the dialect speakers, were somewhat selfconsciously ranged as commenting chorus. But they too gained force from the solidity of the country setting. Their similitudes were continuous with it, a descant of delight in its good things from 'taters and gravy to plentiful linen, diversely modulated according to their respective foibles, from the self-confidence of Mrs Poyser to the querulous anxieties of Lisbeth Bede. It was a pity that they should have so little to do with the action once it began to move, but were haled in and out at the demands of the author, not the story.

The presence of a Methodist preacher, Dinah Morris, made an alien element in all this pastoral largess. Living 'like the birds o' the air . . . nobody knows how', coming from a far country where people 'live on the naked hills like poultry a-scratching on a gravel bank', she was securely established by such contrasts, so that her pallor and her superhuman patience seemed right. Her function would have resembled that of the intruders into the cohesive social microcosms of Mrs Gaskell if her advent had effected anything. But her urgent pleading on behalf of sinners, moving enough in itself, came to nothing: her converts had the merest walk-on parts. Her only essential work in the story was to secure Hetty's confession—an important matter to the author, from the ministrations of Mr Tryan in 'Janet's Repentance' right on until something like half of her immense last novel had to be devoted to the development of a confessor adequate for her heroine.

George Eliot's confessor-figures—whether secular like Deronda or professionals like Dinah Morris, Tryan, Savonarola, or Dr Ken— each bring into the fiction a historically modified version of her own ethical urgency. *Middlemarch* has a curious modernity because the main confessions are received by secular characters, while one of the features which gives the painful plot of *Felix Holt* its affinities with Greek tragedy is the impossibility of confession. In *Adam Bede*, set in the days of Old Leisure, who 'had an easy, jolly conscience', and 'never . . . read *Tracts for the Times* or *Sartor Resartus*', the presence of Dinah Morris marked the difference between the settled, prosperous order of things to which the established church ministered, and a world of lower-class discontents which it failed to touch. It was unnecessary to absorb Dinah into the settled order by a marriage to Adam Bede, since, by the end, both the women who had disturbed that order would do so no longer, Hetty being dead and Dinah being ready to accept the prohibition of female preaching. The marriage fostered a reassurance of rural stability at the expense of an awareness of its underlying forces such as had been stirred by the Wordsworthian tragedy.

From the beginning, work on the next novel, *The Mill on the Floss* (1860), showed a determination to make tragedy the main interest. It was selfconsciously foreshadowed by references in the first half of the book to floods and drowning, and to Greek drama; it was later summed up, in a letter, in Aristotle's terms in the *Poetics*, as the 'presentation of a character essentially noble but liable to great error'. Above all, the attempt was to make the sufferings of Maggie Tulliver not only particular but representative. A large set piece at the beginning of Book IV of *The Mill on the Floss* invited the reader to contemplate the ruins of ordinary houses left by the floods of the Rhone which 'oppress me with the feeling that human life—very much of it—is a narrow, ugly, grovelling existence, which even calamity does not elevate . . .'. This was the life of her Tullivers and Dodsons: the stubborn tenacity of its self-interest was fully exposed in comedy which went far beyond external humours into their very habits of thought. Nevertheless, its tragic potentiality was suggested when a phrase from Sophocles was applied, at the end of Book I, to Mr Tulliver's deeds 'inflicted on him rather than committed by him'. The full consequences were borne by his children. In the scheme of the opening of Book IV, theirs was the suffering, also 'represented . . . by hundreds of obscure hearths', of 'young natures in many generations, that in the onward tendency of human things have risen above the mental level of the generation before them, to which they have been nevertheless tied by the strongest fibres of their hearts'. (The author's own suffering had been of their kind, estranged from the father and brother who could not approve of her own independence of thought and action.) It was characteristic of her refashioning of the social novel that she should so present the social context as to foster understanding of processes which were at once more general than any particular society and more deeply impressed upon the inner life of the individual. And if to represent such suffering as that 'which belongs to every historical advance of mankind' might seem to compare small things with great, she claimed the example of 'natural science'. There, 'there is nothing petty to the mind that has a large vision of relations, and to which every single object suggests a vast sum of conditions. It is surely the same with the observation of human life.' For the representative nature of such suffering it was appropriate that she should claim the support of natural science. Darwin's *The Origin of Species* (1859) had been for her an 'epoch'. It had moreover contained approving quotation of remarks and observations of G. H. Lewes, whose *The Physiology of Common Life* was also just out. George Eliot's assimilation of the latter was plain in details such as the psychological vocabulary of 'forces' finding 'pathway' and 'channel'.

Allusion to Greek tragedy brought further support to the representative quality of the sufferings of the young. Maggie Tulliver's impulsive cutting of her own hair was likened to the madness of Ajax in Sophocles, slaying the sheep which he supposed to be the sons of Atreus; her warring passions when jealous of her cousin would, it was claimed, have made a tragedy but for the lack of the magnitude Aristotle required. The disabled Philip Wakem, who first engaged Maggie's interest, told the story of Philoctetes and exemplified his insight; and Tom Tulliver, whose lack of imagination was the source of his lack of sympathy with his sister and with Philip, became, like Philoctetes, wounded in the foot: this was in the course of the only action prompted by imagination in his whole career, his play-acting with a sword; his tragedy was that he recovered without the slightest effect.

The part of the action in which Maggie became tied by the strongest fibres of her heart to her native place, to the generation before her, and to the brother who concentrated the practical virtues of his mother's people in himself, was developed, like the rural idyll in *Adam Bede*, with delectable leisure. The marvellous 'sunshine undimmed by remembered cares' took a whole volume; the forgiveness exchanged by brother and sister and the reasons for it, up to the time of their father's death, occupied all the second. So only one volume was left for the preparation of the catastrophe. This, as the author recognized, was not enough. For the insistent suggestions of tragedy were to be fulfilled in the experience of a heroine still young. The impulsiveness she had inherited from her father, which had been likened to that of

Oedipus, led her to love rather than (in his case) to hate, and from this arose her 'great error'. The difficulty was that she was still young enough for errors with such a cause to be remediable, inseparable as they were from the process of growing up.

The process was movingly presented in her early splendid physical maturity and her attraction to Philip when, as he later recognized, 'she hardly knew what she felt'. Ominously, in their exchange of declarations of love near the end of the second volume, the touch of 'real happiness' came with the notion of sacrifice, 'a moment of belief that, if there were sacrifices in this love, it was all the richer and more satisfying'; when her brother forbade the relationship, she was 'conscious of a certain dim background of relief in the forced separation', and when her cousin spoke of helping them to marry, 'Maggie tried to smile, but shivered, as if she felt a sudden chill'. Her earlier premature attempts at religious self-sacrifice could be seen continuing in this relationship with Philip. They continued into that with Stephen, who stirred her more completely: again, insidiously, renunciation intensified 'rapture'. And the strength of their attachment, which to the author as moralist increased the virtue of their renunciation, only increased the reader's doubts about the concealing of it. The rightness of doing so seemed to be called into question when a surprise meeting with Stephen caused Maggie 'a beating at head and heart' like 'the sudden leaping to life of a savage enemy', and a similar energy on his side was imaged in 'the flanks and neck' of his horse 'streaked black with fast riding'. Their belief that this degree of disturbance could and should be concealed was what started the calamitous chain of events which led the very two people whom Maggie and Stephen were protecting from pain to send them off in a boat alone together. Not only did the artist emphasize the connection of the resulting 'temptation' with the concealment, but she also underlined the fact that, by the time Maggie refused Stephen's final appeal to marry him, both of the other two people involved had forgiven her. Moreover, although Maggie believed Stephen would return to his former love, this appeared entirely unlikely; Philip had in any case decisively accepted that she would not return to him. The 'great error' appeared to be the attempt to behave, in both the concealment and the refusal, as if complete renunciation were the best thing.

In the face of these facts Maggie was presented as moved by the power of 'the long past . . . the fountains of self-renouncing pity and affection, of faithfulness and resolve'. But even the author's clergyman-mouthpiece emphasized that 'moral judgements' need 'perpetual reference to the special circumstances that mark the individual lot', and such 'reference' here by no means reinforced the claims of the past. Maggie's own past, seen with great distinctness from her own point of view in the first volume, had in any case hardly been marked by the faithfulness and resolve which she found there on looking back; it had been more rebellious and more interesting.

The power of the past to the author, however, had been from the start overwhelming. It had caused her to dwell, in the passages which moved Proust to tears, on the 'deep immovable roots in memory' of 'the loves and sanctities of our life'. It was Wordsworth, assimilated over many years, who enabled her to assert so confidently the connection between the distant past and peremptory duty in the present. But to explore all this as part of 'the great problem of the shifting relation between passion and duty', there was insufficient space in the third volume, so that the drowning which returned Maggie to the arms of her brother seemed an afterthought rather than, as the author had planned, a tragic culmination. Even in holding to the point of honour, the determination not to make her happiness out of another's misery, Maggie seemed to be, as she had been in the past, betrayed by the immaturity of her good intentions; in so far as she might yet mature, her final predicament could hardly be regarded as one to be resolved only by death. It was plain nevertheless that the story fulfilled the sort of aim Wordsworth had set himself in *The Prelude*, Book XIII to present

> . . . miserable love, that is not pain
> To hear of, for the glory that redounds
> Therefrom to human kind, and what we are.

Whatever their outcome and however their presentation might be marred by the author's championing of Maggie, there was a subtlety in the presentation of her predicament from within, which enlarged the reader's awareness of human possibilities. Wordsworth had triumphed over Aristotle.

The strenuous pursuit of high seriousness continued in the preparations for a challenge to her other idol, Scott, in *Romola*. Then, early in 1861, *Silas Marner* 'came *across* my other plans by a sudden inspiration' from 'the merest millet-seed of thought', a childhood glimpse of 'a linen-weaver with a bag on his back'. 'I should not have believed that any one would have been interested in it but myself (since William Wordsworth is dead) . . .' The epigraph directed the reader to Wordsworth's Michael, whose child brought him 'hope . . . and forward-looking thoughts'. The tale developed two symmetrical stories, one of the solitary Marner who gained this blessing, the other of the more sociable egoist, Godfrey Cass, who refused it. Each endured the consequences of self-centredness for some fifteen years before being awakened, Marner by the miracle which endowed him with Cass's child, Cass by his increasing discontent at the failure of his more respectable second marriage to yield him another. The child brought Marner back into a community; the child's own choice, to stay with him rather than go to her natural father once he had owned her, left Cass to rebuild his marriage, 'resigned . . . to the lot that's been given us'. The first of the two stories was 'legendary', encouraging hope in both the character and the reader; the second, based on Cass's being forced 'to accept the consequence of his deeds', was realistic but brought its own kind of muted hope, the kind which George Eliot's best work was to yield. The humour of characterizing rustic talk, in the liveliest vein of the two preceding novels, made the realistic milieu an integral part of the 'legendary tale' by impressing, equally strongly with Marner's own experience, the community to which he became joined.

A certain lightness of heart in the presentation of the spokesmen of this community, the parish clerk Macey, and Dolly Winthrop who took the place of an interested aunt to the child, relieved moral admonition of half its strenuousness: the pathos of Marner's self-enclosed life could be unselfconsciously touched with hope once Macey spoke of him to his face as 'allays a staring, white-faced creatur, partly like a bald-faced calf'. Cass was seen as unfortunate in having no one like this to try, however bunglingly, to set him right except the author, whose readiness to do so appeared with a dry good humour, veiling the threat of retribution. 'When we are treated well, we naturally begin to think that we are not altogether unmeritorious, and that it is only just we should treat ourselves well, and not mar our own good fortune.' The good fortune was solidly present to the reader in the traditional virtues of the Lammeter sisters, one of whom he was now enabled to marry, and in the freedom he gained from his father's threats, which had been memorably dramatized. The author's comment about his merit conveyed sympathy by seeming to share a joke about it with the reader. It was a kind of tact and economy which proved less easy to achieve when her plans, '*across*' which *Silas Marner* came, were more ambitious. The contrast between smaller aims perfectly realized in *Silas Marner* and the large ambitions inadequately fulfilled in *Romola* has been plain ever since their publication.

In her elaborate preparations for the writing of *Romola* (1862–3) George Eliot was seeking an action equal to the range of her mature mind. Looking back later, she recognized that 'it is my way, (rather too much perhaps) to urge the human sanctities through tragedy': in this case it had demanded a 'necessary idealization' which 'could only be attained by adopting the clothing of the past'. She recognized also that, in relation to her heroine, the idealization had been imperfectly controlled: 'the various *strands* of thought I had to work out forced me into a more ideal treatment of Romola than I had foreseen . . .'. In both her interior changes and her outward actions the heroine was curiously abstract, a sort of Maggie Tulliver without either her subtlety or her sensuality. Like Maggie separated from her brother and losing the infirm (in this case blind) father to whom she had been devoted, Romola rose, with her brother, 'above the mental level of the generation before them' in being stirred by the religion their father despised in his devotion to the new learning. The claims of this religion were dramatized in the brother's deathbed visions, later picturesquely fulfilled, and in the insight into her predicament of the Dominican 'dictator of Florence', Savonarola. When religion appeared to fail her, her attempted suicide, 'Drifting Away' (the chapter heading) from the seashore, curiously combined Maggie's enchanted voyage away from other people's claims upon her and her death by water; Romola's drifting away, and her salvation, 'belonged to my earliest vision of the story and were by deliberate forecast adopted as romantic and symbolical elements'.

Much of the plot was in the vein of the current form of romance, the sensation novel. A villain-hero, at first of mysterious origin, and later suspected of bigamy, a spy and counter-spy, under constant threat of murder by an avenging foster-father he had publicly disowned, seemed curiously calculated for the reigning popular taste. When George Eliot 'ran through' Bulwer's *Rienzi*, 'to examine his treatment of an historical subject', she would see an acknowledged forerunner of the sensation novelists at work, and on Italian city politics not unlike those of *Romola*—an aristocratic party against a party of lower clergy, merchants, and artisans, and all the excitement of spying, attempted murder, and papal interference. Scott's *The Pirate* was also reread, no doubt for its prophesies and vigorous adventures, but Bulwer's effect was dominant, appearing in the laborious historical accuracy no less than in the sensational events. George Eliot has been given insufficient credit for the way in which, especially in the third book, starting from Romola's return to her husband, her plot offered suspense and surprise. There was no attempt as in *Villette*, which George Eliot admired, to develop the contrast of romance with reality; rather they were identified with each other by the devices of fulfilled visions and symbolic artefacts—rings, paintings, and a box to conceal the crucifix of Romola's brother beneath a picture of Bacchus and Ariadne. Sterner overtones sounded when the painter of the picture saw Romola not as Ariadne but Antigone, daughter of the blinded Oedipus and devoted to the memory of her dead brother; but these were not developed in the slick story.

There was something more than slickness in the account of the villain-hero's decline, with its insight into the 'reiterated choice of good or evil that gradually determines character'. Explanation predominated over event, however, and the sense of the character's distinctive life which might have been achieved by starting from his position in society—as with Arthur Donnithorne—was unavoidably lacking: Tito Melema was an egotist succeeding with difficulty in a world to which he belonged only by the sharpness of his wits in mastering its intricacies. He was bound to it by neither of his two marriages, even though the mock-marriage produced children. Nor did either marriage touch his senses in the fashion so fully apparent in the case of Maggie Tulliver. He was a wonderfully intricate piece of machinery, though by the author's own standards inadequate. Her pursuit of 'as full a vision of the medium in which the character moves as of the character itself' could hardly succeed when the society was as foreign as fifteenth-century Florence. The best of the reviews, R. H. Hutton's, saw in the story 'the conflict between liberal culture and the more passionate form of the Christian faith', with Savonarola as an 'Italian Luther', and it was at this level of generality that the historical element appealed. Nevertheless, the fullness of 'vision' at which the author aimed was a valuable, perhaps a necessary, preparation for presenting a social 'medium' equally extensive, but more intimately known, in *Middlemarch*.

Felix Holt the Radical (1866), which took a further step in the same direction, marked the beginning of the major phase, prompted by an awareness of the radical changes which had been taking place in the government of the country during the long political career of Lord Palmerston. His prominence, beginning when he became foreign secretary in 1830, had constituted the period of transition between aristocracy and democracy. As prime minister from 1855, with one brief break, he had been supported by conservatives for fear of the radicals and by the radicals to keep out the conservatives. With his death in October 1865, reform of the franchise, which had for long been discussed and which he had always resisted, became a possibility. George Eliot, now a liberal conservative with a strong attachment to the past, had shown the direction of her own interests by the critical light she threw on the sudden political changes which formed the historical part of the action in *Romola*—Savonarola 'withstanding the vicious tyrannies which stifle the life of Italy' after the seventy years' subjection of Florence to the Medici. Of the three novels of the major phase two were set in the thirties, when the decline in the political power of the aristocracy began, and one in the later sixties, at the end of the period of transition to democracy.

In *Felix Holt the Radical* the social 'medium' of the characters became that of the county she knew best, 'that central plain watered at one extremity by the Avon, at the other by the Trent', imagined in the period of her own childhood. This place and period and a subject like that of *Silas Marner* solicited her in the midst of more strenuous pursuits, a heroic tragedy in verse no less, *The Spanish Gypsy*. This time, however, the preparations for the more ambitious undertaking continued to influence the more congenial. The reading in Greek tragedy undertaken for the poem affected the form of the novel, which displayed within a limited time-scheme the results of long-antecedent actions off-stage. The popular success of the sensation novel, at the other end of the spectrum, reinforced this example. So the subject of *Silas Marner*, the gaining and losing of a child, was turned into a story of secrets gradually revealed. Rufus Lyon, a dissenting clergyman, was placed in Marner's position of foster-father to a girl, Esther, who, when given the opportunity to leave him for a more comfortable life, rejected it. Mrs Transome, whose family the girl was legally entitled to disinherit, was presented, in the strongest part of the novel, suffering the inward effects of her adultery, years before, and of the long absence of the illegitimate son from whom she hoped for a return of love. The symmetrical contrast of her case with Lyon's appeared when he secured the child's affection by disclosing the past, while Mrs Transome's specific refusal to do so led to tragedy.

The secrets connected with Esther's ancestry and the imprisonment of her father were revealed with a plodding exactitude in conformity with conventions of the sensation novel. George Eliot took careful advice about the facts and reasons which established Esther's claim and even inserted into chapter 35, without change, the summary of legal opinion that Frederic Harrison, then a young lawyer, had given her. On the other hand the secrets of Harold Transome's ancestry and their discovery were dignified by being treated as a strictly Aristotelian tragedy for him, and for his mother as tragedy in a freer mode. Her neglected house, her husband who 'was born old, I think', her own 'anxious and eager face', were strongly impressed in the first chapter, supplemented in the ninth by her bitterness at having to suppress all signs of 'the degradation she inwardly felt'. In the manner of Greek tragedy she was fixed in this posture of torment, her inner life seen vividly in its essentials rather than studied in its changes, although the possibility of happiness seemed repeatedly to be opening before her. Within these limits she was capable of actions, positive in accepting Esther's claim and welcoming her to the house, negative in refusing to allow her own son to be told that Jermyn, the family lawyer, was his father. And these were of central importance, the positive actions augmenting Esther's temptation, for they made everything seem so easy, the negative forming the ground of the development of the tragedy according to the principles of Aristotle's *Poetics*, which George Eliot had been rereading. For it was

Harold Transome's ignorance of his true ancestry which led him to the 'great error' of digging up the past, in order to restore some of the family fortune and his own position, by prosecuting Jermyn. The result, a peripeteia, a reversal in which his own action recoiled upon him by the public discovery of his paternity, formed a change from ignorance to knowledge of just the kind Aristotle had discussed in his eleventh chapter.

The classical neatness of this part of the action, however, failed to engage the author's sympathies. To the student of Shakespeare and Wordsworth, Mrs Transome was the imaginative centre. It was her suffering which the Introduction powerfully foreshadowed in its references to the pain which found a voice only in the dolorous, enchanted forest of Virgil, Dante, and Spenser, as well as to the classical 'pity and terror' evoked by 'some tragic mark of kinship in the one brief life to the far-stretching life that went before'. It was in the chapters which stressed her sufferings that the epigraphs most movingly furnished tragic parallels. The epigraphs from Shakespeare looked forward: in chapter 9 Richard II's queen foreseeing his downfall and Constance in *King John*, the mother of Arthur, the heir who was never to succeed, expressing her fears; each of them, like Mrs Transome as she faced Jermyn, was speaking to a man distrusted. The classical epigraphs referred in the other direction, to the characters' past: in chapter 42 (where Jermyn suggested the mother should herself undeceive the son) the parallel was with Electra throwing back at Clytemnestra the shame of 'the ugly deed that made these ugly words' and Tecmessa urging upon Ajax the claims of their son and their shared memories which only an ignoble man would forfeit—already echoed, in chapter 35, in Mrs Transome's 'Men have no memories in their hearts.' This vision of 'the passions of the past . . . living in her dread' was set against the wish-fulfilling romance of those who freed themselves to look to the future, either by breaking with the past—like Holt in his decision to live not by selling his father's nostrums, but by educating the people who would eventually gain political power—or by choosing to live in touch with the past, as Esther did in returning to the class to which her foster-father, and Holt, belonged.

Esther's choice of a past of her own (the same choice as Eppie had made in *Silas Marner*, except that this time the foster-father had a smaller part in it) meant more to the author than was transmitted through the character. A fuller sense of the past came in the opening conspectus of Loamshire, the county of *Adam Bede*, filled, like the opening of *The Mill on the Floss*, with the pleasure of departed things. Here they were surveyed more impersonally, to show the phases of English life, in both deep-rutted lanes and rattling paved streets, 'the busy scenes of the shuttle and the wheel, of the roaring furnace, of the shaft and the pulley', making 'crowded nests in the midst of the large-spaced, slow-moving life of homesteads and far-away cottages and oak-sheltered parks'. Although the first half of this Introduction was suffused with a sublimated nostalgia, the time whose passing it registered was the time of social history, its scale established by reference to science, technology, and religion: the 'innovating farmer who talked of Sir Humphry Davy . . . driven out by popular dislike', the coachman embittered by 'the recent initiation of Railways', 'the breath of the manufacturing town' fostering Dissent in 'multitudinous men and women' who felt they were better than their rulers and 'if better, might alter many things which now made the world perhaps more painful than it need be and certainly more sinful'. The coachman on an imagined journey led these promising lines of interest to the Reform Bill of 1832 and the disputed Transome Estate, with its suggestions of an upper class losing its grip. The historical perspective lengthened when the coachman was likened first to Virgil conducting Dante through an underworld where medieval Italian politics met their judgement, and then to the Wanderer in *The Excursion* who had guided Wordsworth through the evidence of the delusive hopes raised by the French Revolution to a vision of Freedom on the throne of England

> Whose steps are equity, whose seat is law.

The seventeenth-century English revolution was to come into view with Rufus Lyon's politics of Independent dissent and their overtones from Milton, Richard Baxter, and the Puritans of Scott's Civil War Novels, *Old Mortality* and *Woodstock*.

The continuation of this Introduction in chapter 3 moved more specifically towards the political subject by developing the difference between 'the old-fashioned, grazing, brewing, wool-packing, cheese-loading life of Treby Magna' and the 'more complex' life brought to it by the canal and the mining of coal. Even the private matters of the story, it was claimed, had been directed by these 'social changes': 'there is no private life which has not been determined by a wider public life'. It was perhaps a weakness that this principle could be applied by the reader only to the precipitating events in the present time of the story, not to those long antecedent which in fact 'determined' its main personal relationships, the Durfeys' gaining of the Transome name and inheritance by purchase, and Mrs Transome's attraction to a 'soft-eyed slim young Jermyn (with a touch of sentiment)' who thereafter abused his power. Although, in the long perspective of the Introduction, public life might well have affected these events, the novel did not make plain how this had occurred (as it did in the case of the separation of Esther's parents and her father's exchange of names with another prisoner of war). In so far as the antecedent events concerned the abuse of the law they could be loosely traced to 'comparatively public matters' which, within the longer perspective insistently suggested in both Introduction and narrative, had been of great importance to the development of English liberty. Certainly, behind these events of the plot, there rose, even in the mind of Mrs Transome, 'the burning mount and the tables of the law', while in Rufus Lyon's vision, with its overtones of both Milton and Jeremiah, 'true liberty' led to 'the millenial reign, when . . . one law shall be written on all hearts'. But it is doubtful whether such matters unify the novel for the reader as the practices of a salient section of the law do in *Bleak House*. Dickens too used a lengthened time-scale, giving emphasis to connections between the seventeenth and nineteenth centuries, and centred his plot upon a mystery about the parentage of a girl called Esther. George Eliot's rigorous methods of direct realism, however, meant that the reader required details before he was prepared to accept that a similar determinism had been active in the antecedent events, as in those in the foreground. The suggestions of a declining upper class and of the effect upon Mrs Transome of the fashions set by Byron and the Regency were not quite enough. The model of Greek tragedy, which required the plot to show the results, within a limited sequence of present time, of facts and decisions from the past, was not reconciled—though no doubt it could have been, given more space—with the novel of public life 'determining' private, which George Eliot was trying at the same time to write.

The political context within which the tragedy was eventually worked out was full of topical suggestions at the time of writing. These seem to have entered George Eliot's preparations after Robert Lowe began, in May 1865, his famous series of speeches in the House of Commons against extending the franchise. John Stuart Mill, whose attitude to parliamentary reform was more temperate, stood for election in July. It is in Mill's *Political Economy*, from which she made extracts in her notes, that the germ of the contrast between the radicalism of Felix Holt and that of Harold Transome may be found. Like Holt, Mill conceded that 'the too early attainment of political franchises by the least educated class might retard, instead of promoting, their improvement'. Like Holt and like Wordsworth's Wanderer, he relied for this 'improvement' in their lot upon education. The view which Mill regarded as obsolete, that their 'lot . . . should be regulated *for* them, not *by* them', lay behind the Radicalism of Harold Transome. This view, Mill believed, idealized the willingness of the powerful to guide and protect the powerless, whereas 'all privileged and powerful classes, as such, have used their power in the interest of their own selfishness'. Esther saw this, with 'the keen bright eye of a woman', when considering Transome as a suitor: 'the utmost enjoyment of his own advantages was the solvent that blended pride in his family and position, with the adhesion to changes that were to obliterate tradition . . .'. George Eliot was too ready to go to the other extreme and idealize Holt in 'his renunciation of selfish claims, his habitual preoccupation with large thoughts . . .'. The result risked falling under the author's own strictures, in the famous letter of 15 August 1866, by ceasing to be 'aesthetic teaching' and lapsing 'from the picture to the diagram'. The riot accompanying an election, however, cast an ironic light both upon Harold Transome's political innocence and upon Holt's ambition to guide the masses, which, like Alton Locke's at just the same point, two-thirds of the way through the novel, landed him in gaol.

Felix Holt would be much more valued if someone else had written it and it did not stand under the shadow of *Middlemarch*. Certainly its social vision, although sketchier, resembled that of the greater novel. Its approach to character, however, was quite different, dealing with most of its people in their consistency, as a matter not of 'process' and 'unfolding' but of typical behaviour repeated once their main features had been established. The one character presented undergoing inner change was Esther and upon her changes the whole plot turned—her choice of Felix Holt rather than Harold Transome and of her foster-father rather than Mrs Transome. She was presented as the kind of woman, 'verging neither towards the saint nor the angel', whose perfection 'must be in marriage', but there was far more of Romola in her than of Maggie Tulliver, and her debate with herself about her love for Holt was of more interest as coming from the author than from the character: 'A supreme love, a motive that gives a sublime rhythm to a woman's life, and exalts habit into partnership with the soul's highest needs ...'. Even her rejection of Transome Court was framed largely in high-minded abstractions: 'dulness ... absence of high demand ... the higher ambition ... nullified'. Adjectives might enliven them—'a silken bondage that arrested all motive and was nothing better than a well-cushioned despair'—but they did not concentrate the reader's own experience with anything like the force of the similar expressions applied to Mrs Transome's unchanging inner life 'in the midst of desecrated sanctities, and of honours that looked tarnished in the light of monotonous and weary suns'; these formed the summary of matters the reader had been able to share, since he had seen for himself the quality of the Jermyn she had formerly loved, the son he had given her, and the husband she had rejected.

Without this experiment in parallel private lives 'determined by a wider public life', George Eliot would hardly have attempted the multiple plot of *Middlemarch* (1871–2), one story lighting up another and all of them subject to decisive influence from their common social medium. Moreover, with the completion and moderate success of her verse tragedy *The Spanish Gypsy* (1868), she seemed to have worked out her interest in subjects involving 'a great dramatic motive of the same class as those used by the Greek dramatists' and could return to the problems and situations of what was to be her major phase. Her preoccupations were clarified in an 'Address to Working Men by Felix Holt' which opened *Blackwood's Magazine* for January 1868. Circumstances had by now made *Felix Holt* topical; there had been disorder in July 1866 when, after Russell's government had fallen, through being unable to get a Reform Bill accepted, a large public gathering was refused admission to Hyde Park; Disraeli had succeeded in having a

Reform Bill passed which his chief called, in a phrase from Thomas Hobbes, 'a great leap in the dark'. In these circumstances, the Address articulated the ideas which *Middlemarch* would develop.

Holt was imagined looking back on his experience of an election 'after the Reform Bill of 1832' and urging 'the care, the precaution, with which we should go about making things better, so that . . . no fatal shock may be given to this society of ours, this living body in which our lives are bound up'. Among the 'old institutions' of this 'slow-growing' society, the address gave most attention to 'the various distinctions and inherited advantages of classes,' which must 'be used as they are until new and better have been prepared'. Above all, there was 'the common estate of society', the 'treasure of knowledge, science, poetry, refinement of thought, feeling, and manners, great memories, and the interpretation of great records'. To hamper 'the classes who hold the treasures of knowledge' would be to 'do something as short-sighted as the acts of France' against the Huguenots or of Spain against the Jews:

> . . . let us watch carefully, lest we do anything to lessen this treasure which is held in the minds of men, while we exert ourselves first of all, and to the very utmost, that we and our children may share in all its benefits. Yes; exert ourselves to the utmost, to break the yoke of ignorance.

The yoke must be broken for the 'future masters of the country . . . to get the chief power into the hands of the wisest, which means to get our life regulated according to the truest principles mankind is in possession of'. Men could be made to see and act on such principles only by 'the slow stupendous teaching of the world's events', an 'outside wisdom which lies in the supreme unalterable nature of things'.

> Selfishness, stupidity, sloth, persist in trying to adapt the world to their desires, till a time comes when the world manifests itself as too decidedly inconvenient to them. Wisdom stands outside of man and urges itself upon him, like the marks of the changing seasons . . .

The *Middlemarch* that we know took shape, it may be suggested, when these considerations enabled her to see the connection between two works she had in hand, one about 'Miss Brooke', and the other about the family of a provincial mayor, Vincy: both were about members of 'the endowed classes' who either held or believed themselves to hold or wished to hold 'the treasure of knowledge, science, poetry, refinement of thought, feeling, and manners'. In varying degrees they were under illusion about their position in this respect or about the responsibilities it entailed. At worst, through 'selfishness, stupidity, sloth', they persisted 'in trying to adapt the world to their desires'.

Out of the originally separate Middlemarch story developed the selfishness of Vincy's daughter Rosamond, 'the flower of Mrs Lemon's school', and the sloth of his son Fred, university-educated to little purpose but saved by his attachment to the Garth family, the morally stable centre of this part of the novel. Garth himself, with his 'reverential soul' and 'strong practical intelligence', like Adam Bede's, brought into it the recognition of 'this outside wisdom which lies in the supreme unalterable nature of things'; his daughter Mary, 'having early had reason to believe that things were not likely to be arranged for her peculiar satisfaction', stood out for her ability to view as comic the illusions of other people in this regard. But the Garths and Fred Vincy became displaced from the centre of the novel once the author's interest moved to the more obviously 'endowed classes', who, at the beginning of the transition to democracy, stood equally in need of 'this outside wisdom'.

The genuine 'treasure of ... science' was in the hands of Lydgate, a medical reformer newly come to Middlemarch. In another inversion of Mrs Gaskell's characteristic subject, the town 'counted on swallowing Lydgate and assimilating him very comfortably'. Its defeat of his ambitions, through the allies he had to choose, and through his marriage to Miss Vincy, appears to have been the development of his story opened out to the author by the comparison with 'Miss Brooke'. Dorothea Brooke, marrying a man more than twice her age out of sheer eagerness to learn from him, suffered a similar reversal. Not only was she given reason to be critical of her husband's 'method' in 'the interpretation of great records', but she was shut out from informed participation herself by his self-centred suspicion, 'the certainty that she judged him' with 'a power of comparison by which himself and his doings were seen too luminously as a part of things in general'. Lydgate, strongly aware of the system of 'things in general' in so far as they related to his scientific interests, was yet in matters of 'refinement of thought, feeling, and manners' where women were concerned, prone to an 'unreflecting egoism', like the 'stupidity in a man of genius if you take him unawares on the wrong subject'. Dorothea, although she 'had early begun to emerge' from the 'moral stupidity' of unreflecting egoism, stood nevertheless in need of a knowledge that was subtler both as to its object—that her husband had his 'equivalent centre of self'—and as to its form, 'that distinctness which is no longer reflection but feeling—an idea wrought back to the directness of sense, like the solidity of objects'. The distinction here was equally applicable to the author's own transcendent skill at its best, and suggested the importance of her own art in 'the common estate of society', contemplation of which was generating this supreme novel. Certainly ideas bearing on 'the treasure of knowledge' were being wrought back to the directness of sense—most wonderfully in the case of the characters whose understanding of that 'common estate' was most limited, like Casaubon carrying his 'dim taper', 'lost amid small closets and winding stairs', and constraining Dorothea to give up 'the large vistas and wide fresh air which she had dreamed of', and Rosamund Vincy holding to her own ideas of 'refinement of thought, feeling, and manners' but reacting with the quick, unexamined responses of a plant. The triumphant result fulfilled George Eliot's ideal expressed in a letter of 1866 'to get breathing, individual forms, and group them in the needful relations, so that the presentation will lay hold on the emotions as human experience'.

The 'needful relations' required work on a larger scale than anything of hers before this. She had hoped that 'nothing . . . will be seen to be irrelevant to my design, which is to show the gradual action of ordinary causes rather than exceptional'—not, as in the previous novels, adultery, murder, or bigamy, whether suspected or actual, or the strange circumstances in which a child might be forfeited or gained, but the commoner experiences concerned with marriage and with money such as might afford 'the deeper insight . . . into the causes of human trouble', into 'the wonderful slow-growing system of things', and into 'the dependence of men on each other' foreshadowed in these words from the 'Address to Working Men'. Setting the action back several decades into the time before and during the passing of the Reform Bill of 1831–2, George Eliot developed a more suggestive set of relations between public and private life than in *Felix Holt* because she now had in view the role, intermediate between public and private life, of 'the common estate of society', especially the state of knowledge. Dorothea's first husband, Casaubon, combined prejudiced conservatism with faulty scholarship. Her second marriage (forfeiting wealth, like Esther's marriage to Felix Holt) was to Casaubon's liberal-radical nephew Ladislaw, whose activities he had tried to veto as the 'sciolism', the superficial knowledge, 'of literary or political adventurers'. It was not a mere matter of approval of the liberal and depreciation of the conservative. While 'sciolism' was a palpable hit at the superficiality, delightful to the reader, of Ladislaw's employer Mr Brooke, the term also applied to Ladislaw who, when criticizing Casaubon, had pretended to rather more knowledge than he possessed, 'being not at all deep himself in German writers; but very little achievement is required in order to pity another man's shortcomings'. Further adverse light was thrown on Ladislaw (despite the author's indulgence) by his opinions. To Mr Brooke he appeared 'a sort of Burke with a leaven of Shelley', but comparison between the opinions of 'Felix Holt' in the 'Address' and Ladislaw's told a different story: 'Wait for wisdom and conscience in public agents—fiddlestick! The only conscience we can trust to is the massive sense of wrong in a class . . .' In terms of the 'Address' this was a strongly partial view, taking up only 'the claims of the unendowed multitude'. Where that multitude came directly into view in this novel, it was with reference either to their ignorance (as in the figure of Dagley, Mr Brooke's neglected tenant, whose idea of Reform was that 'them landlords as never done the right thing by their tenants 'ull . . . hev to scuttle off') or to the view taken of them in literature, as in Dorothea's vision after the vigil of her despair, a vision echoing not only *The Prelude* but also Eliot's own *Silas Marner*, the action of which had been impelled by the very figures Dorothea saw, 'a man with a bundle on his back and a woman carrying her baby'.

Emphasis on 'the common estate' made the correlation of the main events in the provincial foreground with the fate of the Reform Bill in London seem not only natural but significant. Casaubon and Brooke, Tory and Whig, could be seen to belong to the class whose power was on the wane. The first passing of the Bill by the lower house took place near the time of Casaubon's death; the dissolution of parliament followed, and Mr Brooke's farcical candidature and withdrawal, which dismissed him from the novel until near the end. New interests took their place, centred upon the limited religion of the banker Bulstrode, who favoured Reform, but found that in his private life it was more difficult to separate himself from the abuses of the past. The eventual passing of the Bill was outside the main action, along with Dorothea's marriage: in each there was an element of generous illusion, suggested by the glimpse of subsequent politics in which Ladislaw 'became an ardent public man, working well in those times when reforms were begun with a young hopefulness of immediate good which has been much checked in our days'. The 'unalterable nature of things' was not to be so easily evaded, and provincial people were not entirely misguided in taking more interest in the betterment of the conditions of life—farming, cottages, gates, the special interest of Caleb Garth but not at all of Mr Brooke, the 'sciolist' of obscure parliamentary ambitions. The author was bringing her narrative to bear upon the attitudes of her first readers towards the period of transition initiated in the 1830s, in order to prompt questions. Steadily she implied that, in their 'old', 'slowgrowing' society, events of forty years ago were still of importance; the contrary view was quietly mocked in remarks like, 'In those days the world in general was more ignorant of good and evil by forty years than it is at present.'

Such considerations, associating *Middlemarch* with *Felix Holt*, are obviously of less intrinsic importance than the new level of subtlety in the drama of inner life which developed out of them—the recurrent shocks of pain and forgiveness in the Casaubon and Bulstrode marriages, the mastering of Lydgate by a wife who would understand nothing but her own 'reticent self-justification', the final triumph over personal resentment which secured Dorothea's happiness. It was a subtlety dependent on the author's direction of the reader's sympathies by methods varying between the extremes of open partisanship and the reticence of the mere presenter. In the process, traditional literary forms, instead of being taken, like tragedy in *The Mill on the Floss* or *Felix Holt*, as the height to which the author might aspire, were more allusively handled, mixing them one with another. Lydgate's story might be introduced with an epigraph from Ben Jonson about the kind of comedy which shows a contemporary 'image' and makes 'sport with human follies not with crimes'; no doubt George Eliot appreciated, too, the connection of the latter words with Aristotle's brief view of comedy in chapter 5 of the *Poetics*, which she had recently reread; but Lydgate's actual position in allowing himself the folly of fascination by Miss Vincy was from the first treated with a characteristically thoughtful irony, mingling suggestions of both comedy and tragedy: he was shown comparing her with Dorothea, who 'did not look at things from the proper feminine angle', so that a man could hardly relax after his work in her 'too earnest' society; but an allusion to *King Lear* glanced at the grave consequences which would lead him to 'modify his opinion as to the most excellent things in woman'.

The 'excellent things' in Dorothea, her earnestness among them, were not exempt from a similar irony, the overtones of which formed a highly original blend. The novel's Prelude glimpsed with amusement the 'epic life' demanded by an 'ideal nature' which, 'fed from within, soared after an illimitable satisfaction', but was reduced to accepting a determinate duty. The acceptance was called an 'epos', the unwritten rudimentary form contrasting with finished epic in George Grote's *History of Greece* (recorded amongst the preparatory reading for the novel). Those 'later-born Theresas' who, in the absence of a 'coherent social faith and order which could perform the function of knowledge', found for themselves no epic life of even this rudimentary form were left to alternate 'between a vague ideal and the common yearning of womanhood'. Once Dorothea had given up the first alternative, the second transformed 'her epos' into 'the home epic' of the Finale, a more complete marriage than her first. 'Epic' here subsumed both comedy and tragedy in so far as its subject was 'the gradual conquest or irremediable loss of . . . complete union'. The modification 'home epic' was offered with a smile at the Eden which is lost to achieve it and at the fact that its subject, marriage, 'has been the bourne of so many narratives'—a reference back to the Prelude, where 'the favourite love-stories in prose and verse' were said to have set narrow 'limits of variation' to 'female indefiniteness'. Between Prelude and Finale those limits had been widened in an action which mixed the traditional categories of comedy, tragedy, and epic. The indefiniteness, when it reappeared in chapter 3, took the form of a 'haze' produced in Dorothea's mind by her inadequate education. The lack of the 'coherent social faith and order' which might 'perform the function of knowledge' for her was presented as tolerant comedy in which her distinctive actions, though pointing towards a painful first marriage, stirred the other characters only to disapproval, whether with pain or mockery—the mockery spirited and amusing, the pain easily assuaged. When the reader was admitted to an inner view more favourable to Dorothea's aspirations, the pathos, even at its most poignant, as in her bewilderment on her wedding journey, rested upon the premiss of her limited self-knowledge and so formed a continuum with the comedy at the expense of the same limitation, right from the time when, feeling that she enjoyed riding 'in a pagan sensuous way', she 'always looked forward to renouncing it'.

It was necessary to be aware of this continuum, this mutual interpenetration of suggestions of both comedy and tragedy, before Dorothea's development could be understood. When her unconscious attraction to Ladislaw by way of his grandmother's miniature, 'so like a living face that she knew', was presented without authorial comment, the reader was to notice the demands of her own nature appearing without her being aware of them. A movement similar to Maggie Tulliver's had begun, from an unequal relationship founded on intellectual interests to another which the heroine at first understood even less, but now both the fears of the reader for the heroine and the author's compassion for her were touched with the comedy of her limited self-knowledge. The same applied to Dorothea's eventual second marriage. The author obviously liked Ladislaw too much, but his inadequacies were not concealed, neither his superficiality of knowledge nor his less than iron resolve ('his own resolve . . . was by no means an iron barrier, but simply a state of mind liable to melt into a minuet with other states of mind, and to find itself bowing, smiling, and giving place with polite facility'). Despite the author's obvious enthusiasm for Dorothea's romantic fulfilment of her own nature, the character's own comment, once she had found herself, set the second marriage in a more complex light: 'I might have done something better, if I had been better.' There was an echo of her sympathy with Lydgate as one 'who had meant to lead a higher life than common, and to find out better ways', and of the author's sympathy with Bulstrode, 'who had longed for years to be better than he was'. At the same time the context of Dorothea's comment lightened its tone without subtracting from its seriousness: it was made in conversation with the sister whose commonsense had been a steady source of comedy, but now the roles were reversed and it was Celia who expected Dorothea to play the tragedy queen.

The tragicomedy in *Middlemarch* of the limited, ordinary people who did not understand their own actions and the extraordinary people who came to understand their own actions all too clearly was continued in *Daniel Deronda* (1876). It was complicated, however, by a didactic preoccupation with correcting common misapprehensions about the Jews. The result showed a strange contrast between the story of Gwendolen Harleth, with its detailed sequence of inner cause and effect, and Deronda's story, a romance depending on coincidences—that a friend of his unknown Jewish grandfather should see and recognize him on a brief visit to Frankfurt, that a girl Deronda saved from suicide should also be Jewish, and that he should find, by chance, her lost brother and be taken by him for the deliverer of his race.

Deronda's story, like that of *The Spanish Gypsy* and Esther's in *Felix Holt*, projected on to a larger scene the subject of *Silas Marner*—a central character brought up among one set of people but belonging to another by birth and making a choice between them once the truth was known. In the two earlier novels it was the foster-parent and his people who were chosen, in *The Spanish Gypsy* the people of the heroine's birth. Deronda repeated the second, more doctrinaire choice. In the attempt to endow him with an inner life there was also much that was doctrinaire, the product of the author's determined wish-fulfilment. His 'sense of injury' at the suspicion of being illegitimate bred 'not the will to inflict injuries' but 'a hatred of all injury', 'a meditative interest in learning how human miseries are wrought'. He was given an inner history modelled on the more abstract and explanatory parts of *The Prelude* in his 'ceasing to care for knowledge . . . unless . . . gathered up into one current with his emotions', and in the 'meditative numbness' he fell into without 'some way of keeping emotion . . . substantial and strong in the face of a reflectiveness that threatened to nullify all differences'. Although he reached a meditative, almost mystical altruism, his will remained paralysed—'forgetting everything else in a half-speculative, half-involuntary identification of himself with the objects he was looking at, thinking how far it might be possible habitually to shift his centre till his own personality would be no less outside him than the landscape'—until his rescue of Mirah from suicide. He was prey to a similar paralysis just before encountering his grandfather's friend: 'the habit of seeing things as they probably appeared to others' had made 'strong partisanship . . . an insincerity for him', so that sympathy fell 'into one current with that reflective analysis which tends to neutralize sympathy'. He was like the poet at the beginning of Book XIII of *The Prelude* when ready to seek his remedy 'In man, and in the frame of social life'. But his state tended to the abstract in its expression and to the vague ('how miseries are wrought', '. . . nullify all differences'), and his remedy came from pure coincidences, events which did not need to be understood in their causes.

Gwendolen Harleth's history, on the other hand, was specific in content, vigorous in language, and startling in the kind of dramatization, surpassing even the best in *Middlemarch*, by which inner causes were comprehended. An egoist like Rosamond Vincy, Gwendolen was not presented as pitiable in her limitations but as spirited and energetic, full of 'a *naïve* delight in her fortunate self, which any but the harshest saintliness will have some indulgence for in a girl who had every day seen a pleasant reflection of that self in her friends' flattery as well as in the looking-glass'. The danger in which she stood was conveyed by the control of the narrative, presenting her loss of fortune first, and only then the fortunate self in which she had formerly delighted. The circumstances which had given that self play, and yet allowed the girl to glimpse within it depths which she did not understand, were then developed in twelve chapters of flashback—nearly three of the eight books, a month apart, in which the novel was published. And within these chapters dramatization preceded explanation: Gwendolen's remarkable proneness to 'fits of spiritual dread' appeared first in her terror, at a moment of theatrical triumph, when the picture of a dead face suddenly sprang into her view on a movable wall-panel. It was quite a different procedure from that used in Deronda's case: interest in the character and what she might do came first, explanations afterwards. The development of this dread, which could suddenly give 'an undefined feeling of immeasurable existence . . . in the midst of which she was helplessly incapable of asserting herself', stirred George Eliot to a collaboration of ethical insight with imaginative power remarkable even in her strange line of novels, each revealing unexpected differences from the others. Looking back just before the first instalment of *Deronda* was published, she called them 'simply a set of experiments in life—an endeavour to see what our thought and emotion may be capable of . . . I become more and more timid—with less daring to adopt any formula which does not get itself clothed for me in some human figure and individual experience . . .'

The contrast in this novel between Gwendolen and the title character, the woman fully known as an individual, the man seen mainly in abstract terms as typical of a certain predicament, that of the solitary, well-disposed intellectual, can, of course, easily be exaggerated. No less than Deronda's, Gwendolen's case depended upon the author's assimilation of Wordsworth with his emphasis on 'unknown modes of being', his revulsion against those who value the 'universe', and even the 'human soul',

> No more than as a mirror that reflects
> To proud Self-love her own intelligence;
> That one, poor, finite object, in the abyss
> Of infinite Being, twinkling restlessly!

Gwendolen's sense of 'immeasurable existence' was felt in 'solitude in any wide scene', without the Coleridgean refinements of Deronda's identification of subject and object. Nevertheless, the common source threw into relief the subtlety with which the logic of her inner development was followed, once she was forced to confront everything from which her fortunate self had been protected. There was melodrama in both stories; but in Deronda's it resulted from chance, in Gwendolen's from her own choice once she had accepted a suitor's advances for reasons other than the more usual inner reasons connected with love. She had already recognized with distress, to her mother, that 'I can't love people . . . I can't bear any one to be very near me but you.' The result was a marvellous dramatization of ineluctable responsibility in moral choice, in a person apparently 'determined' not only by outward circumstances but by her own disposition.

Even before circumstances oppressed her she had felt, when faced with apparent freedom, an unexpected terror at the 'subjection to a possible self, a self not to be absolutely predicted about', so that 'her favourite key of life—doing as she liked—seemed to fail her'. When, after she and her family had been 'totally ruined' financially, the suitor, Grandcourt, again presented himself, 'her native terror' reappeared: 'again she seemed to be getting a sort of empire over her own life. But how to use it? Here came the terror. Quick, quick, like pictures in a book beaten open with a sense of hurry, came back vividly, yet in fragments, all that she had gone through in relation to Grandcourt.' Even while she seemed to herself not to be choosing but to be drifting, the very things amongst which choice must be made presented themselves. The thorough imagining of mind and sensibility in action, 'the dark seed-growths of consciousness', marked the completeness of the transformation into art, 'clothed . . . in individual experience', of insights that had appeared as generalizations in *Romola* without that transforming power. What the earlier novel had paraphrased from sources English and Greek—'all those inarticulate sensibilities which are our deepest life' (chapter 37), 'that dread which has been erroneously decried as if it were nothing higher than a man's animal care for his own skin: that . . . fear at anything which is called wrong doing . . . fear . . . as the guardian of the soul' (chapter 11)—now became the very substance of an individual inner drama as satisfying in its suspense as in its discernment of preconscious states. The terms from *Romola* reappeared in Deronda's advice, 'Take your fear as a safeguard.' But the inner drama stirred the reader's concern for the outcome more keenly than in any situation George Eliot had yet created. Gwendolen's terrors, becoming clearer to her in the 'new consciousness . . . awaked' at her inability to forget the prior rights of her fiancé's children and their mother, the 'new soul, which had better, but also worse, possibilities than her former poise of crude self-confidence', generated extraordinary expectations once she had married and her will, at first as strong as her husband's and one ground of the affinity between them, was forced to give way to his.

The steady menace of the husband, Grandcourt, conveyed with a contrasting restriction of the inner view, preserved suspense as to his moves. His naked negative will appeared less stagey than it otherwise might, through the sheer force of the contrast with Gwendolen's more complex state of mind and growing self-awareness. The reader's suspicions that her former energy of will could hardly remain in complete abeyance sharpened the suspense when her only apparent help came from Deronda's moral advice and she seemed to be exposing herself to a charge of vulgar intrigue. In these dangerous circumstances his altruism became more real for the reader than in all its careful explanation and romantic course hitherto; because Gwendolen was known so thoroughly there seemed to be substance, as the novel approached its climax, in his 'nervous consciousness that there was something to guard against not only on her account but on his own'.

At this climax the presence in the novel of even a story as far-fetched as his found its justification because, like that of Dinah Morris in *Adam Bede*, it allowed to appear, as a familiar acquaintance of the reader's, a character who could elicit confession without condemning. It was a more sophisticated confession than Hetty's, befitting Gwendolen's greater self-awareness, though it was made while carrying the weight of an equally unbearable recollection which seemed as if it would never change: 'a dead face—I shall never get away from it. . . . It can never be altered.' Its melodrama was mitigated by juxtaposition with the simpler sensation novel of Deronda's quest, with its coincidences, its Secret, and its suspicions. As the means of decisively separating her from him once she was tending towards dependence on him, his vague mission served well the purposes of George Eliot's characteristic kind of tragedy, the climax of which is not death but the necessity to go on living. Gwendolen's resolve, in the penultimate chapter, 'I mean to live', was made in the midst of hysterical fits of shrieking which suggested a painful continuation of the 'circle of punishment' brought into the narrative by quotation from Dante.

It was clear she was to be left painfully to climb the purgatorial mountain by her unaided efforts as, 'like one who had visited the spirit world', she sat through conversations about the Church-Rate Abolition Bill and the telegraphic cable. The limitations of the country gentry amongst whom she moved had all along been quietly stressed by the contrast with her own experience, 'so far as pastoral care and religious fellowship were concerned, in as complete a solitude as a man in a lighthouse'. Grandcourt's acceptability to these people, who could not look beyond his 'faded aspect of perfect distinction' and see his negative 'will like that of a crab or a boa-constrictor', conveyed its comment upon them. The author reinforced it with steady, unobtrusive satire of their insensitivity, sharpening on occasion into pointed humour—the father who, to protest against his daughter's engagement to a distinguished musician, 'took his cigar from his mouth and rose to the occasion by saying, "This will never do, Cath."'

Again the 'coherent social faith and order which could perform the function of knowledge' for the inexperienced was lacking, as in the Prelude to *Middlemarch*. The consequent ignorance in which a young woman could be left was no less in the 1860s than in the 1830s. Equally plain was the enclosed self-satisfaction common on the verge of the aristocracy, 'the world—I mean Mr Gascoigne and all the families worth speaking of within visiting distance . . .'. The sort of politician they approved of entered with 'the expectant peer, Mr Bult, an esteemed party man', astonished at an 'outburst . . . on the lack of idealism in English politics' from a musician whom 'he hardly regarded in the light of a serious human being who ought to have a vote'. The musician, Klesmer, spoke for creative artists and for the author (despite her awareness of comedy in his vehemence): 'We are not ingenious puppets, sir, who live in a box and look out on the world only when it is gaping for amusement. We help to rule the nations and make the age as much as any other public men.'

George Eliot's own making of the age in her major phase had come from the understanding she promoted of the people whose personal lives, near the time of the two Reform Bills of 1832 and 1867, seemed to her symptomatic of what was wrong with their society. To her these were times when politics, at first national, then international, had made even more urgent the conquest of ignorance. Her heroines had been young women in search of, or ready to welcome, knowledge, her heroes men who, with all their faults, were ready to furnish it. The experiment with the life of Gwendolen Harleth showed the process in a setting of what George Eliot called in a letter 'the intellectual narrowness . . . which is still the average mark of our culture'. The aim to 'widen the English vision a little' with reference to Jewish life and history resulted merely in recommending it to readers, not in experimenting with it, but she had 'meant everything in the book to be related to everything else there'. The elements of romance in Deronda's story, by comparison with the realism of cause and effect in Gwendolen's, may have made it difficult to accept the predicament of the Jews as the characterizing feature of the age which it was intended to be, but there was no doubt about the dangerous national and international setting of the action. Gwendolen and her family lost their money through the 'reckless' speculation of those who had charge of it. This 'recklessness' and 'lawlessness' came under the same condemnation as her own gambling and, as part of 'the last commercial panic', was one of the 'political and social movements' specifically said, at the beginning of chapter 48, to be ignored by Grandcourt, indifferent to everything but his own rent-roll. Of the others, 'the policy of Bismarck, trade-unions, household suffrage', the first was not only the subject of scattered allusions to establish the novel's time scheme but furnished the analogy for Grandcourt's own iron will and watchfulness, in exercising within his small sphere 'qualities which have entered into triumphal diplomacy of the widest continental sort'. As typical issues, trade-unions and household suffrage were well selected; their interconnection and their relation to the position of the aristocracy at the end of the period of transition to democracy are independently noticed in the Oxford History of England, volume XIII:

> The attack on the legal position of trade-unions [regarded as beyond the scope of the Friendly Societies Act and as associations 'in restraint of trade'] made the leaders of the working class realize the importance of direct representation in parliament, and the economic crisis of 1866 revived the almost forgotten agitation against aristocratic misgovernment.

For all that George Eliot had inflected with amusement her words at the end of chapter 8, 'I like to mark the time, and connect the course of individual lives with the historic stream', it was clear that they applied to other matters than the crinolines in question there. The 'historic stream' was more specific in the passing reference to Bismarck's victories than in the fairy-tale of Deronda's lineage, marriage, and vague political plans under the influence of a minor character's portentous rhetoric. But it was through these plans that 'the larger destinies of mankind' entered the experience of a heroine who had been presented, like most of the other characters, and especially her husband, as insulated from them.

In the novels of the major phase political and social issues impinged on the lives of her heroines in this way, as representative of the general life which might discipline the ego. Although individual character was the main focus of attention, the author's ironic and compassionate vision comprehended not only the connection of individual choice and individual retribution but also their relation, whether specified or only suggested, with what she called in the last novel 'the fermenting political and social leaven which was making a difference in the history of the world'. The element of 'panorama' to which James objected was essential to such a vision, though it is doubtful whether 'panorama' was the right word for the social medium seen in process of change, slow but offering suggestive resemblances to the presentation of individual character as 'a process and an unfolding'.

The hypothesis she adopted for the purpose, that the main issues were matters of ignorance and knowledge, curiously resembled Dickens's in *Bleak House* twenty years earlier. Despite the difference in treatment, such a hypothesis was obviously of value to both novelists in their attempt to clarify the essentials. It led George Eliot, in *Middlemarch*, into subjects more congenial to her as an intellectual than any she had yet taken up, reaching to the kind of knowledge which was her own province as an artist—ideas 'wrought back to the directness of sense, like the solidity of objects'. No such effect resulted from the more limited kind of ignorance—about the culture and expectations of a particular people—which she took up in *Daniel Deronda*, except in so far as it was an example of the ignorance of the general life which had marked her heroine. But that was sufficient to bring together the final two novels as the climactic examples of her mind and sensibility. It was a sensibility which had all along accepted, as her characters were required to do, the discipline of the mind. In the case of the characters this meant a process of coming fully to understand what unusual sensibility—whether unusual for its range or its limitations, its subtlety or its very obtuseness—had revealed. In the case of the author it meant the acceptance of a discipline 'which even in my doubting mind is never shaken by a doubt, . . . my conviction as to the relative goodness and nobleness of human dispositions and motives'. Such apparent ethical assurance, sometimes felt as ponderous from its manner of expression, made her work uncongenial to many in the two or three generations of readers following her death. A century later her statement of conviction, though strange, is more likely to be read with emphasis on the activity of comparing implied in 'relative'. This stimulated her imagination where it is strongest, where, as she said in a late essay, its work is 'based on . . . a keen consciousness of what *is* and carries the store of definite knowledge as material for the construction of its inward visions'. The results still command the attention, if not of such a great variety of readers as Dickens, yet of a very large number for whom, at the peaks of Victorian achievement, it is only the inessential that is old-fashioned.

9. Trollope

'A novel should give a picture of common life enlivened by humour and sweetened by pathos. To make that picture worthy of attention, the canvas should be crowded with real portraits, not of individuals known to the world or to the author, but of created personages impregnated with traits of character which are known. To my thinking, the plot is but the vehicle for all this ... ' Trollope's[1] own view of the novel in his *Autobiography*, written when most of his work was done, was well exemplified in the novels with which he began. They offered a picture of common life in Ireland, at first in set pieces reminiscent of Scott. In *The Macdermots of Ballycloran* (1847) the humour was incidental to the story, and the pathos, from the failure of a son's efforts to prevent the family decline, was not of a kind to sweeten it. The portraits, by meeting Trollope's later requirements, showed what was lacking in them; for known Irish 'traits' were vigorously presented, but without giving much sense of created individuality. Thady Macdermot, his sister Feemy, and her seducer were more than lay figures—the first especially, in his troubled tactlessness—but they were far from the successes which Henry James summed up after Trollope's death: 'We care what happens to people only in proportion as we know what people are. Trollope's great apprehension of the real, which was what made him so interesting, came to him from his desire to satisfy us on this point—to tell us what certain people were and what they did in consequence of being so.'

In this respect the second novel, *The Kellys and the O'Kellys* (1848), made a considerable advance, despite its stage villain, Barry Lynch. The more moderately presented wickedness of Lord Kilcullen bore the marks of individuality in his brisk effrontery and cheerful acceptance of his own shortcomings, capped by an incongruous ability to be sorry for the heiress whom he hoped to

[1] Anthony Trollope, 1815–82, fourth son of Frances Trollope, was largely self-educated while a Post Office clerk in London, despite having been both to Harrow and Winchester. After serving the Post Office with distinction in Ireland, Wales, and the English provinces, and on official visits to Egypt and the West Indies, he retired in 1867 to edit a new magazine, *St Paul's*. Unsuccessful Liberal candidate for Beverley in 1868, he resigned the editorship in 1870 and travelled (writing all the time) in the antipodes, the United States, Ceylon, and South Africa.

marry, even to the extent of feeling 'strongly tempted to act a generous part; to give her up, and . . . to deserve at any rate well of her, and leave all other things to chance'. The plot, made up of two independent love-stories, might seem inferior to the single intensifying action of the first novel, but it was a step in the direction of the most common form which Trollope's fiction was to assume. The very disjointedness of this early attempt may have taught him something about the necessity, noted in the *Autobiography*, for subsidiary plots to 'tend to the elucidation of the main story', even though it took him twenty years to master the method of bringing this about.

In both novels the social importance of religious divisions focused his attention upon the differing influence and abilities of clergymen. Important situations in both were resolved by their intervention and guidance, and it was far clearer in their case than in the case of the nominal heroes and heroines that what they could do was determined by what they were. The relation between the religious standing of such characters and their social role was to become a source of comedy in his first successes. His attempt directly to follow Scott in *La Vendée* (1850) gave a militant clergyman, Father Jerome, a prominent place, his crucifix brandished in battle and stained with blood, his long grey coat patched and darned after violent wear. Other personages were singled out for the contrast between their humble status in times of peace and their commanding roles in war. The ostensible heroes, young aristocrats, were as null as such figures in Scott. Their love-stories, set into the course of the 1793 revolt in La Vendée, illustrated the acceptance of social levelling in times of crisis. (The hero's unease about this exactly paralleled that of Lord Hampstead, in *Marion Fay* thirty years later, who felt 'that associations which were good for himself might not be so good for his sister'.) Demonstration of a practical absence of snobbery among the royalists (except for the villain Denot, who, when reformed, had to accept a master-baker as his second-in-command) went with denunciation of doctrinaire republicanism. Robespierre was introduced, in an unnecessary chapter, as the man who 'seems almost to have been sent into the world to prove the inefficacy of human reason to effect human happiness'. The next novel, *The Warden*, was to speak of a reasonable reformer as criticizing 'time-honoured practices with the violence of a French Jacobin'. The conclusion of

La Vendée looked for a return of the monarchy, now that 'France is again a Republic'. For the 1874 reprint, hastily done (prompted perhaps by the appearance that year of Hugo's novel of La Vendée, *Quatre-vingt-treize*), two sentences were added to lament that 'France must again wait till the legitimate heir of the old family shall be willing to reign as a constitutional sovereign'.

The author's conservative liberalism would find expression in the future through less vigorous action than this story of civil war. Already he seemed to acknowledge the lack of what he later called Scott's '20-horse-power of vivacity' by choosing to show the results of the culminating battle as observed from a sick-room. His pattern was to be *The Kellys and the O'Kellys*, which a reviewer in 1848 recommended to 'readers who dislike to be wrought up to a pitch of strong sensation'. Work of this quieter kind made his name in *The Warden* (1855). But the simple history of Mr Harding's conscience there was surrounded with heroic echoes of martial crusade. The wardenship which Mr Harding felt bound to resign was to Archdeacon Grantly part of 'the holy of holies' to be defended 'from the touch of the profane'; 'to guard the citadel' was 'to put on his good armour in the best of fights; . . . part of a never-ending battle against a never-conquered foe—that of the church against its enemies'. The ambiguity of the word 'church' allowed the enemies of the external organization to be taken as those of the community of the faithful, and mockery of an excess of zeal to become almost, but not quite, satire at the expense of a radical incongruity. It became natural, and all the funnier for that, that the archdeacon should also misapply the language of personal religion, calling the retention of the comfortable wardenship an 'affliction' which Mr Harding should 'bear', when in fact he was resigning it with pain for reasons of conscience.

Trollope later claimed to have been exposing 'two opposite evils', the misapplication of Church endowments and 'the undeserved severity of the newspapers'. With the former, but not with the latter, criticism took second place to comedy. When writing on clergymen in the *Pall Mall Gazette* in 1865–6, and claiming the viewpoint of 'all men who care for the Church of England', he pointed out how 'it is so picturesque and well-beloved in its old-fashioned garments' that 'we can put up with anomalies which elsewhere would be unendurable'. In *The Warden*, affection for these mingled respect with ridicule as Mrs

Gaskell had in *Cranford*. Since *The Warden* was begun while *Cranford* was appearing in *Household Words*, there may even be a direct connection. At any rate, the language of religion in *The Warden* led to a mock aggrandizement like that when Mrs Gaskell called her inoffensive ladies 'Amazons'. At the centre of each book was a character who resisted the militant mores of the place, a small provincial community suddenly disturbed by forces emanating from a large modern city. Instances outside the novels—in Trollope's case the mishandling of the finances of a Winchester hospital and a Rochester school—gave both writers a foothold in contemporary problems from which they moved to the defence of the old ways which could withstand the disturbance. But the life of the characters, not the merits of the case, became the central matter—Archdeacon Grantly, 'fond of his own way and not too scrupulous in his manner of achieving it', and Mr Harding, the Warden, firm only in following the demands of his conscience. The central action was one to which Trollope repeatedly recurred—the giving and taking of advice. Mr Harding was offered a great deal, from the officially impeccable sources of the church and the law, and rejected it all. This action was supplemented, unfortunately, with heavy satire of *The Times*, ('Britons have but to read, to obey, and be blessed') and parody of the methods of the two chief literary counsellors of the nation, Carlyle and Dickens. Uncertain of his own powers, Trollope seemed not to recognize that in Mr Harding he had a character like Goldsmith's Dr Primrose whose guilelessness was a sufficient triumph in itself, when contrasted with the worldly wisdom of those who attempted to dissuade him.

The order of composition, which the present discussion follows, is not always that of publication. The next work to be written was not published till 1972, *The New Zealander* (1972), a book of social criticism. Comprehensive as Carlyle's *Latter-Day Pamphlets* (1850) in scope, it was, luckily, rejected for publication. Trollope could then turn to developing, in fiction, the resilience of his small community threatened with change from outside. *Barchester Towers* (1857) extended the community to include the ultra-conservative squirearchy, the Thornes of Ullathorne, and the threat to include, in the main action, the low-church Proudies and, in the background, the Tractarian movement and the continued activities of the Ecclesiastical Commission which the reformed parliament had set up in 1836. The latter more topical matters were kept subordinate to controversies and mock-controversies of longer standing. The old ways, gently and affectionately mocked in Miss Thorne's domestic arrangements, her 'breakfast on the lawn' with Elizabethan games, and her early eighteenth-century tastes in literature, had again something of the flavour, if little of the delicacy, of *Cranford*. As in *Cranford*, too, comedy included a comforting glance back at the more turbulent times past, the steward who had carried in his boot-heel letters for the royalists 'when the directory held dominion in Paris' corresponding to Miss Matty's recollected fears of French invasion.

The low-church invasion of Barchester became lively farce, suggestive in its paradoxes. The most outrageous of the outsiders, the Stanhopes, were part of the old order, the children of a non-resident pluralist towards whom 'poor dear old Bishop Grantly had . . . been too lenient', but their presence in Barchester was part of the new order instituted by a low-church bishop, who summoned absentees home under threat of having their names 'submitted to the councils of the nation'. So the independence of thought and action of Bertie Stanhope and his sisters, Charlotte and the Signora Madeline Neroni, had grown up under the old dispensation but disturbed Barsetshire under the new. Outfacing the Countess De Courcy, patronizing Bishop Proudie, and, in the person of the signora, implausibly gaining the instant amorous attention of clergymen of both low and high church as well as of Mr Thorne with his Grandisonian compliments, their fooling avoided puppetry by being touched with a range of individual feeling, from the signora, with her interest in the variety of her admirers, to the countess, who became 'excited, happy, and merciless' on hearing how 'abominable' was the signora's conduct. The Stanhopes' insouciant extension of the range of things that could be said in Barchester company constituted a lively critique of it, supported by a muted continuation of the author's needle-points of ecclesiastical mockery from *The Warden*. The farce concerned with the low-church party depended as much on human oddities as on ludicrousness of situation. The active partnership at first was between a large masculine bishopess, Mrs Proudie, and her husband's chaplain, Mr Slope, her fondness for whom was so well known that his attentions to her eldest daughter had been concealed from her by all. The destruction of this comfortable patronage by Slope's court to the signora, who had laughed when her brother knelt to disentangle the debris of Mrs Proudie's dress from the sofa he had propelled over it, held improper overtones, rendered more outrageous by the clerical setting. They reached their climax in the dismissal of Slope, officially by the bishop but actually by his wife.

> 'Do you think I have not heard of your kneelings at that creature's feet—that is if she has any feet—and of your constant slobbering over her hand? I advise you to beware, Mr Slope, of what you do and say.'

Trollope next reverted to a more derivative kind of comedy in *The Three Clerks* (1858), conforming at the same time more closely to the model of the social novel than in *The Warden*. He exposed the ineffectiveness of the 'loathed scheme of competitive examination' for positions in the civil service and made fun, in Thackeray's manner, of the practice of doing so in novels. *Pendennis* had contained the precedents for the young author among the three clerks and for commenting on current fashions in fiction by parody of the kinds of novel he tried to write, even of the social novel, with its 'slap at some of the iniquities of the times'. The high spirits of the attack on the editor's requirements, for exciting incidents, especially at the beginning, rather than descriptions or the 'history of who's who', or for variety rather than unity—'our game is to stick in a good bit whenever we can get it'—reflected Trollope's own concerns. Open journalism of his own, lifted from, or based upon, the revised manuscript of *The New Zealander*, appeared in unassimilated chunks. The jocular attitude to parliament taken over from the satire—'The pursuit is certainly full of interest, but it is somewhat deficient in dignity'—was to persist until the author tried to become a member himself. The reviewer who found the book 'straggling' had some cause. It was an experiment in multiple structure and only just managed to give sufficient prominence to the downfall of Alaric Tudor in spite of his success in examinations. It was also a London novel, of which there were to be many more, equally lacking the sense of comforting sameness and stability that made Barsetshire attractive at the time and spurred Trollope's revival in the succeeding century.

In returning to Barsetshire in *Dr Thorne* (1858), however, he still had his eye on the social thesis novel. His imaginary county became a more definite geographical and social entity, making more specific the suggestions of *Barchester Towers* and widening their range so as to bring in other classes than the county and clerical aristocracy. More emphasis fell on the historical setting in the generation after the coming of the railways and the passing of the Reform Bill, which divided the county into two. So the main features of Barsetshire were those characteristic of the whole country when the old ways were accommodating themselves to the new. The heroine was heiress to the property of a railway contractor, Sir Roger Scatcherd, who had himself gained titular admission to the gentry, though noted for 'the violence of his democratic opposition' to them. Her marriage to Frank Gresham restored the fortunes of an old Tory family, for long in debt to Scatcherd, and for long insistent that Frank should clear the debt by marrying money. Without the same desperate need, the Whig magnates, the De Courcys, had the same aim, and in its pursuit the same disregard for 'blood'. The book began with the premiss that England might still 'be called feudal . . . or chivalrous' rather than 'commercial', the owners of land still 'the true aristocracy, . . . trusted as being best and fittest to rule'; it proceeded to show them in pursuit of the wealth generated by commerce, Greshams and De Courcys in turn pursuing an alliance with the daughter of an ointment manufacturer, Miss Dunstable.

The paradoxical attitudes of the characters, and, in some degree, of the narrator, to these matters were the subject of a good deal of mechanical comedy where the De Courcys were concerned. Trollope's own kind, based on the unexpectedness of human character and human prejudice, appeared in his 'hero' Dr Thorne, coming of an ancient family, the Thornes of Ullathorne, and offending no rank 'by boasts of his own equality' although 'there was that in his manner that told it', hating 'a lord at first sight' although at heart a thorough Conservative, friend both of the unpresentable Scatcherd and of the squire of Greshambury. Mary Thorne, orphaned and illegitimate, was presented as 'at war with herself', believing in her own equality with anyone in inner worth, yet loving the prejudice in favour of 'blood'. The fairy-story which made her in manners fit to associate with the Greshams and, unknown to herself, of wealth sufficient to allow her to become one of them was paid out very, very slowly with considerable repetition and an amazing absence of dullness. Despite the routine thoughts and actions of Frank's mother, herself a De Courcy, and the pathetic blackguardliness of Scatcherd's heir who also courted Mary—implausibly enough when he stood to gain nothing by it—Trollope kept going the threadbare plot which his brother had given him. The very sameness curiously led to expectation of change, especially in Frank, devoted as he was to his dogs and his horses. The secret was in the managing of a number of small passages of argument in dialogue, each taking up a single point bearing upon the fate of Mary Thorne. The main lines of the plot were those which directly affected her prospects; the sub-plots—Mr Moffat, the wealthy son of a tailor, horsewhipped by Frank for jilting his sister Augusta, the Lady Amelia de Courcy advising Augusta not to marry a fashionable London lawyer, and then marrying him herself—offered parallels with the marriage of wealth to 'blood' which Mary's happy outcome was to exemplify. The characterization offered small surprises of individuality—Lady Scatcherd keeping all her rough loyalty and affection, Miss Dunstable retaining a 'heart' despite her plain speaking.

At the time *Dr Thorne* would have seemed one of a sequence of social novels, for it was followed by *The Bertrams* (1859).

> To him that hath shall be given; and from him that hath not shall be taken even that which he hath. This is the special text that we delight to follow, and success is the god that we delight to worship.

The modern competitive spirit, political opportunism, and legal chicanery illustrated by this sombre, tendentious tale gave it 'magic' for Harold Laski. And when Trollope was invited to write a serial for the newly established *Cornhill* he was at work on *Castle Richmond* (1860), counterpointing the Irish famine of 1846–7 against blackmail and bigamy among the upper classes. Something of a primitive sensation novel, with a lawyer detective following one bigamous marriage back to its source to reveal another, it brought its chastened hero to offer, as a 'thankoffering', all his 'devoted attention to the interest of the poor around him'. Less lugubrious than *The Bertrams*, it was equally improving.

Thackeray, editor of the *Cornhill*, hoped for 'as pleasant a story' as *The Three Clerks*, which he admired, but, for *Framley Parsonage* (1860–1), Trollope chose his own model and reworked the central situation of *Dr Thorne*, looking further into the characters. Lady Lufton's opposition to her son's marrying beneath him became the means of revealing more than the routine snobbery of

Lady Arabella Gresham, while the feelings of Lucy Robarts were presented with a greater individuality than those of Mary Thorne. For all the spirited dialogue of the scenes in which Mary had defended her position, her sorrows had been those of the princess in hiding, extreme for purposes of eventual contrast: 'I sometimes think I was born to be unhappy, and that unhappiness agrees with me best.' Lucy's self-disgust at finding herself in love made no such direct appeal for sympathy:

> 'I'll tell you what he has: he has fine straight legs, and a smooth forehead, and a good-humoured eye, and white teeth. Was it possible to see such a catalogue of perfections, and not fall down, stricken to the very bone? . . . Here, standing here, on this very spot—on that flower of the carpet—he begged me a dozen times to be his wife. I wonder whether you and Mark would let me cut it out and keep it.'

Her ability sharply to take the 'injudicious' initiative when summoned into Lady Lufton's presence was not merely comical but stamped with the interest of a love which the girl had not meant to 'learn' but which, once confirmed, gave her the assurance that she could marry Lord Lufton 'feeling that I was doing my duty by him'. The moral terms were lighted up and given content by individual emotion. The resolution of the impasse, not by the romance-writer's wand as in *Dr Thorne*, but by the comedy of the comparison of Miss Robarts with the beautiful, vacuous Griselda Grantly, threw the emphasis on to something good: 'After all, love was the food chiefly necessary for the nourishment of Lady Lufton . . .'. The scene in which she asked Lucy to be her son's wife was moving in the simple way characteristic of Trollope at his best. Its climax in the mother's admission, 'I shall be wretched, Lucy, if I cannot teach you to love me', and the girl's independent 'Tell him—tell him—he won't want you to tell him anything' showed Trollope's power, after so much explanation, to make the scene contain the simplest feelings and to measure everything else against them. Griselda's insensibility therefore needed no comment, whether she were about to be jilted—'Then, mamma, I had better give them orders not to go on with the marking'—or leaving on her honeymoon—'I suppose Jane can put her hand at once on the moire antique when we reach Dover?'

There was some continuity with *Barchester Towers* in incidental episodes of farce and in the concerns of Lucy's brother Mark, an ambitious young parson; he owed his position to Tory patronage but needed the support of the ruling Whigs if he was to advance himself as Bishop Proudie had done. Parliament, which was to displace clerical Barsetshire as the main element of continuity in Trollope's *comédie humaine*, had already an important role. As yet, it was not a matter of systematic satire that politicians should influence ecclesiastical appointments, or fight, merely for party advantage, their battle over a bill to create more bishops. Behind the politicians, and subject to the same kind of distancing jocularity, stood the press, credited with the power 'to overturn the government and throw the whole country into dismay', and claiming, quite unjustly, the credit for Robarts's resignation of the prebendal stall to which he had been appointed after helping a politician in need of money. The fresh, uncomplicated feelings which had furnished the standard by which things were judged in the love-plot found their counterpart here in the ill-paid necessary labour of the perpetual curate of Hogglestock, Mr Crawley. Comical in his tactlessness with Lady Ludlow and even in his pride—elaborate subterfuges were needed to circumvent it if his family were to be fed—he brought in a touch of stiff-necked integrity to match the independence of Lucy Robarts.

Framley Parsonage sealed Trollope's fame for making entertainment out of the goodness of his characters without concealing their faults. He at once began a quite different kind of novel with a stronger centre to it, *Orley Farm* (1861–2). It was to be called by the *Spectator* reviewer 'the nearest approach he has made to the depth and force of tragedy'. At its centre stood a woman, Lady Mason, who likened herself to Rebekah, prepared to incur guilt in order to secure her son's welfare—'I never loved anything but him . . . I care more for his soul than for my own'—a strong idea, though the embodiment of it fell considerably short of tragedy. Even at the high point of her self-recognition her language had very little individual stamp compared with that of Lucy Robarts, and its content had to be taken on trust, corroborated as it was by very little that the reader had been shown. The narrator's own language was of the most undistinguished: 'The lines of her face were terrible to be seen as she thus spoke, and an agony of anguish loaded her brow . . .'. With characteristic simple realism, he was willing to see her, after confessing her guilt, 'huddled up in the corner of the sofa, with her face hidden, and all those feminine graces forgotten which had long stood her in truth so royally'. But he repeatedly relied on the more effusive gestures of the stage, equally marked whether Lady Mason was confessing or concealing her guilt. The rub was in the concealment; for Trollope was also managing, in his own way, a mystery story, its plot, he said later, 'probably the best I have ever made'. He may have been stimulated to this by the fact that, when he began to write, in mid-1860, *The Woman in White* was coming to the end of its run in *All the Year Round*. In the *Autobiography* he found Wilkie Collins's construction 'most minute and most wonderful', though he could 'never lose the taste of the construction . . . One is constrained by mysteries and hemmed in by difficulties . . .' In admitting that his own plot in *Orley Farm* had been marred by 'declaring itself, and thus coming to an end, too early', he implied an acceptance of the demand of the sensation novel for a Secret the divulging of which should be kept back as long as possible.

In the main plot of *Orley Farm* he skilfully reconciled this demand with his own freer kind of plotting, by making the actual mystery slighter and less dependent on a multiplicity of facts and reasons, but more revelatory of the strangeness of human life and emotion. In the process he showed how completely he had penetrated the difference between *The Woman in White* and the mere crime novel of incidents arousing strong feeling. Collins's novel gained as much force from threatened as from actual crimes; Trollope's gained as much from the prospect of secrets as it did from the one divulged. When Lady Mason confessed, halfway through the novel, the author came forward to claim that if the reader felt surprise, 'I must have told my tale badly'. But the hints made in the course of the telling, and even the specific evidence afforded, would have been taken by readers of the new sensation genre as mere sauce to whet the appetite for a more exact and more surprising resolution of the mystery. It is hard to believe that the author was not aware of this. He was already skilled in keeping expectation alive in a sequence the conclusion of which was, at least in part, known. When, at the end of chapter 15 of *Barchester Towers*, he had ventured 'to reprobate that system' which violates 'all proper confidence between the author and his readers, by maintaining nearly to the end of the third volume a mystery', the disclaimer had been disingenuous; he had divulged merely the suitors whom the heroine would not marry, leaving open the questions whether she would and, if she did, whom. About these questions he clearly expected some curiosity. It was only inferior examples like Florence Marryat's *Woman against Woman* or Le Fanu's *Wylder's Hand* which violated the reader's confidence by withholding, in the interests of a mystery, information needed to make sense of the narrative.

It is a question, however, whether in *Orley Farm* Trollope was not violating such confidence himself by concealing crucial aspects of Lady Mason's inner life. 'All that passed through her brain on that night I may not now tell,' he claimed—in order to keep going the expectation that he was playing a game of mystery. Especially was this so in her stagey appeal for support in chapter 26—'You do not then think that I have been guilty of this thing?'—which, read as implying a protestation of innocence, outweighed a great deal of innuendo. It was a game stimulated (as in *Lady Audley's Secret*, which began its serial appearance in July 1861, the month of the fifth number of *Orley Farm*) by *Bleak House*: again a lady of rank endured the threat of disclosures about her past, again she felt their effect most poignantly in relation to her only child, again the reader was assumed to be interested in the imperfect operation of the law. Like Dickens, Trollope was less successful in presenting Lady Mason's sufferings from within than in demonstrating their outward, visible aspect, in her case the inadequacy, as a man and as a son, of the child on whom her affections were centred. In the power to make the reader regard the operation of the law as characteristic of the working of society in general, Trollope fell far behind Dickens, despite some intention, in the sub-plots, to show, in other reaches of society, the effects of and the analogues for the law's activity.

The ability of comedy in the sub-plots to relieve the prevailing sombreness in the main story was put to the test here. No doubt the more casual requirements of the serial reader were well served; but the reader of the volume form of the novel was offered insufficient reward for farcical interference with the main line of interest. In *The Small House at Allington* (1862–4) the comic sub-plots had a more specific relation to the main story of Crosbie's jilting of Lily Dale in favour of a De Courcy daughter. The common fictional situations of a man caught between two women, or a woman beween two men, were replicated at differing levels of seriousness and of social position. In the entanglement of Johnny

Eames, a humbler civil servant than Crosbie, with the landlady's daughter, the author's affectionate placing of Eames's immature foolishness and his strength of devotion to Lily Dale combined the pathos with the comedy of perversity in love. At the other extreme, farce in high society showed, with diverting artificiality, Mr Palliser endangering his whole social position by contemplating an affair with Griselda Grantly, now Lady Dumbello: 'making up his mind to forget himself' with her was the paradoxical phrase for a lover's irrationality, in the artificial ambience of Griselda, with her 'unmeaning, unreal grace'. The sufferings of Lily Dale, completely bound to the lover who had deserted her, and of Eames, putting his pursuit of her in jeopardy by succumbing to other attractions, not only contrasted with the ease with which Mr Palliser could accept his dismissal, but suggested a perversity parallel to his, though stronger. Written to amuse, the stories were interwoven so as to prompt thought to a far greater extent than the more pretentious subject of *Orley Farm.*

Nevertheless, the book effortlessly induced the reader to accept that this was life and not the arbitrary fancy of the author. Lily's clear-sightedness in seeing the inconvenience marriage would be to Crosbie and offering him release from the engagement, her moving gaiety when he persisted in it, and her determined devotion when he formed another, were individualized by her mode of speaking, more self-aware than that of other heroines in its pertness, its occasional portentousness, and its honest, humorous expression of the determined will to tyrannize which grew out of her deprivation. There was obviously a conventional pastoral element in the sufferings of a heroine aptly named for rural virtue and seclusion. Yet, like the virtue, the seclusion was not idealized. The Dale inhabitants of the Small House, with all the promise of an idyllic life, failed to fulfil it because of their uneasy dependence on the Great House. The squire there, unable to express his affection for his nieces and their mother, oppressed by the dullness of his solitude, was a fine small realistic portrait: the importance of unspoken attachment subsisting beneath unpromising appearances could be seen movingly in his ability to be hurt by their attitude towards him, as well as in the insight into his needs and nature possessed by his curmudgeonly gardener. The bachelor Earl de Guest, who looked and lived like a farmer, took a sincere pleasure in befriending Johnny Eames, and on occasion acted up to his own position 'somewhat grandly', told a similar unemphatic tale of the limitations of country life, his jerky speech catching his small self-assertions, his dull round of existence well suggested. The charm and clarity of small domestic matters—packing to move house and unpacking again when plans changed—did not match the female novelists in this line, but nevertheless gave opportunities for practical talk to make substantial to the reader the everyday continuance of the emotional lives of the remaining inhabitants of the Small House. The contrast was no doubt too marked between this and Crosbie's misery, without one single alleviation, in his aristocratic marriage. No personage like Dickens's Sir Leicester Dedlock mitigated the satire. It entailed considerable vigour of characterization in impressing the reader with the Earl de Courcy's malice and financial mismanagement and the countess's dodges to deal with it. So Crosbie's marriage, followed soon by separation from his wife, who departed with her mother for Baden, fitted into the chronicle of the De Courcys' decline. The former subject of alliance between the gentry and middle-class wealth continued in the marriage of the second son to a coal-merchant's daughter.

Rachel Ray (1863) made a small comedy, equally suggestive, out of 'the commonest details of commonplace life', Trollope's phrase when sending a copy to George Eliot. Her reply praised their organization 'into a strictly related, well-proportioned whole, natty and complete as a nut on its stem'. The organizing principle came, as in *The Warden*, from the giving and taking of advice. The willingness of Rachel's mother to take advice and put off the girl's suitor, even though her own judgement was in his favour, became echoed in the willingness of her adviser, the rector, Mr Comfort, to take advice in his turn—although protesting against it—from Dr Harford. Harford, at the centre of the story, a former Liberal who had changed his party from dislike of the consequences of the Reform Bill, especially as seen in the activities of the Ecclesiastical Commissioners, had made up his mind that everything was in decay; without questioning the common opinion in the town, he readily gave his judgement against a suitor who would disturb 'the old familiarity of the place'. Common opinion and the commercial plans of the suitor neatly brought into play a wide range of small-town life. Once he was able to take advice himself, from the rector's daughter, who had been prepared to investigate the common talk and find it mistaken, his difficulties disappeared. The plot led to Trollope's kind of simple uncertainty about the characters—whether, in another chain of advice, Rachel's former friends the Tappits would deter their mother from urging Tappit, the suitor's commercial rival, to give in to his terms, and whether Tappit would be able to do so, recognizing his own limitations. His farcical difficulties left scope for the author's manipulation; when things were nevertheless left to take their quieter course the reader's impression of being shown causes and effects in 'common-place life' became strongly reinforced.

The return in *Can You Forgive Her?* (1864–5) to less 'ordinary' people, and to multiple plot in order to extend the import of the novel by internal comparisons, marked the beginning of the development of Palliser's marriage in the sequence which, after a century, outclasses the Barsetshire novels in the minds of most readers. Although, in the *Autobiography*, this series of 'semi-political tales' was said to begin only with *Phineas Finn*, more than two years (and five novels) later, it was in *Can You Forgive Her?* that Trollope became aware of the possibilities of political life as a subject in itself: chapter 42 could use the former jocular imagery of Olympus from *Framley Parsonage*, in order to make a contrast with the actuality, in which 'exciting changes occur which give to the whole thing all the keen interest of a sensational novel'.

As yet, however, the changes were important mainly for their effect upon the Pallisers' private life. Yoking this subject to the plot and sub-plot of a play he had written in 1850, *The Noble Jilt*, gave him two heroines, motherless in contrast to Rachel Ray, but like her troubled by advice about matrimony. One, Lady Glencora, was 'driven like a beast' by advice to safeguard her fortune from the man she loved, Burgo Fitzgerald, and marry Palliser. The other, Alice Vavasour, accepted advice from 'learned ladies' on the importance of the question, 'What should a woman do with her life?' Each heroine was tempted to give up 'safety' for the 'risks' of life with a man of dubious character but undoubted sway. Alice was in no doubt as to the moral superiority of John Grey, whom for most of the novel she rejected; Glencora's hope that if she had been allowed to follow Fitzgerald she would, before he had spent all her money, 'have won even him to care for me' showed a shrewd estimate of him, if not of her own chances. But each was afflicted with the contrast between the safe and predictable course which was within her power and the unknown possibilities of a different kind of life, imagined in terms which recalled the Romantic poets: Fitzgerald suggested Byron in his self-centredness, his personal beauty, and his casual generosity; George Vavasour began his disturbance of Alice's peace with the opinion that

> 'In this world things are beautiful only because they are not quite seen, or not perfectly understood. Poetry is precious chiefly because it suggests more than it declares. ... Now I'm made up of poetry.' After that they began to laugh at him and were very happy.

His opinion seemed to support the feeling which 'tormented' Alice, that in her love for him she 'had had a more full delight' than in the love for Grey 'that had sprung up subsequently'. Her first love, now a 'dream', was not stressed in her attempts to reason with herself, 'How might she best make herself useful'; indeed, her physical revulsion against George, her 'excessive fear' of any 'demand for an outward demonstration of love', was plain; but her strange state appeared in the words 'Bygones will not be bygones. ... One might as well say so to one's body as to one's heart.' This state correlated movingly with 'a special winter's light' on a Westmorland fell, 'where the eye ... sees at the world's end the faint low lines of distant clouds settling themselves upon the horizon'. Although the author endorsed her view that 'it was not her love for the man that prompted her to run so terrible a risk', her dawdling sense of futility, except when repelling George's reasonable demand for some 'outward demonstration of love', or when offering conventional advice to Glencora, suggested a battle with herself below the level of her ambition to be 'the wife of the leader of a Radical opposition'. This 'undefined ambition', which, by making her 'restless without giving her any real food for her mind', appeared slightly ridiculous, was apparently satisfied when the staider man she married, Grey, became a Liberal MP. Yet, despite some obvious didacticism, there was no doubt of the author's sympathy with her state of indecision. It was less intensely imagined than Maggie Tulliver's and the mistakes to which it led were not clarified like hers by the inexorable movement of the plot, which was allowed to resolve itself by melodrama. But, although unable to get the girl quite into focus—

or the George who disturbed her—Trollope had moved a considerable distance beyond the 'complete appreciation of the usual' which Henry James thought his 'great merit'. Of course, things concluded in the triumph of the 'usual' over the dream, in the comforts of Grey's flat Cambridgeshire patrimony, yet there were oddities in this vision of the usual which would have borne elaboration: to Grey, for instance, Alice was 'all he had ever loved with the perfect love of equality'.

Glencora's toying with the idea of adultery—well sustained, like the other 'sensation' motifs which Trollope, in novel after novel, brought to the bar of the usual—yielded characters more fully understood and, in consequence, a better plot. Politics were more completely assimilated into it than ever again: Palliser acknowledged his love by giving up the Chancellorship which had been the goal of his ambition, and Glencora's actual revolt—as distinct from the elopement which she could 'think of' but not carry out—arose from her anger at being, as she thought, spied upon and not allowed to exercise her own judgement—like the hero of the next political novel, *Phineas Finn*, who rebelled at having, when in office, 'no right even to think of independence'. Palliser's sacrifice was moving, but the emphasis fell on the woman's emergence from the impasse with pain—able both to dismiss her lover 'Because I am a man's wife, and because I care for his honour, if not for my own', and to deny that there was any gain in this: 'There is no room for thankfulness in any of it;—either in the love or in the loss. It is all wretchedness from first to last!' Although Trollope lavished praise, in the *Autobiography*, on Charlotte Brontë rather than Emily, he appears to have understood one aspect of *Wuthering Heights*, which may also have suggested Alice's strange moorland epiphany.

Before enlarging two of the constituents from this development of *Wuthering Heights*—'sensation' in *The Claverings* and the stronger sense of place in *The Belton Estate*—Trollope attempted a drab and direct realism in *Miss Mackenzie* (1865), presenting a woman who, in middle life, found herself suddenly with independence and suitors. The economical individuality with which two of her suitors were invested allowed the reader pleasurably to share her doubts about them. The resolution came only after introducing, on the one hand, some of the shop-soiled devices of inferior fiction, and on the other, the master-stroke from *Pride and*

Prejudice, when a mother's efforts to forestall her son's attachment to Miss Mackenzie in fact revealed it to her.

The Claverings (1866–7) offered again two heroines, one ill used in the same way as Lily Dale, the other making a resolute marriage of convenience like Catherine in *Wuthering Heights*, though to a man who left her a prey to his disreputable associates, like the heroine in *The Tenant of Wildfell Hall*. The Lady Julia who attracted the hero away from a placid engagement suffered from the attentions of Mme Gordeloup and Count Pateroff, both of them apparently possessed of secrets which might decisively alter her situation and the reader's view of her. By the time it was discovered that they had no such secrets, pleasure had already arisen from the mere presence of this pair. It was characteristic of the book that whatever appeared to be touched with cheap romance should be brought relentlessly within the rules of the everyday world. Mme Gordeloup, although talked of as a dangerous spy, was able to outwit only the most delightfully foolish of clubmen and the witless Archie Clavering, who, on the model of Shakespeare's Andrew Aguecheek, had believed she would act as intermediary in his suit to Julia.

Much of Julia's story used the breathless repetitions of the sensation novel—'She had been mean enough, base enough, vile enough, to sell herself to that wretched lord'—but the self-laceration was followed by her simpler refusal 'to pretend that she mourned the man as a wife mourns'; and if not a wife, then ... She could gain dignity when forming the facts into the simplest language:

> 'I would not marry him because he was poor; ... it does not make me love him the less now ...'

Her attempt to rehabilitate herself in the real world was even accomplished with the stiffness of a Post Office report: 'I am prepared to take any step that may be most conducive to the happiness of the man whom I once injured ...' There was an over-decisive emotional shorthand for the easy magnification of the result—'she would immolate herself without hesitation', or see her rival 'destroyed without a twinge of remorse'—but the author intervened to question whether the man was worth the passion, at the same time as he pointed out that 'the passion, nevertheless, was there, and the woman was honest in what she was saying'.

The young man, Harry Clavering, had been presented, when losing Julia, as 'somewhat proud to have had occasion to break his heart'; going to meet her for the first time as a widow, he 'liked to think of himself as one to whom things happened which were out of the ordinary course'. But it was the ordinary course all the same and it left him humiliated and wretched, while the novelist drew the professional conclusion: 'if he attempt to paint from nature how little that is heroic should he describe!' The conventional happy ending, brought about by an accident, had more than a hint of mockery.

The sub-plot gave the book some title to inclusion in the Barsetshire 'series': an easy, conventional churchman in Bishop Proudie's diocese, the Revd Henry Clavering, was exposed to criticism by the behaviour of the son he had raised and by the contrast between himself and the curate whom he rejected as a suitor for his daughter. As in the first Barchester novels, parody and farce dulled the edge of the satire upon English life. Mme Gordeloup, for instance, spoke in obvious parody of Dickens:

> 'Look at the street out there. Though it is summer, I shiver when I look out at its blackness. It is the ugliest nation! ... You see that black building,—the workhouse. ... The great fat beadles swell about like turkey-cocks inside.'

The beginning of this passage glanced directly at the number of *Our Mutual Friend* for the month in which Trollope began to write, August 1864—

> It was not summer yet, but ... nipping spring with an easterly wind ... Such a black shrill city, ... such a gritty city ...

The reference to the workhouse repeated the terms of the parody in *The Warden*, where the villain of Mr Popular Sentiment's novel had 'a chin which swelled out into solid substance, like a turkey-cock's comb'. The solid presentation of English heartlessness in its strength as well as its weakness came not in such words, which Trollope casually half-admitted to be those of a spy, but in the less obtrusive simplicities of Sir Hugh Clavering—in his logical self-limitation to his own direct personal interest, his refusal to comfort his wife and his rejection of consolation for himself when their only child died. The novel contained hints in its parts which made the reader wish that Trollope had allowed himself time to consider more carefully how they might be arranged so as really to tell.

In *The Belton Estate* (1865–6) he turned the simplest of his plots, a girl's choosing between suitors, into an experiment in the relation between character and place. The opening of *The Claverings* had established a connection, prompted by *Our Mutual Friend*, between Clavering Park and its harsh owner—'the large, square, sombre-looking stone mansion . . . the parterres, beds, and bits of lawn . . . dry . . . from the effects of a long drought'; such a connection was now developed into the contrasts of a whole plot. The losing suitor, conventional in his feelings and his calculations, came from a house 'with a portico of Ionic columns which looked as though it hardly belonged of right to the edifice, and stretched itself out grandly'; the successful suitor was a farmer of land not noted for its beauty, and the force of his obsession with the girl was made plain in the account of his pursuits: 'As he watched the furrow, as his men and horses would drive it straight and deep through the ground, he was thinking of her . . .'. Both his rejection and his acceptance took place beside 'a high rock which stood boldly out of the ground, from whence could be seen the sea on one side, and on the other a far track of country almost away to the moors'. Similar contrasts present in *Can You Forgive Her?* were of less importance to the whole novel than in the simpler structure of *The Belton Estate*, devoid of sub-plots, except for the slight threat of mystery hanging over the hero at the end of chapter 12, though never carried out.

In *The Last Chronicle of Barset* (1866–7) every ounce of menace was wrung from a small mystery, to make it the centre of Trollope's largest structure to date. Although he later thought of it as his best novel, 'taking it as a whole', he 'was never quite satisfied with the development of the plot . . . I have never been capable of constructing with complete success the intricacies of a plot that required to be unravelled.' The statement implied that the kind of unravelling characteristic of the sensation novel was his aim. Amusingly, Mr Toogood, the attorney, brought forward to play the detective, was allowed, by chapter 42, exactly in the centre of the novel, to happen on most of the clues to the mystery while the author was powerless to reveal that they *were* clues without giving away the solution. Although it was the intricacies of the character of the supposed culprit that interested him, he assumed that these would only be set in satisfying order by the external details of what had actually happened and why. In his own development of the genre he so managed the revelation of the central character, Mr Crawley, as to show him in doubt precisely about these himself.

Josiah Crawley was developed from the 'strict, stern, unpleasant man' of *Framley Parsonage*, after the eighth of Trollope's clerical sketches, 'The Curate in a populous Parish', had appeared in the *Pall Mall Gazette* on 20 January 1866. The disappointment, there, of the ill-paid curate, 'a soured and an injured man', and his 'irrevocable misery and distress' if he should marry, became in *The Last Chronicle* the near-madness of a brooding and resentful enemy of himself and his family. The whole imbroglio about a missing cheque arose from the extremes of his pride and his humility; the denouement came in the discovery that these extremes were indeed the cause of all. Pride, which had hurried him away with 'no time for explanatory words' about the money he so reluctantly requested and accepted, had prevented him from learning what money it was; later, out of humility, he had believed the word of the Dean about this rather than his 'own memory'. As with such situations in actuality, behaviour which was quite comprehensible when close to the event became less so on looking back.

This plot strained to the limit, however, the organizing principle which Trollope claimed, at the end, for all of the novels concerned with the diocese of Barchester, namely the presentation of 'the social and not the professional lives of clergymen'. The case of Mr Crawley was so intimately involved with his professional life as to involve also his personal religion. Where George Eliot might have seized the opportunity, Trollope walked round it, though with some skill. Crawley's humility and pride drove him, movingly, to blame himself and also 'to pay a tribute to himself for the greatness of his own actions'. He tended to see his powerlessness and his 'memory of former strength and former aspirations' mainly in terms of Classical and Old Testament heroism—and the second of these as interpreted in the light of the first, by Milton—with only passing reference to the hardships of St Paul and none at all to intermediate examples which might have been thought even more authoritative for him. In a comedy his eventual reward might have been appropriate, a living of small duties which had been adequately carried out by the ageing Mr Harding and would allow his successor plenty of time for Pindar. Such a resolution was harder to absorb into a sequence so near to tragedy. The oblique light it cast on old Barchester, which could so accommodate Crawley, represented a reversion to the mode critical of the professional as well as the social life of the clergy with which *The Warden* had begun.

The plots disturbingly woven in with the main one concerned strange emotional complications, some of them expected, like the flirtatious fidelity of Johnny Eames, some completely unexpected, like the development into pathos of materials which had earlier been those of farce: the marital relation of the Proudies now showed new inner possibilities, in the bishop's listless humiliation, Mrs Proudie's sufferings at his exclusion of her, and her apparently wilful shutting of herself up, 'her heart . . . too full for speech', when convinced she was hated. Even her death from a physical 'heart complaint' threw new light upon her: 'We suspected it, sir, though nobody knew it. She was very shy of talking about herself.' Other characters from earlier novels appeared in a fresh light; the archdeacon was moved by the sight of the bishop, at Mr Harding's funeral, 'looking old and worn,—almost as though he were unconscious of what he was doing'—to resolve 'that there should be peace in his heart, if peace might be possible'. He had already appeared a less simple authoritarian when won over by a moment's acquaintance with Grace Crawley, whose engagement to his son he had relentlessly opposed. At the end he was left uncertain whether, compared with Mr Harding, whom he credited in death with 'the spirit of a hero', he himself lacked guile or even feared God. This was turning the spotlight on to his professional life indeed. The penultimate chapter returned to his social position in the old, apparently unchangeable Barchester, to 'the temporalities of the Church' as the reason why 'the Church was beautiful to him', and to his subjugation of Mr Crawley with the assertion that 'We are both gentlemen'—and a volume of the sermons of the previous bishop whose benevolent, lax rule had been the strongest single cause of the action in the first Barchester chronicles. Neither Ecclesiastical Commissioners nor evangelical appointments had made more than a temporary stir.

The incongruous City underplot was detached completely from Barchester, but it had a similar increased concern with what Meredith called the inner 'machinery'. When presented as farce, the result bore a curious resemblance to *Sandra Belloni* (published three years earlier)—for instance, in the case of Mrs Dobbs

Broughton, neurotic wife of an alcoholic money-lender, talking and thinking endlessly about herself while pretending both love for another man and self-sacrifice in his interests. Trollope could see more than he was prepared to work to assimilate into this curious book.

In the decade from 1867 to 1876 he was to attempt a clearer focus in each novel. In most of them this meant an emphasis on the political and social issues foreshadowed in *The Bertrams*, *Dr Thorne*, and the sub-plot to *The Last Chronicle*. The time of the writing of the latter, the first nine months of 1866, had been a time of great political excitement—the defeat of Russell's Reform Bill and the resignation of his government in June, the big Hyde Park demonstration in July and other demonstrations in large cities, addressed by John Bright, the radical member for Birmingham. Trollope accepted the editorship of a new monthly magazine, *St Paul's*, which, according to his introduction to the first number in October 1867, would be political, since the study of politics was of all studies 'the first and the finest', and would contain fiction because 'the preaching of the day is done by the novelist'. For this purpose he produced *Phineas Finn* (1867–9). While he was writing it, from November 1866 to May 1867, the new government brought in and passed its Reform Bill. He was not stimulated, like George Eliot when contemplating this period, by the long perspective of change, but by the immediate foreground—not only the question of the franchise but the opposition of a group of Liberals, the 'Adullamites', which had prevented Russell's Reform Bill from passing, and the matter of Russell's successor in any future Liberal government. At the time of writing, the question of the franchise was still *sub judice*; yet it was unlikely to remain so and the novel would have to appear after it was settled. There was also the problem that immediately topical detail would risk turning the novel into a pamphlet. Meredith and George Eliot gained attention for more general issues by means of a setting in a past time which resembled the present in certain salient features. Trollope's solution was to set the action in the time of writing, to keep something like the alignment of parties then obtaining, and to leave the leaders recognizable while altering their names. He chose, skilfully, to keep the issue of Reform going in the background, but to place in the foreground as the chief concern of his hero a matter of importance to himself in *The New Zealander*, the independence of the individual member in the service of truth and his country rather than of his party. This was no matter of scruples far removed from actual politics. The Adullamites had recently exemplified it, and in any case, in the 1860s as much as half the time of any one session might be given to non-government business which did not demand party loyalty.

Trollope had much quiet amusement at the expense of franchise reform. Phineas Finn was devoted to it, while himself standing successively for pocket boroughs which were abolished in the course of legislation, but he developed in political understanding as a result not of this inconsistency but of his attempt to take an independent stand on an Irish issue. As criticism of parliament, this part of the story, while high-minded, was hardly spell-binding. The basic excitement arising from the fact that the concern of politics is *power* seemed missing. Trollope saw the House of Commons as a most exclusive club. There, unlike at the Garrick or the Athenaeum, he was an onlooker. His descriptions of debates in the House, adequate for a reporter, lacked the immediacy demanded by the reader of novels; Mr Monk, the hero's principal mentor, was a grave sermonizer, of little interest in himself.

Finn's social life, upon which the plot relied for most of its movement, was so handled as to reduce both his personal and his political stature. He was in love with four women, not quite one after the other. Unavoidable moments of emotional double-dealing were exposed in a man at first apparently unaware of his own mixed motives. In love, for instance, with Lady Laura Standish, and recognizing that, as a man without means, he would need to be in office if he were to be able to support a wife (for members were not paid), he 'then'

> remembered that Lady Laura was related to almost everybody who was anybody among the high Whigs. . . . he thought of it because he loved her; honestly because he loved her. He swore to that half a dozen times, for his own satisfaction.

The steady flatness of the style reduced the inflections of irony, especially when the characters' naïve cogitations were not easily distinguished from the author's commentary. To the extent that Laura became Finn's political guide she extended the mockery: her finding him a second parliamentary seat on the loss of his first underlined how little he was doing and how much was done for him. The matter on which Meredith was to base *Beauchamp's Career* ten years later, the contrast between the sexes in political power and in the ability to use such as they had, appeared in Trollope's contrast of the hero's uncertainties with the women's decisiveness, Laura in making a marriage of convenience, Mme Goesler in her own offer of her hand. But the near-nullity of the hero, which no doubt sufficed for the male side of the contrast, was inadequate to serve as the focus of female certitude. Both women had a confidence in his potential which the reader could not share.

When Finn eventually decided to marry an Irish girl and to stand up for Irish tenant-right against his own party, he seemed not to be finding his feet but helping the writer to a conclusion. For despite Finn's occasional feeling that his life in England was a mere 'pretence', the women he met there made it the strongest part of the novel, and his life in Ireland had not been substantial enough for his feeling of being a 'cheat' in London to be adequate reason for his obvious hollowness. At the end all he had left was his Irish loyalty, insisted upon, though of little interest as fiction: his political 'independence' had not even freed him from his party, since he received at their hands the Irish appointment which enabled him to marry. Trollope's avoidance of sub-plots here (except in his development of the main women beyond the demands of their immediate relation to Finn) had thrown into relief the problem of making a more unified plot tell an unconfused tale.

The moving presentation of the tyranny of Lady Laura's husband led to the close observation in *He Knew He Was Right* (1868–9) of a quarrel in which both husband and wife might find it impossible to give way. There was considerable monotony and repetition in the demonstration, but something approaching tragedy in the concluding spectacle. Trollope regarded the book as a failure (in the *Autobiography*) because he had not fulfilled his intention of creating sympathy for the husband, led astray by mere 'unwillingness to submit his judgement to the opinion of others'. This strange view of the novel's purpose explicitly recognized the importance for the author of the giving and receiving of advice as a subject. Significantly, in this novel he at last allowed himself favourable presentation of a newspaperman—even though ready 'to instruct the British public of tomorrow on any subject, as per order'—and made fun of scepticism about the parallel function of the writer of novels—'Who would ever think of learning to live out of an English novel?' It was true in this case that the friends of the husband, Trevelyan, were ready to urge that his quarrel had no substance. But his obsession was more active than a mere refusal to listen to them; the force of its presentation, though wavering, was strong and its analysis often subtle. It reached its climax in the perversity of wishing for what he feared, proof of his wife's infidelity. He 'was continually telling himself that further life would be impossible to him, if he, and she, and that child of theirs, should be thus disgraced;—and yet he expected it, believed it, and, after a fashion, he almost hoped it'. This 'hope of the insane man, who loves to feed his grievance, even though the grief should be his death', when yoked to the self-righteous intransigence of his wife, formed a bleak and barren situation, hardly made more tolerable by selfconscious Shakespearean echoes in Trevelyan's speech, merging into claptrap about 'the terrible hand of irresistible Fate' in his thoughts. Some of the most memorable things came in the imagining of the man from the outside, the ease with which he gave up his child after vehemently defending his own possession, his appearance, unshaven and pale, in loose slippers and an old dressing-gown in the Italian heat, his solitary observation of the dragonflies by a muddy stream. It was remarkable, though prolonged beyond the reader's patience.

In the comedy and farce of interwoven stories, other characters who 'knew' themselves to be right were contrasted in so far as they had the ability to love and to give in—strong-minded provincial spinsters like Miss Jemima Stanbury and her niece Priscilla, or a theoretical American feminist who approved of the *Saturday Review*'s opinions, no doubt about 'The Girl of the Period'. Some of the farce led to parody of 'sensation'—a parson's jilted fiancée secreting sharp knives and displaying one as she sat on the lid of her trousseau while he declared he 'had better take poison and have done with it'.

Nevertheless Trollope was not above seriously employing conventions of the sensation novel and of the thesis novel to which it was nearly related. In *The Vicar of Bullhampton* (1869–70) a murder was mysteriously connected with the members of a family on whom centred a thesis about what the Preface called 'castaway' women. The *Autobiography* made 'raising a feeling of forgiveness for such in the minds of other women' the chief object of the novel. Sentimental means to this end were rather too apparent—'the lost one lay asleep there, with her soft ringlets all loose upon the pillow ...'—but it was characteristic of Trollope not to blink the envy tinging the affection of her plain sister or the attraction towards the 'castaway' of the Vicar himself. He was hardly a representative clergyman, though Trollope obviously wished he were, in his enjoyment of a fight, his accusing the marquis who traduced him of 'malice', and his replying, when asked if 'malice' were not a very strong word, 'I hope so.' His cheerful toleration of dissenters turned the content of chapter 6 of *The New Zealander* into vigorous comedy.

The tendentious commentary on the action claimed not only that clergymen are men but that 'We are all men', with natures and demands which must be acknowledged. The part of the novel concerned with Mary Lowther, which repeated the plot of *The Belton Estate*, allowed the physical delight of her love to appear; the commentary emphasized that 'Nature prompts the desire, the world acknowledges its ubiquity ... but it is required that the person most concerned should falsely repudiate it, in order that a mock modesty may be maintained ...'. 'Saturday Reviewers' who attacked 'the forward indelicacy' of 'the girl of the period' were blamed. The young women whose conduct was in question were, however, very slightly individualized, the outcast indeed hardly at all. But the latter's father, whose stubborn silence and aggressive manners marked not impassibility but affection, was memorably glimpsed, even though Trollope could not always find the right language for him.

The succeeding short novel, *Sir Harry Hotspur of Humblethwaite* (1870–1) took up the possibility that the kind of love exemplified in *The Belton Estate* could become a trap. The physical presence of the man, his voice heard as 'music', his touch felt 'as had never been the touch of any other human being', proved calamitous when the story showed him to have no other sort of compatibility with the heroine.

From now on Trollope had increased recourse for his plots to sixteenth- and seventeenth-century drama, in which he was unusually well read. For *Ralph the Heir* (1870–1) he combined suggestions from Shakespeare's *King Lear* (directly misquoted) and Middleton's *Michaelmas Term*, the first for Squire Newton's preference of his illegitimate son, the second for the story of the legitimate heir. These were supplemented with a livelier political plot based on the author's experience as a Liberal candidate in 1868. It was a temporary and unsuccessful return to the earlier mode of domestic realism from the sensational materials which were to characterize the big novels of the rest of his career.

In *The Eustace Diamonds* (1871–3), the plot was managed like a sensation novel, as if it contained a mystery to be resolved. Although Trollope repeatedly expressed his determination not to depend on such a plan—'He who recounts these details has scorned to have a secret between himself and his readers'—he always left the reader uninformed about how what was divulged had come about. But facts and reasons about the status and fate of the diamonds also dramatized Lady Eustace's life of systematic lying. External events again and again showed her lies coming true. Before her first marriage, in order to secure some jewellery, she lied about marriage having been proposed to her, but soon it was. She lied too about her age; that she soon ceased to be a minor was not mere commonplace, for it was carefully underlined that the time during which she attained her majority and began her course of lying was the time which had deprived her of her 'heart'. 'It had become petrified during those lessons of early craft ...' This was the 'one truth she could not see', even when, while coveting the attentions of Frank Greystock, she recognized the virtues of Lucy Morris, his fiancée. And repeatedly she was presented as drawn comically near to recognizing what was true by her very pursuit of what was false. She admired Byron's Corsair, no doubt for his devotion to Medora and his daring in 'a thousand crimes', but there was no sign of her appreciating the passage Trollope had had in mind when writing of her heart as 'petrified':

> His heart was form'd for softness—warp'd to wrong;
> Betray'd too early, and beguiled too long;
> Each feeling pure ...
> ... sunk, and chill'd, and petrified at last.

Her hilarious misunderstanding of Shelley's *Queen Mab* (helped by his confused syntax) missed the emphasis, a little further on than she read, upon virtue and 'the unfailing consciences of men'. She took a vague pleasure in Tennyson, but Greystock had to point out that the 'heart' for which she admired Lancelot and praised himself might be no more than 'a talent for ... running away with other men's wives'—or, he implied, fiancées, for she was engaged to Lord Fawn. Moreover, in *Elaine* where she had been reading of Lancelot, she could also have read about his 'nine-years-fought-for diamonds' which were 'the kingdom's not the king's—For public use ...'. To Trollope, in chapter 35, this avoidance of the truth resembled that of a novelist who would 'depict a hero ... absolutely stainless, perfect as an Arthur, ... above all, faithful in love', rather than 'tell of the man who is one hour good and the next bad', like Greystock. Moreover, he believed this 'superlative vein' of idealizing fiction to derive from the language of commerce: 'name the thing ... as it is, and the market is closed against you. ... No assurance short of A1 betokens even a pretence to merit.' Yet the ability to 'name the thing as it is' and so present 'the true picture of life as it is' was essential if the moral aims of the novel-writer were to be fulfilled: only if he could 'show men what they are' could he show 'how they might rise, not, indeed to perfection, but one step first, and then another on the ladder'.

Lizzie Eustace was betrayed by 'the desire to make things seem to be other than they were. To be always acting a part rather than living her own life was to her everything.' Her life, beginning as the precariously bedizened daughter of a spendthrift admiral, had taught her to like lies, 'thinking them to be more beautiful than truth'. Much comedy was extracted from this, but also much pathos:

> ... she did perceive, in some dark way, that, good as her acting was, it was not quite good enough. Lucy held her ground because she was real. You may knock about a diamond, and not even scratch it; whereas paste in rough usage betrays itself. Lizzie, with all her self-assuring protestations, knew that she was paste ...

The comic story growing from her tenacious hold upon actual diamonds appeared, then, as the pathetic analogue of her half-acknowledged hunger for what was genuinely valuable.

There was some forcing on the author's part to show 'as it is' the kind of life which her 'guiding motive' yielded her; it was not presented as inevitable that she should pick up with quite such a dreadful crew of adventurers as Lord George Carruthers, Mrs Carbuncle, and her 'niece' Lucinda. They seemed to belong to the more mordant sections of *The Last Chronicle*, but now they were enveloped in the atmosphere of lies, making every statement about them seem doubtful. The savage farce of Lucinda's hatred of lies, her stubborn refusal at the last minute to follow through to marriage the lie she told at her engagement, and the effect on her associates, after all the bargaining over the presents, reached its climax just as Lizzie Eustace faced the police court. The Jonsonian falling out of rogues (by 1876 Trollope was writing in his copy of Jonson, 'The Alchemist is certainly a very great work') was played out against the indefatigable romantic innocence of Lizzie's hopes that even at the very last minute Lord George might 'be the Corsair still'. As a result the worst of the adventurers, Mr Emilius, could come forward to claim her—not a Corsair who would act out Byron's verse, but a fake clergyman who could read it so well as to convince her 'for the moment that after all, poetry was life and life poetry'. Trollope denounced him with unnecessary iteration: his ludicrous, flowery proposal was enough; her only reservation arose from doubt whether he was quoting *Don Juan* or the Bible. The sustained comment on the substitution, in life and in literature, of beautiful lies for actuality gives the novel connections both with Thackeray (Trollope called Lizzie Eustace, in chapter 3, 'that opulent and aristocratic Becky Sharp') and with the M. E. Braddon of *The Doctor's Wife*. *An Eye for an Eye*, which transposed false romanticism into the key of melodrama, for it depended on the central character's romantic view of Ireland, was written next, though not published till 1878–9.

Materials taken over from the sensation novel served an even more didactic purpose when the political disillusionment of Phineas Finn was continued in *Phineas Redux* (1873–4). Bigamy and murder by Mr Emilius now reflected on the onlookers' ability to remain unmoved by the horror of the events which attracted and scandalized them. Phineas Finn, brought back (*redux*) from Ireland into Westminster politics, was again relatively passive, this time in the face of an attempt on his life, his arrest and trial on a charge of murdering a political rival, and the continued attentions of two of the women from the earlier novel. Again the emphasis fell on his ability to react to the initiatives of others rather than to act decisively himself. In parliament he had little to do but observe and feel slighted; in the trial the more active party was Mme Goesler, who secured his heart by seeking in Bohemia the crucial evidence for his acquittal. Lady Laura's continued love for him, breaking out with a greater intensity than anything else in the book, further reduced, by comparision, the force of his emotions, on which, in default of his actions, the plot depended. Had the novel been written in the first person this could have been taken as modesty and the story might have developed into that of the hero's self-recognition, perhaps long after the events. But Trollope was content to keep attention awake with uncertainties about the outcome without attempting to penetrate very far into Finn's character. Nevertheless, his experience as the victim of rumour and circumstantial evidence, culminating in psychological illness on release from prison and trial, was remarkably imagined—his doubt of his own identity, his attraction towards the scene of the murder, his approach to tears at being well treated. But the illness was temporary, its effect on his view of politics curiously bathetic.

Bathos might have appeared much sooner had not Finn's reactions to politics in the first half of the book been overshadowed by the author's own extraordinary creation of parliament at the time of Daubeny's minority Conservative government. The circumstances in which Disraeli, governing without a majority, had had his Reform Bill passed in 1867 were amusingly echoed in Daubeny's presenting a bill to disestablish the Church of England. The emphasis fell on the conjuring of this Cagliostro. Daubeny's clinging to power, 'indifferent alike to the Constitution, to his party, and to the country', exaggerated only a little the situation just before Trollope himself had stood unsuccessfully for election as a Liberal in late 1868; there was less exaggeration in Gladstone-Gresham's 'earnestness' and 'ferocity', Daubeny's 'air of affected indifference', and the beautiful parody of Disraelian eloquence:

> 'The period of our history is one in which it becomes essential for us to renew those inquiries which have prevailed since man first woke to his destiny . . .'

Such things were much more vigorously done than in *Phineas Finn*, but they hardly helped the hero's stature when his own speech in the House merely echoed his chief and he was soon found echoing the author's words about Daubeny as conjuror. It was a remarkable *tour de force*, but it failed to make dramatic Finn's disillusionment with politics for other reasons than his exclusion from office (once scandalous talk about him 'had injured his prospects with his party'). In his strange state after the trial he imputed to himself an earlier faith in the idealism of the Liberals for which there had been no evidence at the time, in order to contrast his present feeling: 'It has all come now to one common level of poor human interests.' If this change had been established, it would surely have been the foundation for an open-eyed political career, balancing patriotism and party. Finn's resignation and re-election, the offer of office under Mr Gresham and his refusal of it, and the final suggestion that, now that he had accepted Mme Goesler and her money, he might eventually also accept office 'if opportunity offers ... under better auspices' did not make a sequence which was firm in its connections. The reader was apparently to sympathize with Finn's disillusionment with party to the extent that only a coalition under the completely high-minded Palliser, now Duke of Omnium, would furnish 'better auspices'. It was still a question, however, whether Finn was substantial enough for the role; he had suffered much but acted little and, what was worse, he seemed to have learned little about the realities of politics , though the authentic feel of them had been communicated , despite the satire, in the first half of the book.

It was typical of Trollope's fecundity that, before embarking on his novel centred on the prime minister (of a coalition government) committed to a Liberalism 'of lessening distances—of bringing the coachman and the Duke nearer together', he should write his own version of *Felix Holt the Radical* in *Lady Anna* and his version of *Our Mutual Friend* in *The Way We Live Now*, taking two months over the first and only six months over the second. *Lady Anna* (1873–4) was set back, like *Felix Holt*, into the past—'fifty years ago' to George Eliot's thirty. The radicalism important to its plot was of the abstract kind associated with the Lake poets—the young radical tailor engaged to Lady Anna was brought actually to consult the ageing Wordsworth. Like Felix Holt's Esther (the parallel was noticed at the time in the *Saturday Review*), Anna refused a match in accordance with her 'lineage'. The tailor, whose 'grand political theory' was 'to diminish the distances, not only between the rich and the poor, but between the high and the low', appeared patently more magnanimous than the Tories and the aristocrats ranged against him. It was characteristic of Trollope, however, to describe him as 'sullen and tyrannical', hating the domination of others, but 'prone to domineer himself', and his enemy, 'a violent Tory', as 'thoroughly humane and charitable'. Typically, too, actual events qualified even these descriptions. Anna and her husband were sent to Australia, where 'he became perhaps a wiser man'.

The Way We Live Now (1874–5) started from the concern of the social novelist with the over-valuing of success in *The Bertrams* and of commerce in *Dr Thorne*. There his comment had been, 'Buying and selling is good and necessary; . . . let us hope that it may not in our time be esteemed the noblest work of an Englishman.' By the seventies 'it seemed that there was but one virtue in the world, commercial enterprise'. Setting up house in Montagu Square, in that region 'between Portland-place and Bryanstone-square' inhabited by Mr Dombey, had apparently moved Trollope to take up the topic of financial speculation which had been an important strand in one Dickens novel in each of the three preceding decades. On the model of *Dr Thorne*, the gentry were shown eager to share the fruits of commerce offered by a pair of sharpers from abroad, Fisker from California and Melmotte from New York via the Continent. The promotion of a 'grand' railway into Mexico, an enterprise such as was normal beyond the Atlantic, caused a comical stir in what another visiting American called the 'stupid tranquillity' of England.

Trollope's extant plan shows he had intended his 'chief character' to be the novelist, Lady Carbury, an admirer of Melmotte and 'thoroughly unprincipled from want of knowledge of honesty'. As presented in the novel, her work resembled his own in being carried out 'from day to day . . . with a firm resolve that so many lines should be always forthcoming', and subjected to the same warning as he had once received from a publisher, 'Your historical novel is not worth a—.' In her fiction, however, 'she considered that she could best deal with rapid action and strange coincidences' and she courted the praise of the press, which gave or withheld it, like its praise and dispraise of Melmotte, for dishonest reasons of its own. So the danger that dishonesty might become accepted as the norm was underlined by reference to the corruption of two important public organs of advice, fiction and the newspapers.

Although in the event this became a subsidiary theme as Melmotte himself took more and more attention, comedy still arose from the effect that he and his admirers exerted on characters and plots which Trollope, as the honest workman, had handled before and from allusions to the 'rapid action' of the kind of novel Lady Carbury aspired to write. A weak young man, for instance, caught, as in *The Claverings*, between the love of two women, one young, the other older and experienced, became Paul Montague disentangling himself not only from Melmotte but from Mrs Hurtle, an American experienced in the violence of Oregon and Texas, and an admirer of Napoleon, to whom she likened Melmotte. Her frustrated wild-cat vehemence, her 'silken hair, almost black', the mystery about her married state, and her handiness with weapons—the table knife upon which she clenched her fist when Montague broke with her, the pistol which 'luckily for his comfort she had left . . . in her bedroom', the horsewhip she threatened to take to him—recalled Aurora Floyd. The mysteries of Mrs Hurtle were parallel to those of Melmotte's origins and previous activities: was he French, or the son of an Irish coiner in New York, Melmody? Was his daughter, whom English aristocrats were keen to marry as his heiress, illegitimate? His reputed large plans resembled those of Count Fosco, of equally shady foreign origins.

The importation of violent short cuts to prosperity by characters like these, whose domestic arrangements were anomalous, was the sort of subject that had interested Trollope since *Barchester Towers*. Now they disturbed the peace of mind of a group of the gentry in London, Lady Carbury, her dissipated son Sir Felix, Lord Nidderdale, and Dolly Longstaffe, son and heir of an impoverished squire 'specially proud of his aristocratic bearing'. As the foreigners preyed upon these typical personages, so these in turn battened upon their own compatriots outside London: Sir Felix, the farcical Byronic misleader of a country girl, was parasitic, with his mother, upon the head of the family, Roger Carbury, the half-comical representative of those who had 'been true to their acres and their acres true to them through . . . Commonwealth and Revolution'. Like the chinese-box construction of *The Woman in White*, the plot recoiled from the country back upon London and so back upon the foreign intruders: Melmotte's downfall began with a country property which he tried, by forging Dolly's signature, to filch from him. The whole was managed as Jonsonian comedy, complete with multiple ways in and out of the thieves' den in the City. The magniloquence of their boasting and of their victims' hopes had its analogue in the advertising of Melmotte's success by rumour, the press, and his own elaborate entertaining of English and foreign royalty.

The satire extended to politics, which Melmotte, in his anxiety to become thoroughly assimilated, aspired to enter. His election as 'the head of the great Conservative mercantile interests of Great Britain', introduced into the house by the leader of the Conservative party himself, turned some of the satire obliquely against Disraeli, who was in power by the time monthly parts of the novel began to appear. Melmotte's candidacy was connected not only with 'unprecedented commercial greatness' but with vaguely imperialist plans 'to open up new worlds, to afford relief to the oppressed nationalities of the over-populated old countries', with his 'fleet of emigrant ships', his 'contemplated line from ocean to ocean across British America', his company for 'a submarine wire' so that 'England need be dependent on no other country for its communications with India', and his 'philanthropic scheme' for a concession of British territory 'on the great African lakes'. The suggestions of this all pointed towards the Disraeli who had spoken at the Crystal Palace, in mid-1872, in favour of 'reconstructing as much as possible our Colonial Empire, and of responding to those distant sympathies which may become the source of incalculable strength and happiness to this land'. Against this the Liberals in the novel could produce only the results of their investigation of Melmotte's dubious character and career. The election took place against a background of their charges and his own loss of prudence as 'crisis heaped itself upon crisis' in his proliferating financial affairs. The working classes were behind him—'Whom had he robbed? Not the poor.' After his election there was an idea that he might become

> as it were a Conservative tribune of the people,—that he might be the realization of that hitherto hazy mixture of Radicalism and old-fogyism, of which we have lately heard from a political master, whose eloquence has been employed in teaching us that progress can only be expected from those whose declared purpose is to stand still.

As much as any of the more obviously political novels, this one had some title to belong to that series.

But a curious sympathy with Melmotte in his increasing isolation made him more than the butt of satire and even admitted the reader to his rudimentary inner life. After his death (a last-minute decision on the author's part) the book modulated to a blander form of comedy, at some risk to its satirical bite. 'All his proved liabilities' were met, suggesting he had lacked not assets but time. The narrative ceased to support the anti-Semitic prejudices against his associates: Mr Brehgert, especially, accepted without resentment both financial misfortune and the refusal of Georgiana Longstaffe to honour her engagement to him; the demonstration of his good sense and good humour suggested that the loss was hers. But despite the lessening of the attack in the last two instalments (ten chapters), the main charge stood that dishonesty was 'no longer odious'. The charge had already been made at the end of *The New Zealander* and related to the then recent suicide of John Sadleir, the swindler (out of whose 'precious rascality' Dickens admitted he had 'shaped Mr Merdle'). The happy ending, once the dubious foreigners were either dead or gone to America, seemed to support the opinion the bishop offered when Roger Carbury quoted Horace's fourth Epode against Melmotte: 'The world perhaps is managed more justly than you think, Mr Carbury.' To Trollope (in a letter) this made the satirist's activity more not less urgent, for he warned of dangers in their beginning:

> It is the proclivity and not the depth of sin which he handles. No satirist will redeem a man from the lowest pit,—only the man who is going thitherwards.

After the general election of February 1874 had made Disraeli prime minister, Trollope began a novel in which parliament would be exposed to satire by the unfitness for its manoeuvres of his ideal gentleman. As head of a coalition government, Palliser showed how his probity could be mistaken for a 'pretentious love of virtue', like that for which Emily Lopez, in the sub-plot, could easily be found irritating by the reader. Each of these two was married to a person whose conduct they did not approve but could not openly deplore. The one marriage threw a strange light upon the other, not only from such parallels but from their more obvious links through the Duchess's sympathy for Emily's personable but unscrupulous husband. The Palliser marriage gained from having been contemplated over the ten years since *Can You Forgive Her?* Although it had not been prominent enough to weld the political novels into one, Trollope had kept it in view—in the Duchess's adoption of advanced political opinions ('making men and women all equal'), in her fame for doing 'exactly as she ... pleased', in the 'caprice' of her support for attractive young men like Phineas Finn. *The Prime Minister* (1875–6) showed the marriage under strain from her liveliness and ambition. Her personal voice irradiated with individual humour the criticism both of parliament and of women's exclusion from it.

> 'They should have made me Prime Minister, and let him be Chancellor of the Exchequer ... I could dole out secretaryships and lordships, and never a one without getting something in return. I could brazen out a job ... make myself popular with my party, and do the high-flowing patriotic talk for the benefit of the Provinces. A man at a regular office has to work. That's what Plantagenet is fit for. He wants always to be doing something that shall really be useful, and a man has to toil at that and really to know things. But a Prime Minister should never go beyond generalities about commerce, agriculture, peace ...'

Her clarity about her own position, in her own idiom, put blood into the more abstract issues, as Phineas had signally failed to do, by constituting a story in themselves, touching, and ironical when connected with her earlier history. Her attachment to Palliser was still romantic, but in an older sense:

> '... there is a dash of chivalry about him worthy of the old poets. ... And he has a much higher chivalry ... He is all trust, even when he knows that he is being deceived. He is honour complete from head to foot.'

Yet her reply to his claim that 'A man's wife should be talked about by no one', had been, 'That's high-foluting, Plantagenet.' And she could see his over-scrupulous sensitivity as 'simply a disease'.

As with Mr Crawley's religion, Trollope eloquently side-stepped the details of the Duke's politics. Issues which should have been made more specific by the character's profession and experience remained abstract. As Crawley had exulted in ancient exemplars of tragic resistance, so the Duke, when opening his mind to Finn about politics, recurred not to his experience as a Chancellor of the Exchequer but to more distant matters. These would have been characteristic of the radical tailor in *Lady Anna* if the qualification had not quite removed them from practicability:

'You are a Liberal because you . . . would still march on to some nearer approach to equality; though the thing itself is so great, so glorious, so godlike,—nay so absolutely divine,—that you have been disgusted by the very promise of it, because its perfection is unattainable.'

The Duchess's mockery when told they had been talking politics—'Of course. What other amusement was possible? But what business have you to indulge in idle talk . . . ?'— led them from 'cloudland' back to 'the realities of the world'. But, 'in the realities of the world', the Duke resisted having any distinct policy when in power; decimal currency, the one issue upon which he had taken action himself, had been mildly mocked by the novelist, whose sympathies were obviously more with Palmerston ('Oh, there is really nothing to be done') than with his successors. Only a very young man, Emily's brother, thought he knew what had to be done—and he was satirized as believing 'that he had gone rather deep into politics, and . . . had the great question of labour, and all that refers to unions, strikes, and lock-outs, quite at his fingers' ends'. This was all topical enough before the 1874 election, when Gladstone's government was insecure and Disraeli was making his name with the catchwords of social reform as well as of Empire. The Duke's coalition, of 'politicians . . . really anxious for the country' and prepared in its interests to forgo party, represented one possible solution to the impasse of 1873, when Gladstone had resigned, since the House was evenly divided on his Irish University Bill, but Disraeli had refused to form a government. Trollope's point, 'A plague on both your houses', does not appear to have been taken—understandably enough, by a public no longer inclined to believe that there was 'nothing to be done'. The book's poor press, said a note to the *Autobiography*, 'seemed to tell me that my work as a novelist should be brought to a close'.

Nevertheless, the novels of his last eight years, although they tended to ring the changes on earlier themes, formed a vigorous and varied group. The first to be written, *Is He Popenjoy?* (1877–8), returned to obviously earlier interests, to a county which was brother to Barsetshire, and to a London where the heroine's behaviour could resemble that of Glencora with Burgo Fitzgerald; other earlier constituents, such as satire of feminism and a background of Italianized aristocratic villainy, had less point here than in their former appearances in *He Knew He Was Right* and *Lady Anna*. In *The American Senator* (1876–7), a visitor's rational criticism of English life was intended to be refuted by the close view of one undistinguished rural society. The skill shown in the close view, especially when focused on hunting, so far exceeded that shown in the presentation of the Senator and his opinions as to render them unnecessary to the reader's pleasure. Trollope was much more stimulated, in *The Duke's Children* (1879–80), by the development of the argument of the political novels, patriotism versus party loyalty, in a situation where memories of the past made the present into a test for the chief character. Palliser was imagined as widowed when the desire of his children to marry out of their rank brought into collision his two convictions, that 'the highest duty of those in high rank was to . . . elevate those beneath them', and that, as a pre-condition, they were 'required . . . to maintain their own position'. His daughter's love for a commoner brought reminders of his own marriage which made the matter a test of his own inner quality as well as of his understanding of these two duties. Attempting to reconcile them too simply, he backed a pair of possible fiancés for his children who seemed to come out of previous books and were all the more unsuitable for that—farcically in the one case, Lord Popplecourt for his daughter, and pathetically in the other, Lady Mabel Grex for his son. For Lady Mabel, the character of greater interest than either of the children, Trollope re-created the opening situation of *The Claverings*, while making her both cleverer than Julia and more given to melancholy, more capable of duplicity but finally unable to deceive herself. Her inability, at first unrecognized, to conquer the love she had tried to give up inhibited her from carrying through any of her plans to marry for wealth and position. In particular, her own perverse forestalling of her plan to marry Silverbridge, the Duke's heir, revealed her as 'a manufactured article', lacking spontaneity by comparison with an American, Isabel Boncassen. The honesty and vitality of Miss Boncassen, triumphing over the Duke's disapproval of her as a bride for his heir, reproduced the subject of *Framley Parsonage*. This prevented a less derivative action in which the girl might have had to stand by the declaration, made to her family, though not to her lover: 'If he loves me well enough to show that he is in earnest, I shall not disappoint him for the sake of pleasing his father.' In consequence her transatlantic vitality was left to be illuminated less by her interaction with the heir than by her brusque speech, her tireless tennis, and some of Trollope's best humour of the purely local kind in her exchanges with Dolly Longstaffe. Looking back to his first big success, Trollope missed his chance of vying with Henry James, who at this time devised much more active roles for the American girl in Europe. He was looking back, too, when siting the more emphatic small events in places which held strong associations from earlier novels. The ruins at the Matching estate, for instance, where Glencora in *Can You Forgive Her?* had confessed to a friend her continuing love for the man she had given up in order to marry, were the setting when Silverbridge confirmed to Miss Boncassen that he would resist his father's plans to marry him more suitably.

The political element in the novel carried the reader back to the 'high-minded, proud, self-denying nobility' of *Phineas Redux* and to Finn's acquaintance there with 'the patriotism of certain families'. In obedience to this principle the Duke returned to cabinet office, though believing himself 'unfit' for 'parliamentary strategy'. Trollope's old simplification of politics, making patriotism incompatible with party loyalty, reappeared in another 'conjuror', Sir Timothy Beeswax. In contrast to this mere opportunism, Plantagenet Palliser was still, as in *Can You Forgive Her?*, the kind of politician to give the country 'that exquisite combination of conservatism and progress which is her present strength and best security for the future'. (Some sixty years later the adequacy of this ideal was to be tested when Lord Halifax, a dedicated representative of Palliser's order, faced the total opportunism of the Nazis with imperfect comprehension—in contrast to a son born to the marriage, quite recent at the time Trollope was writing, of Lord Randolph Churchill with an American.)

This conclusion to the 'semi-political tales' was followed by Trollope's most straightforwardly sensational novel, *John Caldigate* (1878–9). The bigamy and blackmail plot was carried through with unflagging zest and assurance, born of a complete mastery in the unspectacular narration of spectacular events. They culminated in a vigilant clerk's overturning of the evidence for bigamy, late in the story, with an exactitude on which Wilkie Collins could hardly have improved. By contrast, the gentle, romantic farce of *Ayala's Angel* (1881) offered amused variations on earlier amorous themes. Trollope added, at one end of the romantic scale, Ayala

Dormer's expectation of a perfect lover like an Angel of Light and, at the other, Gertrude Tringle's determination to elope with one or other of her successive suitors. The old stories led now to everyday conclusions: Gertrude, starving herself when broken-hearted, made 'sly visits to the larder'; lovers who had decided, as in *The Claverings* and *The Duke's Children*, that they were for ever parted by poverty, accepted the help of a maiden aunt, and married. One of this pair could even contrast his behaviour with that 'required for a three-volume novel'. The smile at the practices of novelists, less broad than in *The Three Clerks*, was equally self-aware. But it was an indulgent rather than a critical smile, and for the feelings as well as for the fiction: although the heroine might think that her 'dreams had all been idle . . . the Ayala whom her lover had loved would not have been an Ayala to be loved by him, but for the dreams'.

The case of conscience at the centre of *Cousin Henry* (1879) was more serious. It offered a fresh development of the story of Sir Harry Hotspur's matchmaking between his daughter and his heir, the daughter now unattracted and the heir unattractive. Small cowardly decisions of which everyone recognizes the possibility in themselves were exposed to a quiet, steady gaze. Solitary choices in the entire absence of advice made it a nightmare case. The result, unpredictable, and therefore constant in its interest despite the very narrow range of possibilities, showed vividly how far Trollope had come from the loud villainies of Barry Lynch in *The Kellys and the O'Kellys*.

The doctor of *Dr Wortle's School* (1880–1), on the other hand, whose 'rule of life' was 'to act . . . entirely on his own will', found himself in a situation where 'he could not endure the responsibility of acting by himself'. Encumbered with advice, resentful of it even when he had requested it and, even when accepting it, taking his own line of action, he generated a plot which combined the threat of violence in Chicago with quiet English comedy in pointed dialogue. It was a small triumph compared with the deployment of more various materials to no particular end in *Marion Fay* (1881–2). There the evoking of sensational expectancy only to defeat it made a dubious parallel to arguments and fanciful events which led nowhere, though they concerned religion, and social rank in relation to friendship and marriage. It was entertainment such as the *Autobiography* took most of his works to be, sufficient for a single reading, but offering little material for the kind of reflection which had seemed, in his best work, to be reflection on life itself because of his awareness of strange complications in apparently simple characters. There was little of such material, either, in the two shorter things, *Kept in the Dark* (1882) and *The Fixed Period* (1881–2), which preceded his last long novel. The first returned to a marriage broken for reasons almost as insubstantial as those in *He Knew He Was Right*. The second, set in 1980, satirized doctrinaire reformers in Mr Neverbend, its first-person narrator. Trollope's earliest interests reappeared also in the Irish problems, now more desperate and expounded at greater length, of the unfinished *The Land Leaguers*.

The last long work, *Mr Scarborough's Family* (1882–3), however, resembled a Wilkie Collins novel more than any other of Trollope's. Its mysteries, unresolved until near the end, grew out of the unconventional deeds of a hero who despised honest folk and made an ass of the law.

> For the conventionalities of the law he entertained a supreme contempt, but he did wish so to arrange matters with which he was himself concerned as to do what justice demanded. Whether he succeeded in the last year of his life the reader may judge.

For himself Mr Scarborough claimed to seek his 'duty . . . beyond the conventionalities of the world', that is, of the law, of religion, and of morality. 'All virtue and all vice were comprised by him in the words "good-nature" and "ill-nature".' To the local doctor, watching over him in his fatal illness, he had 'a capacity for love, and an unselfishness, which almost atones for his dishonesty'. These virtues were less apparent to the reader than his vitality, ingenuity, and implausible articulateness when mortally ill. It was clear that his determination not to tyrannize over his sons had left them without guidance, so that the 'good-nature' of the one led him to compulsive gambling and the 'ill-nature' of the other to self-centred intrigue. The reader, moved to expect some further light on the riddle before the finish, only saw Scarborough die respected, even loved, by his doctor, his sister, and his heir. Contemporary pieties seemed as much outraged as in the Jacobean plays, especially, again, Ben Jonson's, which may have prompted the creation of such a figure. But the old-fashioned lawyer acting for Scarborough, Mr Grey, who believed 'that the law and justice may be made to run on all-fours', was driven into premature retirement. His partner survived by accommodating himself to dishonesty as easily as the people in *The Way We Live Now*; he admired Scarborough as 'the best lawyer he ever knew'. He summed up a great deal of Trollope's social criticism ever since the admonitions, in *The New Zealander*, about 'lying . . . gradually becoming not abhorrent to our minds'. But now Scarborough's character seemed to leave the moral rather less clear.

It made an arresting and equivocal climax to a long series of novels marked by the careful adaption of means to ends. The reader's way had been smoothed by an unpretentious lucidity of language, by dialogue which gave steady if relatively shallow illumination of character and its development into plot, and by plot which progressed slowly, each small stage marked by repetitive discussion, prolonging the reader's pleasure in wishing to know the outcome, and offering no profounder analysis of motive or event than would promote that pleasure. Nevertheless Trollope had believed he could not avoid 'teaching': 'The writer of stories must please, or he will be nothing. And he must teach whether he wish to teach or no.' He had taken a special interest, humorous, satirical, or sympathetic, in people who assumed that same obligation when giving advice—clergymen, journalists, parents, friends—and in those who deferred to them or resisted. In pursuing this interest, his success had come from so managing plot and sub-plot as to appeal to the reader's powers of judgement. The very activity of writing to entertain, even the length demanded by readers—'short novels are not popular,' he noted in the *Autobiography*—stimulated Trollope to consider and compare predicaments, characters, motives, and by so doing repeatedly to achieve elucidation which was far from commonplace, for all his concern to give 'a picture of common life'.

10. Meredith

Few of Meredith's[1] readers have been undisturbed by his cultivation of an eccentric gentleman's freedom of language, both in the bread and butter of the prose—the odd placing of modifiers or the carelessness about the ambiguity of pronouns—and in 'the fine flavour of analogy' and metaphor. These are the signs of a selfconscious literary ambition also evidenced in George Eliot and even in Trollope, though less idiosyncratically. Meredith's unusual range of allusion, both to fiction and other genres, English and Continental, contemporary and antecedent, his appropriation from them of techniques for concealing the highly personal nature of his material by means of the structure of the work, and his management within it of a running critical debate about its aims and achievement, eventually found their fit admirer and emulator in James Joyce. Meredith, however, wanted to be popular, and with some reason. He was, with it all, as much concerned as any novelist to flash upon the inward eye individual character and immediate event, and he often succeeded, though more often with minor than with major characters and in episodes than in longer sequences. There was often a subtlety in these aspects of his work to match the over-eager subtleties of literary aspiration, though the latter ambitions tended to pull him in one direction while the wish to satisfy the demands of the common reader of novels pulled him in another. Uncommon readers overpraised him for the range of his literary sophistication or depreciated him for not rather choosing to refine the methods peculiar to prose fiction. A 'pure' artist like Henry James could eventually speak of his 'lack of aesthetic curiosity'. But Meredith was concerned with the novel in the older, looser sense. The form to him was still that of Fielding, Thackeray, and the Brontës, not yet finally separated, with its own

[1] George Meredith, 1828–1909, grandson of 'the Great Mel' (i.e. Melchizedek Meredith, tailor, of Portsea), was educated at the Moravian school in the Rhineland principality of Neuwied and articled to a solicitor on his return to London, 1846. He married the widowed daughter of Thomas Love Peacock, 1849. The marriage failed and he was left in 1858 with the care of a son, while working as a journalist, and as publisher's reader for Chapman and Hall, 1860–95, in which capacity he gave valuable advice to, among others, Thomas Hardy and George Gissing. Meredith remarried, 1864, and from 1868 lived on Box Hill, which, as his fame increased, became a place of pilgrimage for his admirers. The newly created Order of Merit was conferred upon him in 1905.

distinctive rules, from epic and romance and the drama, and still drawing sustenance, as in George Eliot, from allusion to as well as contrast with these other genres. Within these wider limits Meredith was restlessly experimental.

The results tended at first to be odd rather than original or satisfying. Fantasy out of the *Arabian Nights* in his first book, *The Shaving of Shagpat* (1856), was praised by George Eliot, in a review, for 'significant humour' as well as for 'picturesque wildness of incident'; but there was not enough of the first to make the second satisfying to the mind. Nor was the story sufficiently clear to make the incidents all seem part of a single conception, although it has proved perfectly possible to extract from it what George Eliot called 'deep meaning . . . for those who cannot be satisfied without deep meanings'. In much of the work following this, the parts similarly gave the impression of meaning more than they said, and the whole too often failed to clarify what this was, even though a governing 'idea' might be glimpsed; 'significant' humour was frequent, but its nods and nudges only increased the reader's bewilderment when he could not see how the attitudes and emotions of one episode were to be related to those of another. Thackeray at his best had worked to induce in the parts of the novel the critical sympathy or amusement which would serve to complete the whole in the reader's mind, once reading was done. Although Thackeray when not at his best came short of such an ideal, it is by this standard that Meredith deserves to be judged.

Few readers have been happy with the mixture of literary modes in his second long fiction, *The Ordeal of Richard Feverel* (1859). It depended, like *Shagpat*, on literary allusion for the arousing of expectation and the control of tone, particularly in the first thirty-two chapters, (that is, the first thirty-three, before the 1896 revision). After that, life itself was intended increasingly to take over, until the culmination in a catastrophe which fitted into none of the literary categories at all. The strategy which carried on longest was an apparent inversion of that in *Vanity Fair*: far from subtitling his work 'a Novel without a Hero', Meredith called Richard Feverel 'Hero' repeatedly and showed him playing up to the title while failing to meet the obvious demands of his actual situation—crucially, in forcing himself to fight a duel when he was married and had a child. Before the marriage, there had been allusion to New Comedy in so many words, when Richard's friend looked forward to the spectacle, as in Terence, of a father trying to keep a pair of lovers apart, while the subject with which the novel began, a father's new-fangled System of moral education, bore a suggestive relation to Old Comedy in its satire, farce, and fantasy, all focused, as in *The Clouds* of Aristophanes, upon the disastrous results, for the son, of these new ideas. Meredith appears to have become uneasy with the headstrong fantasy and burlesque in this first third of the novel. After the revisions of 1875 and 1896, the father's System of education occupied a far smaller place, while some broad Dickensian comedy disappeared entirely—the chapter, for instance, in which the father 'exchanged Systems' with a lady who had a System for girls, Mrs Caroline Grandison, 'the petticoated image of her admirable ancestor', Sir Charles. In what remained, Meredith reduced the verbal exuberance which had led reviewers to connect the novel with *Tristram Shandy*. His first version had invited and indeed could stand the comparison. And he openly enjoyed the mixing of allusions: as *Shagpat* had crossed the *Arabian Nights* with Ariosto and E. T. A. Hoffmann, so *Feverel* supplemented Old Comedy with Sterne's Tristopaedia, and New Comedy with Shakespearian romance at the lovers' lyrical first meeting. It is not surprising that Joyce should have admired the novel.

Even in its revised form the mixture becomes violently incongruous for most readers by the end. Meredith gave what he took to be adequate warning in chapter 33, 'Nursing the Devil', when the comedy of the unacknowledged motives of the maker of Systems ('not so much in love for his son as in wrath at his wife,' Meredith explained in a letter) modulated sharply and the evil of his self-regard appeared: he not only saw the son's marriage as something to be forgiven, rather than as the triumph of what was good in his System of education, but also refused to forgive. Unfortunately, however, it was hard to connect this with Richard's separation from his wife and temptation by a vivid and plausible modern Circe. And if the link was weak between the ordeal of the father and that of his son, the catastrophe, when the less complicated ordeal of the son's wife became simply too much for her, was disturbing for an opposite reason, its very connection with the place she had consistently filled. Exempt from the author's irony, she had been presented as in no need of trial by ordeal; that only she should fail to survive it seemed a gratuitous assault on the feelings she had been allowed to evoke, one more incongruity in a book so knowing and contrived.

The parallel to Rousseau's *Emile*, with its sequel in which a careful education proved to be no safeguard against calamity, occurred to one early reader, Justin McCarthy, and is unlikely to have been absent from Meredith's mind. It adds to the volume and variety of literary echo at one extreme, as allusion to the sensation novel does at the other. Expectations of fear were raised by means of devices from the nascent sub-genre; the hero's mother, for instance, twice appeared with all the paraphernalia of a ghost's visitation. Skilfully, Meredith put such devices to his own use, in this case as a pointer to the real inward source of the calamity in the resentments of the estranged husband; that she was in fact haunting the house could be seen in his solitary, unforgiving life spent brooding over his System. So began the stimulating game with the canons of popular fiction which Meredith was still pursuing in his last complete work, *The Amazing Marriage*, in 1895; there he allowed much of the story to be told by Dame Gossip, whose 'idea of animation is to have her *dramatis personae* in violent motion, . . . to make them credible, for the wind they raise and the succession of collisions'. In *Richard Feverel* he was already conscious of offering what appeared to be less than the reader avid for sensation wanted, though he himself believed it to be much more; in the chapter 'In which the Hero takes a Step', he made a direct appeal to a more discerning audience:

> At present, I am aware, an audience impatient for blood and glory scorns the stress I am putting on incidents so minute, a picture so little imposing. An audience will come to whom it will be given to see the elementary machinery at work: who, as it were, from some slight hint of the straws, will feel the winds of March when they do not blow.

For this novel such an audience not only came but has in some degree been retained. At the least it is the one work of Meredith's that more people have heard of than any other; at the most it is the one in which a perennial subject, 'A History of a Father and Son' (the subtitle, which might equally serve for Joyce's *Ulysses*), survives the author's curious inspection of it by means of an arch mixture of styles and the unexpected convergences and divergences of opinion about it among the characters.

The book gained a reputation for impropriety which restricted its sale. By the critics it was treated fairly, on the whole, though its tone and drift were found puzzling. Few people at the time could know of the personal complications which made the author himself the prototype not only for both the protective father and the enamoured son but also for the mocking 'wise youth', Adrian Harley. Meredith's own unresolved feelings in his treatment of first love were apparent in the title of his most lyrical chapter, 'A Diversion played on a Penny-Whistle'. Probing his own wound, he might well be uncertain of finding readers who would attend to the involved narrative of the small causes of large painful effects.

The probing of personal wounds has been thought to continue in *Evan Harrington* (1860–1). Certainly the hero was, like Meredith, descended from a tailor and troubled by relatives who apparently wished to conceal the fact. The action was even set back into a period similar to that of *Vanity Fair* in order to fit the time of the author's aunts who were recognizable in the novel, and that of his own grandfather, the Great Mel, a tailor who appeared in it under that name and as he had been in life, a tailor of social gifts which made him at home with and welcome to people of all ranks. But there was an absence of rancour in the comedy of the efforts of Harrington's aunt, now the Countess de Saldar, to conceal his origins. Her boldness, assurance, and ingenuity emerged as a source not only of amusement but of admiration. The tendency of the tale was no doubt to vindicate Harrington as a gentleman whatever his birth, but this fairly commonplace theme was overshadowed by Meredith's delight in the arch-snob herself.

Writing the story as a serial for *Once a Week*, where the criteria of success were, as his letters show, the 'excitement' and 'tension' of *The Woman in White* which was coming out as he wrote, Meredith laboured hard to produce the required 'animation', restraining his figurative style (to his own disgust) and his indirect methods of narration, in order to engineer an obvious 'succession of collisions'. The point of them all, however, was rather too slight and their succession came too much to resemble a parlour game. Yet if it had been merely a parlour game the intervention of Harrington's mother to tell the truth, in order 'to cast a devil out of the one she best loved' and 'save him from the devils that had ruined his father', would have ended instead of prolonging it. The result was more rounded than mere satire; the brutality of people of birth was not concealed, but neither was their ability to prefer Harrington to the Countess even after his secret was out. It was a first and not unsuccessful attempt at the kind of urbane social comedy with serious overtones which was to become Meredith's characteristic mode. Some imitation of Dickens, in the hope of increasing his serial's popularity, for instance with a pair of brothers who were described as 'roguishly-cosy', showed a tendency which was to be found even in Meredith's last phase. In *Evan Harrington* the incongruity only emphasized his own interest in evoking 'thoughtful laughter', as he later called it, at the spectacle of youth crawling to maturity despite obstacles placed in the way by solicitous elders.

Emilia in England (1864), later *Sandra Belloni*, returned to the less straightforward narrative mode of *Richard Feverel*, if not to quite the same ostentatious concealment of the author's presence and opinions. But he had a subject of much less interest: 'sentimentalism' in a group of middle-class characters he appeared to despise. It grew out of the chapter (24, after 1896) in which Richard Feverel's love which meant business was contrasted with the sentimental dalliance of his father and Lady Emmeline Blandish:

> 'Sentimentalists,' says THE PILGRIM'S SCRIP, 'are they who seek to enjoy without incurring the Immense Debtorship for a thing done.'
> 'It is,' the writer says of Sentimentalism elsewhere, 'a happy pastime and an important science to the timid, the idle, and the heartless; but a damning one to them who have anything to forfeit.'

Now he went on to contrast the happiness of this pastime for the family of a precariously well-to-do merchant, Pole, and its unhappy consequences for the two of them, Caroline and her brother Wilfrid, who, being capable of love, had something to forfeit.

The plot of the previous novels was repeated in their concealment of love from a parent who had other plans for them and in their exposure to a trial or ordeal. This time the reader was easily reassured about the results: in Wilfrid's case, 'you are about to see him grow . . . All of us are weak in the period of growth, and are of small worth before the hour of trial.' Their sentimentalism was presented as characteristic of an England grown suddenly wealthy, the 'natural growth of a fat soil', an almost unavoidable phase for aspiring members of their class. So the Pole family and their adherents, playing at 'the game of Fine Shades and Nice Feelings', were to emerge 'considerably shorn, but purified', as 'examples of one present passage of our civilization'. The action which forced them to give up Fine Shades and Nice Feelings, although it included much that was painful to the characters and much that was embarrassing to the reader, was intended for comedy and narrated like farce. The narrator distanced it as the action of 'little people', and the hint that he reckoned himself among them—'Civilized little people are moved to fulfil their destinies *and to write their histories* as much by distaste as by appetite' (the italics are mine)—was too subdued to affect the tone of superior amusement.

This tone was sustained by comic wrangling with a 'philosopher'. As the opponent of the sentimentalists, this persona owed something to the opinions of Thomas Love Peacock, the author's father-in-law, whose *The Four Ages of Poetry* (1820) had contrasted 'the thinking . . . and philosophical part of the community' with 'that much larger portion of the reading public, whose minds are not awakened to the desire of valuable knowledge', but who are easily moved by 'sentiment, which is canting egotism in the mask of refined feeling'. The narrator of *Sandra Belloni* might complain about the philosopher's 'garrulous, supersubtle' intrusion, while insisting that the story had been conceived 'with his assistance', and must be carried on 'in a sort of partnership'. Now one of them, now the other, took up the role of literary critic. Recalling, no doubt, the conclusion of *Richard Feverel*, the philosopher pointed 'proudly to the fact that our people in this comedy move themselves . . . and that no arbitrary hand has posted them to bring about any event and heap the catastrophe'. The narrator protested (too much) against Thackeray's open acknowledgement of puppetry: 'That showman did ill. But I am not imitating him. I do not wait till after the performance, when it is too late to revive illusion.' The philosopher was blamed for 'making tatters of the puppets' golden robe—illusion' and for 'sucking the blood of their warm humanity out of them'. In its surroundings of an odd kind of comedy, where the narrator tried to remain critical of and detached from his characters, this read like an apology for their obvious lack of blood and the patent fragility of the illusion. Too little effort seemed devoted to directly interesting the reader in what these 'little people' were and did, and too much to discussing their habits, in the belief that 'the primary task is to teach them that they are little people'. Not only was the subtlety of the analysis unmatched by the interest of the narrative, but the tone of both seemed uncertain, difficult to understand in its mocking and its menaces—'Games of this sweet sort are warranted to carry little people as far as they may go swifter than any other invention of lively Satan.' The assurance of the greater exponents of the method, Thackeray and Sterne, was hardly to be expected when writing from a less secure social position and about a society itself in process of change. He was probing, more deeply than in *Evan Harrington*, into the strange 'growth' from a 'fat soil', the emotional risks run by the possessors of the precarious new wealth of the period of the novel, the late 1840s; he was not merely observing but warning his 'civilized little people'; and the narrator who was 'moved . . . to write their histories' was one of themselves. It was not strange that he should give the impression of barely mastering his feelings.

With Emilia, the half-Italian heroine, it was different. Her contrasting directness and naïveté were left to be seen by the reader, unprompted. She was to be the touchstone, like the Italians of E. M. Forster, testing the gold in the English alloy. Yet, once she had acknowledged her love for Wilfrid in a few brief and impressive scenes, her qualities had too little opportunity to exhibit themselves in relation to the Fine Shades and Nice Feelings of the Poles. Most of her opportunities related to a quite separate matter, Italian patriotism, and this cause went uncriticized. The possibility of Emilia's being sentimental about it never arose: it was, as the philosopher said when promising to retire from the sequel (to be written as *Vittoria*), 'a field of action, . . . where life fights for plain issues', and where he was no longer needed to point them out.

'Action' in these terms afforded too simple a contrast with sentiment. It formed the ground of Emilia's final rejection of love and loss of naïveté—'I will not trust my feelings as they come to me now. I judge myself by my acts'—and the ground of the distinction between the English Poles and the Welsh brother and sister who were devoted to the freeing of Italy from Austria—'A sentimental pair likewise, if you please; but these were sentimentalists who served an active deity, and not the arbitrary projection of a subtle selfishness which rules the fairer portion of our fat England.' The same could have been said of the arch-sentimentalist, Wilfrid Pole himself: after all, his turns and changes occupied the interim between two phases of a military career; the first, however, was in India, the second in Italy but in the Austrian army, to which, in *Vittoria*, the British army in India was to be likened. The author's political opinions were taking the place of the internal analysis he claimed to be carrying out. This would have mattered less if the service of 'an active deity' had not been emphasized in the development of Emilia herself and her relation with Merthyr Powys, one of the restrained lovers, acting not protesting, who reappear as Meredith's ideal.

The novel, which gave him great trouble, tended to fall apart, the sections concerned with the Pole sisters becoming less and less connected with those concerned with Wilfrid, Emilia Belloni, and Merthyr Powys, while coherence was further endangered by lacuna and inconsistency in the analysis of the behaviour of the latter group. In particular, the lack of destructive analysis of Emilia to match that of the Poles suggested prejudice, even 'sentimentalism,' on the part of the author. He offered excellent insight in passing, but it was at the abstract level, not carried out in the presented drama: the contrast in chapter 38, for instance, between Emilia's 'surface-self' and 'the submerged self—self in the depths' which 'rarely speaks to the occasions, but lies under calamity quietly apprehending all; willing that the talker overhead should deceive others, and herself likewise, if possible—' was not brought to bear upon the relatively commonplace account of her approach to suicide in the following chapter. The question of Italian 'freedom', by simplifying the issues at the same time as the analysis of the sentimentalists became more and more refined and the mingling of tragedy and farce more original, completed the disintegration of the novel.

The writing of the sequel, *Emilia in Italy*, was interrupted by 'an English novel, of the real story-telling order', *Rhoda Fleming* (1865). It is doubtful whether this is to be identified with *A Woman's Battle* mentioned in a letter of 1861 as the 'next' project: 'I think it will be my best book as yet.' That title would fit the story of *Rhoda Fleming*, but this does not seem to be a story he took up by choice, judging from the one place in which he allowed himself an extended comment on it, in his twelfth chapter. There, a play concerned, like this novel, with a woman as 'one of man's victims . . . had been favoured with a great run' by a public that 'will bear anything, so long as villainy is punished'. The claim that 'classic appeals to the intellect, and passions not purely domestic, have grown obsolete', since 'the mass is lord', turned these into his glum reflections on the story he was now writing. ('Classic comedy' was his phrase, in a letter while *Sandra Belloni* was in progress, for the play he projected, *The Sentimentalists*.) Of course, even in courting popularity with a novel of mainly domestic passions, vying with the notorious bigamy novels of the early sixties, Meredith refused the public much of what it expected. Both the would-be bigamist (marrying under a false name) and the actual bigamist went free, while the first repented and offered valid marriage, only to be rejected. And the reader was made to work hard for his pleasures, since understatement and indirectness often made it difficult to tell what was happening. The plot, like that of some Jacobean plays, the slick tragedies of Shirley, for instance, was contrived to produce violent reversals; but Meredith's confessedly 'uncertain workmanship' lengthened the preparation for them from the projected one volume to two.

Faced with the problems of what he called in chapter 29 'a narrative that is made to revolve more or less upon its own wheels', he seemed to have difficulty in making it reveal its significance without comment from narrator or 'philosopher'. For much of the second volume the direction things were taking was obscure and the pace dawdling, while in the third it quickened unnervingly. The violent swerves as the seducer, after instructing a foolish agent to bribe a boor to marry his victim, tried to stop the marriage, while Rhoda Fleming, the victim's sister, pushed it on by paying the bridegroom (with stolen money), made for suspense of a kind. It was a singularly unpleasant kind when the victim was not only passive and ill but subject to the shock of discovering after the marriage that her lover had been kept from forestalling it—and by Rhoda herself, who had been her only comforter.

A failed attempt at suicide by poisoning, when her husband claimed her, and her freedom when he turned out to be a bigamist combined Jacobean and current motifs. Meredith worked hard to prevent the fault he had found as publisher's reader in *East Lynne*: 'all the incidents forced—that is, not growing out of the characters'. He interested himself in making inner life dramatic by setting in parenthesis the characters' thoughts, at variance with what they said, and by their frenzied self-revelation in monologue, contrasting with brief word and symbolic gesture, like the victim's just before marriage:

> 'Rhoda, I am ready. It is not much.'
> She blew the candle out.

He seized the greatest Jacobean example of all in the allusions to *King Lear* towards the end, repeating Shakespeare's phrase 'the worst', allowing a bemused chorus to call the discovery of the bigamist a miracle 'which had vindicated all one hoped for from Above', and working in the last words of the play: 'You'll never live so long ...'. It reminds one of Shirley's echoing of earlier drama when writing in a tragic vein that was not native to him. The effect upon literary readers at the time was much more favourable—R. L. Stevenson believed that 'if Shakespeare could have read it he would have jumped up and cried, 'Here's a fellow!' By 1897, 'the fiery race of events' in *Rhoda Fleming* was for Arthur Symons, a critic who bridges the nineteenth and twentieth centuries, among the things in Meredith which 'have more of the qualities of poetry than of prose'.

Such verdicts have not been confirmed by later readers. Moreover, the frenetic pursuit of 'animation' obscured the social implications of a plot concerned once more with the strange fruits of the 'fat soil' of wealth. The son of a banker was the seducer; the bigamist was bought off with money which had been stolen for use in speculation; the attractive villainess reformed by marrying the banker; the unheroic hero was the son of 'a yeoman of the old breed' who, at a date before the repeal of the Corn Laws, saw England's 'agricultural day ... doomed' when 'the Yankees ... flood the market ... England was to be a gigantic manufactory, until the Yankees beat us out of that field as well ...' But the plot went off in a different direction, finishing with the victim's appeal, 'Help poor girls.' It is hard to avoid the conclusion that Meredith had underestimated the kind of thought needed for popular success.

In an attempt to succeed by 'vivid narrative' he returned to the sequel to *Emilia in England* on which he had been at work, *Emilia in Italy*, published as *Vittoria* (1866–7). He had spoken of it in mid-1864 as 'all story, ... no Philosopher present: action: excitement: holding of your breath, chilling horror: classic sensation'. Even in the high spirits of a familiar letter this was a large claim. He certainly found it difficult to fulfil, although its fulfilment had become, by the time the novel was nearly finished, a matter of conscience: 'Much of my strength lies in painting morbid emotions and exceptional positions; but my conscience will not let me so waste my time. ... My love is for epical subjects—not for cobwebs in a putrid corner ...' Matthew Arnold had similarly found 'morbid' the situation of his Empedocles, 'in which there is everything to be endured, nothing to be done'; his criticism of mere 'doing' appeared in the articles of 1867–8 which became *Culture and Anarchy*. *Vittoria*, as the sequel to a novel which had defended those who served the 'active deity' of the risorgimento, could be read as criticizing the code which insisted on the strict acceptance of debtorship for things done. Duelling was the mark of this code amongst aristocrats and soldiers; amongst conspirators its mark was mutual loyalty or violent suspicion, and, amongst the women who suffered from the effects of such practices, either devotion or implacable resentment.

Action was presented as especially uncertain and complicated at the beginning, in 1848–9, of the process of Italian unification, under Piedmontese leadership, which had been brought very near to completion only five years before the serialization of the novel started in January 1866. Characteristically for Meredith, it was action in which the decisive part was played by women. He praised Mazzini as an idealist, and made the 'craving for idealistic truths' typical of women and Mazzini himself their champion in a speech when Vittoria was introduced. So in the coherent first half of the novel, everything was focused upon her insistence on playing her part, despite falling under the suspicions of Rizzi, the 'man-devil' of a spy thriller; in the second part things were moved by a female spy in league with a female avenger.

The first part was clear enough, the second overpopulated and much less distinct. Everything in the first twenty-two chapters bore upon Vittoria's giving the signal for rebellion by means of an incendiary aria to be added to an opera and sung in La Scala. After this, lucidity was lost, in a feud which she inherited with her husband, Ammiani, her disagreement with him over the role of the King of Piedmont in the war, and the continued comic pursuit of her by her rich patron, Mr Pericles, 'to cure her of this beastly dream' and return her to her proper career as a *cantatrice*. The action was further complicated by intervention from a former lover of Ammiani and from two lovers of Vittoria, Wilfrid Pole and Merthyr Powys, from the earlier novel. The intrigues connected with a rising in Brescia were meant to form the focus of the second half as those before the rising in Milan had in the first, this time with Vittoria's husband as the possible victim. A sudden suspension of the feud by the woman who had urged it on—'Inveterate when following up her passion for vengeance, she was fanatical in responding to the suggestions of remorse'—opened the surprising possibility that Ammiani, a fugitive once the rising had failed, might be helped to escape by one Austrian in order to defeat the purposes of another. The loyalty amongst the fleeing Italians and Ammiani's insistence on a duel with the Austrian who wanted to save him completed the catastrophe.

It was over-ingenious for 'classic sensation', though the excitements were considerable for those who could master its intricacies. But the pace of events was ill-controlled, the number of cross-lines confusing, the language of activity forced: 'savage eyes reading the fiery pages of the book of hell'; 'A red light stood in his eyeballs, as if upon a fiery answer'; 'Ammiani breathed as one who draws in fire.' This kind of bombast was mixed with an equally mannered brevity of understatement. Seeking an impersonal heroic tone, Meredith showed his ambition to go beyond 'sensation', but the literary artifice was too selfconscious for the events to strike home to the reader. In order to avoid 'morbid emotions' he left the central relationship between Vittoria and Ammiani unexplained except in the stereotypes of patriotism. And their inner natures were not made sufficiently interesting to give substance to the attempt at tragedy in the understanding of themselves which was claimed for each. Ammiani's self-indictment—'Vast numbers admired me. I need not add that I admired myself . . .'—was unprepared for; the inflated feelings with which Vittoria faced the catastrophe seemed hardly her own but rather to be continuous with the picturesque presentation of the events:

> a girdle of steel drove the hunted men back to frosty heights and clouds, the shifting bosom of snows and lightnings. . . . Vittoria read the faces of the mornings . . . the mask upon the secret of God's terrible will; to learn it and to submit, was the spiritual burden of her motherhood . . . ; to be at

peace with a disastrous world for the sake of the dependent life unborn; by such pure efforts she clung to God.

It was a heady mixture, laced with violent rhetoric about Italian servitude and over-resplendent descriptions of the Lombardy background.

The 'love for epical subjects' soon gave way to less weighty projects, judging by the holograph lists of them which survive from just before and just after 1870. Although not all the lists use the headings *Comedies* and *Comediettas in Narrative*, there is some overlap between those that do and those that do not, and all suggest a decisive change of direction, or rather a confirmation of the mode of critical, ironic comedy, as in the novels before *Rhoda Fleming*. The earliest list, which was perhaps made in the mid-sixties and which, remarkably, refers to projects among the last to be published, shows the beginnings of *The Adventures of Harry Richmond* (1870–1) in the note 'Autobiography (with Contrivance Tom)' on which later notes show much work was done before Contrivance Tom became Richmond's inventive father and continuity was secured by exposing the son to the opposing attractions of his father and his grandfather.

In autobiographical romance, *David Copperfield* and *Pendennis* offered precedents for a relaxed kind of plotting, but hardly for these drifting, jejune adventures. The situation of the hero, between his inventive father, who lived by his wits, and his dour, practical grandfather, left the author too much freedom and gave the reader the pleasures of novelty rather than of developing understanding. The strong opening scene dramatized the relationships of Richmond to his father and grandfather pretty completely. Everything was suggested here, even down to the fire which ended the novel, and the heroic love of Richmond's aunt Dorothy; moreover, these matters took a more pointed and economical form than subsequently—the father's courteous effrontery and condescension, though standing as petitioner outside the house, his vague claim to royal rank, his lavishness with words and his twisting of them into slightly unnatural rhetoric ('By no scurrilous epithets from a man I am bound to respect will I be deterred or exasperated') contrasted with the grandfather's actual status ('I'm a magistrate, and I'll commit him'), vigorous language ('I can't think brisk out of my breeches. . . . And here's a scoundrel stinks of villainy . . .'), and wealth with which to wrest the son from his father ('Let go the boy! . . . he shall have Riversley and the best part of my property, if not every bit of it.') The position of the boy between these two, each incapable of understanding the other for reasons which were suggested in their speech, remained unchanged for four-fifths of the novel. His relative nullity, equally unchanged, was clear from the first when his unknown father easily pacified him with sweetmeats and he was able to dream picturesquely, despite the one faintly heard cry after him (of his aunt) and a degree of casually mentioned 'grief'. His subsequent entire absence of nostalgia for or even thought about the home he had suddenly left, even though, later, at the reunion with aunt Dorothy, 'her trembling seized on me like a fire', laid the foundation for the reader's disregard of the boy's reactions, curious though their very slightness was at the time of his mother's death. The unrelenting contrast between the father's engaging trickery and the grandfather's earthbound eye for the boy's more solid prospects carried on into the backing each of the elders gave respectively to a German princess and an English heiress as a match for Richmond. The women relieved the symmetry with their unexpectedness, their selflessness and good sense, and their eventual friendship; all the same it was a little too pat that these two, who apparently met so exactly the respective expectations of Harry Richmond's two patrons, should both fall in love with him.

Meredith staked everything on the climactic situation, as in the two preceding novels. Here it was greatly superior because simpler in itself and in its results. The father, completely broken by the failure of his schemes, and the grandfather, killed by the violence of his own objection, achieved a new definition for the reader and for the titular hero. It was a pity that aunt Dorothy, who had deceived them both, had not been prominent enough in the drifting story to give full value to the eventual revelation of her role in it.

The central character, Richmond's father, showed how much more there was to such figures in Dickens than the exaggeration of a few traits. Dickens not only offered a greater distinctive force and consistency of speech, and a more poetic foolishness, but gave what his characters actually said more substance, sending widening ripples out from the immediate situation. Meredith would have done better not to challenge comparison. His own comments sensibly emphasized local effect—'nimble and overwhelming volubility like a flood advancing', 'a mixture of the ceremonious and the affable such as the people could not withstand'—but the speeches themselves seemed only too easy for the reader to resist. Richmond's changing relation, like Copperfield's, to the woman whom he married at the end also invited comparison with Dickens. The greater power of introspection with which he was credited at the end amounted to little; he claimed to 'stand somewhere between' a fool and a philosopher, but too much had been done for him and too little by him for this to matter. His relation to his father had changed (as had the father himself, in a manner reminiscent of Mr Dorrit), but only by shedding the admiration which the reader had long seen to be unjustified. Both *Copperfield* and *Pendennis* had developed other concerns for which *Harry Richmond* offered no parallel. No doubt German examples, especially that of Goethe, counted for more with Meredith; but his novel had neither the simplicity and solidity of characterization in the episodes (after the first) nor the sense of direction in the whole work to cause readers to connect it with *Wilhelm Meisters Lehrjahre*. Coherence could be made from the criticism of English institutions, especially monarchy and aristocracy, implied in the actions of Harry Richmond's father and debated at length in conversations with his German mentor, as if in a German novel, but this hardly supplied a strong thread in the actual reading.

Nor could the presence of images, like those connected with fire or the sea, bind the work together when they were brought to the reader's mind by incidents which were so arbitrary. He could not be expected to take seriously events which bore so little relation to the agency of the protagonists as the hero's sudden, involuntary sea voyage or the abrupt loss at sea of his rival for the hand of an heiress. Fire stood more chance of reaching the reader's imagination in so far as it arose out of the action as a sleeper's natural expectation at the loud bell-ringing and door-beating of Richmond's father in the opening scene or as the latter's method of preventing a premature discovery of Harry's relation to the Princess Ottilia. It was a fine stroke to have the father die while trying to save the benefactress who had loved and watched over him—and who was in fact not in danger—in the fire caused by his extravagant welcome to Harry and his bride. Nevertheless, the resulting circles of association seemed insufficiently delimited by the ordinary role of this image in the story for them to be constituent elements in an alternative structure.

Beauchamp's Career (1874–6) transposed the story of *Harry Richmond* into terms which brought politics in from the periphery. Again an immature young man was beguiled away from the wealthy relative, in this case his uncle, Everard Romfrey, on whom he depended for his financial prospects, by a striking but Quixotic father-figure, Dr Shrapnel; again the young man evoked the maternal protective love of an older woman and the passion of two younger women, one foreign and one English. As 'The Intellectual Captain', the young man had appeared in the notebooks in the list of *Comedies*. Out of his 'intellectual' interest in politics, under the influence of Carlyle, Meredith made the more curious of the two novels that stand at the apogee of his career. He relished the activities of his friend Captain Frederick Maxse, on whom Beauchamp was modelled, and of Carlyle, 'the greatest of the Britons of his time', without agreeing with either; the two together stimulated his own mind and feelings to greater activity than in any of the previous novels. His narrator now, instead of shadow-boxing with a philosopher, hoped himself to remain 'calm . . . impartial' while keeping 'at blood heat' his characters and the ideas which 'are actually the motives of men in a greater degree than their appetites'—not only the frustrated independence of Romfrey the ex-Whig, the Carlylean radicalism of Beauchamp, and the conservatism of his cousin Blackburn Tuckham (modelled on another friend of the author's, William Hardman), but also the newer radicalism of Seymour Austin, the 'rational Tory gentleman', a 'firm believer in new and higher destinies for women'. A good deal of the irony of the novel and of its standing as comedy depended on this vaunted 'impartiality' of the narrator—in practice, upon his power to prompt a revaluing of the actions of the men from the point of view of the women, as an 'indifferent England . . . reduced in Beauchamp's career the boldest readiness for public action . . . to the . . . influence of our hero's character in the domestic circle'. The domestic circle seemed hardly to offer matter for comedy: Romfrey's newly married wife forcing him to apologize for physically assaulting Shrapnel, Beauchamp losing his life, apparently pointlessly, after forfeiting both of the women to whom he had been passionately attached, Tuckham effortlessly gaining one of the latter, the English heiress, Cecilia, while the Frenchwoman, Renée, after accepting an empty, arranged marriage, proved ready for open rebellion.

Behind the narrator stood the Meredith who was later to express the hope that fiction, 'exposing and illustrating the natural history of man', might help us to 'sustaining roadside gifts' on the way to 'great civilization'. From the ending of *The Shaving of Shagpat* with 'Laws for the protection and upholding of women', right on to the letter written to *The Times* in 1906 when the suffragettes were demonstrating, his view was that such civilization depended upon the 'degree of social equality of the sexes'. This meant, as he said in the letter, that men need to be 'taught that woman is a force to be reckoned with'. The comedy of *Beauchamp's Career* arose from the demonstration of that need in the current circumstances of limited female independence. In a deft small sketch, a womanizer like Lord Palmet, the diverting commentator on Beauchamp's election canvass, became the representative of a traditional belittling attitude towards women. A major character like Romfrey, the old fox fully exhibited in his genial irascibility, presented the chivalrous obverse of Palmet's attitude, assaulting Shrapnel out of a traditional male determination to champion an 'unprotected' woman. (It was of course a more complicated action than that: 'Everard Romfrey never took a step without seeing a combination of objects to be gained by it'; and he himself bore some blame for her vulnerability to 'gossip', besides hating the beguiler of his nephew.) The lady he was defending, Rosamund Culling, could only lament the male 'senselessness ... Is there anything to be hoped of men?' In thinking 'bitterly' of Beauchamp's 'idea of their progress', she was echoing Austin's view that there were 'more certain indications of the reality of progress among women than any at present shown by men'. By contrast, it was possible to imagine that Beauchamp would 'stare, perchance frown conservatively, at a prospect of woman taking council, *in council*, with men upon public affairs, like women in the Germania!'

Beauchamp's politics, while readily comprehensible by a reader in 1874–5 and the subject for direct argument by the characters, entered most actively into the plot in the ironical comedy of his dealings with the women he attracted. The inner view of their sufferings showed vividly the complications he failed to grasp, and the force of the inherited obligations he asked each in turn to desert in order to marry him. The point was in the relative ease with which Beauchamp could put aside the claims of family and class, compared with the extreme difficulty experienced by the women, each of whom recognized in her own way that 'women . . . cannot act independently if they would continue to be admirable in the world's eye'. What the world's eye meant for the young Frenchwoman who refused Beauchamp appeared with the economy of decisive caricature when, after her marriage, he was suddenly summoned to her from the middle of his election canvassing. Her plight, presented as detestable in both its constrictions and its freedoms, contrasted with the conventions, more humane but none the less binding, under which a decision to marry, similar to hers, might be taken by an Englishwoman. With both women Beauchamp failed to recognize the moment at which and the way in which he might have shaken them out of their accustomed ways. With Renée, when 'a word of sharp entreaty would have swung her round to see her situation with his eyes, and detest and shrink from it', he 'committed the capital fault of treating her as his equal in passion and courage'. With Cecilia Halkett, when he 'might have taken her with half a word', he was preoccupied with justifying Shrapnel's politics to her and with the reflection that she 'was not yet so thoroughly mastered as to grant her husband his just prevalence with her, or even indeed his complete independence of action'. Such a reflection was the form assumed by his simpler awareness of having embarrassed her by his eagerness to fly to Renée. The restrained comedy came from the reader's recognition of such processes, not from the art, blessed by what the narrator called 'our elegant literature', of ridiculing 'the comical wretch'.

The reflections of Miss Halkett, which moved her into the position of the chief of the book's heroines, were exposed to a similar kind of sympathetic irony. To praise her as 'a peerless flower of our English civilization' was to recognize English limitations. As a lady, her 'aim at an ideal life' had inclined her to 'self-worship; to which the lady was woman and artist enough to have had no objection, but that therein visibly she discerned the retributive vain longings, in the guise of highly individual superiority and distinction, that had thwarted her with Nevil Beauchamp, never permitting her to love single-mindedly . . .'.

The ability to love and the circumstances which brought politics into relation with it were turned to the light in images of the open air and 'the revelling libertine open sea'. In the strong narrative episode which initiated Beauchamp into his troubled career, a sunrise over the Alps, seen from the Adriatic, for a moment moved Renée to accept his plan of flight from an arranged marriage. The inherited splendours of Venice, to her 'stood for the implacable key of a close and stifling chamber, so different from this brilliant boundless region of air'. Cecilia, who 'loved the sea', was imagined in relation to the craft that conquer it. Her longing, even after rejecting Beauchamp, was 'to embark with him in his little boat on the seas he whipped to frenzy'; the man she had accepted in deference to her father left her in 'an agreeable stupor, like one deposited on a mudbank after buffeting the waves'. 'As the yacht, so the mistress' was Beauchamp's thought as he watched her vessel moving 'through dusky merchant craft, colliers, and trawlers, loosely shaking her towering snow-white sails, unchallenged in her scornful supremacy; an image of refinement of beauty, . . . a visible ideal of grace for the rough world to aim at'. The image brought politics and personal life together in terms which supported the narrator's analysis of her love and the plot of Beauchamp's hesitations.

It was the women of Beauchamp's career who made him most clearly aware of his position, caught between the civilized pleasures for which his taste was fine and the ideal of a more just order which obsessed him once 'he had drunk of the *questioning* cup'. The yacht and the lady 'were precious examples of an accomplished civilization'.

But were they not in too great a profusion in proportion to their utility? That was the question for Nevil Beauchamp. The democratic spirit inhabiting him, temporarily or permanently, asked whether they were not increasing to numbers which were oppressive? And further, whether it was good for the country, the race, ay, the species, that they should be so distinctly removed from the thousands who fought the grand, and the grisly, old battle with nature for bread of life. Those grimy sails of the colliers and fishing-smacks, set them in a great sea, would have beauty for eyes and soul beyond that of elegance and refinement. And do but look on them thoughtfully, the poor are everlastingly, unrelievedly, in the abysses of the great sea. . . . One cannot pursue to conclusions a line of meditation that is half-built on the sensations as well as on the mind. Did

Beauchamp at all desire to have those idle lovely adornments of riches, the Yacht and the Lady, swept away? Oh, dear, no.

Despite the copiousness of incident, the novel relied for the essentials of its action upon the meditations of the characters. As the narrator rather querulously put it,

> It is the clockwork of the brain that they are directed to set in motion, and—poor troop of actors to vacant benches!—the conscience residing in thoughtfulness which they would appeal to . . .

It was characteristic that they should be 'directed' by the narrator to display the working of their brains and that the aim of the exhibition should be an appeal to conscience. Both the directing and the moral appeal were equally plain whether or not the narration was accompanied by commentary. Where there was commentary, it was seldom unmodified by the inner voice of the character; where the latter's inner voice was allowed directly to speak, it was seldom unmixed with open interpretation from the narrator.

It was assumed that the reader's 'thoughtfulness' would be stimulated by the narrator's ironical deference to Carlyle, the favourite author of the 'Intellectual Captain' and the source of his mentor's ideas. Meredith's own mixed atitude to Carlyle is plain throughout his letters: 'an inspired writer' speaking 'from the deep springs of life', but when attempting 'practical dealings . . . impetuous as a tyrant. He seeks the short road to his ends; and the short road is, we know, a bloody one.' The novel subjected Beauchamp's politics to similar praise and blame, the praise through the celebratory images of the sea, the blame through the course of the impetuous career. Even the placing of the career in near-contemporary circumstances had been prompted by Carlyle: the 'impartial' narrator undertook, 'following the counsel of a sage and seer, . . . to paint for you what is, not that which I imagine. This day, this hour, this life, and even politics, the centre and throbbing heart of it (enough, when unburlesqued, to blow the down off the gossamer-stump of fiction at a single breath, I have heard tell) . . .' As in the Proem to the next novel, he could not resist parody while facing the challenge of Carlyle. Here it was the challenge of Carlyle's advice to him (given in a conversation, probably of 1862 and variously reported) 'to turn to history as the repository of facts'. To represent 'this day, this hour', without the limitating associations of direct topicality, he applied the method of *Vittoria* (which was also George Eliot's method) of setting the action back in a time recognizably like that of the novel's composition in its salient features. He chose a period including the Crimean War, a period when, as in the early seventies, the proponents of strong national defence and of franchise reform met the opponents of increased government expenditure and of democracy. Glancing back to the time of Palmerston 'before the liberals were defined', Meredith was looking well forward to the alignment of conservative and liberal conservative against the radical reformers who were concerned about the condition of the common people, 'left to the ebb and flow of the tides of the market, now taken on to work, now cast off to starve . . .'. His hero was endowed with a Carlylean 'faith . . . in working and fighting', but shown as failing in the public arena to become one of the Great Men whom the lectures which first aroused his interest, *On Heroes, Hero-Worship, and the Heroic in History*, had called 'the modellers, patterns, and in a wide sense creators, of whatsoever the general mass of men contrived to do or to attain'.

But Beauchamp's influence in the domestic circle, though mocked as a 'flat result', was of equal importance to an author concerned with 'some degree of social equality of the sexes'. He showed the women who loved Beauchamp disturbing his public career because their position in society hampered or distorted their desire to 'act independently'. Not all their disturbance was negative in effect: Rosamund Culling for instance was able to bring the pressure to bear which allowed Beauchamp 'the great domestic achievement', recounted at the end with laughter, of getting Romfrey down on his knees. Beauchamp's culminating 'domestic achievement' was not, as it would have been in most novels, his marriage and fatherhood: it came when he saved an unknown mother her son, at the cost of his own life. Incomprehensible apart from Carlyle, it was Meredith's concluding challenge to him. '*The end of Man*,' said chapter 6 of *Sartor Resartus*, Book II, '*is an Action, and not a Thought*, though it were the noblest.' The last chapter of the novel showed the Intellectual Captain when his concern for the poor had apparently suffered some commutation, married to a woman of conservative tendencies and engaged on a translation of Plato. ('This philosopher singularly anticipated his ideas,' said the narrator with some irony.) In the concluding scene, among the colliers and trawlers, which, contrasting with Cecilia's yacht, had brought his hero to a sense of his own perhaps false position, Meredith pressed into one act Beauchamp's devotion to people like 'the insignificant bit of mudbank life remaining in the world in the place of him'. The dismissive language for the boy was wholly in Carlyle's manner (as in the phrase from *Past and Present*, Book III, for Mungo Park when saved by an aborigine: 'a horrible White object in the eyes of all'). Beauchamp himself could also be called insignificant by 'the world': 'some few English men and women differed from the world in thinking it had suffered a loss.' As the narrator had planned at the beginning to justify imaginative literature as if to Carlyle, so at the end he appeared to be justifying his hero to the same mentor—who had, after all, claimed in the chapter of *Past and Present* called 'Reward', that

> the brave man has to give his Life away. . . . Give it, like a royal heart; let the price be Nothing: thou *hast* then, in a certain sense, got All for it! The heroic man,—and is not every man, God be thanked, a potential hero?—has to do so, in . . . the most heroic age, as in the most unheroic . . .

Although its tenor was not always clear because the reader was left to supply the relations between its parts for himself, the strange comedy of Beauchamp's guileless impetuosity turned the tide of Meredith's contemporary reputation. He went on in *The Egoist* to a simpler example of the genre after first attempting to clarify the ideas about it which had been implicit in his lists of *Comedies* and *Comediettas in Narrative*. The attempt, in the lecture 'On the Idea of Comedy and of the Uses of the Comic Spirit' (1877), showed him inspecting comic drama in search of another method of dealing imaginatively with the state of an England bloated with wealth but imperfectly civilized. The distinguishing marks of Comedy concerned morality as much as literary criticism, in a circular argument which summarized his own preoccupations up till now. The 'test of true comedy' was that it should 'awaken thoughtful laughter'; its 'flourishing' formed 'one excellent test of the civilization of a country', for it was only in 'a society of cultivated men and women', and one 'where women are on the road to an equal footing with men, in attainments and in liberty', that the thought with which 'true Comedy' informs laughter could be awakened. So 'Philosopher and Comic Poet' were 'of a cousinship in the eye they cast on life', to 'see Folly perpetually sliding into new shapes in a society possessed of wealth and leisure', Folly' being 'the daughter of Unreason and Sentimentalism' in these circumstances. It was clearly the author of *Sandra Belloni* speaking. By 'neglecting the cultivation of the Comic idea', the English were 'losing the aid of a powerful auxiliar'. For the Germans, its 'discipline' was still 'needful for their growth', to 'enliven and irradiate the social intelligence', but the English, he claimed, 'have not yet spiritually comprehended the signification of living in society' at all. The way to comprehend it was to expose 'our conventional life' to this 'first-born of common-sense, the vigilant Comic'.

The doubtful use to him of theorizing about comedy was plain in two shorter fictions appearing at this time, 'The House on the Beach' and 'The Case of General Ople and Lady Camper' . The first, begun in the year of the success of *Great Expectations*, told of a man with something to hide returning rich from Australia, to the embarrassment of an old acquaintance. It was carefully shaped into a comedy in four acts of three chapters each, one for the friends' meeting, one for the Australian's discovery that his friend knew his secret but might keep quiet if married to the daughter he had brought back to England with him, one for the daughter's acquiescence despite 'her wiser intelligence', and one for the storm which unmasked the fiancé as willing nevertheless to expose her father, an 'uncultured Australian who did not seem to be conscious of the dignities and distinctions we come to in our country'. Its form was lucid, the cast small, the place limited, as in a play. Its climax, however, came less in matter for the laughter of an audience than in the 'personal exultation' of the daughter. 'The young lady was ashamed of her laughter; but she was deeply indebted to it, for never was mind made so clear by that beneficent exercise.'

'The Case of General Ople and Lady Camper' was another experiment with the discipline which might be brought to bear upon a character in a comedy. Limiting even further the number of characters and the size of the setting, Meredith produced a mixture of satire and farce which awakened 'derisive' rather than 'thoughtful' laughter; as embodiment of the Comic Spirit, the superior Lady Camper eventually supplied, ready made, all the thought that was needed, in her explanation to her victim.

The mild egoism of General Ople which prevented him from noticing the advances being made to his daughter, like the vanity of the chief occupant of the House on the Beach, was certainly exposed, but it did not follow that the result, though following the theory of the lecture, would be comedy to the reader.

The novel which grew out of these efforts, *The Egoist* (1879), faced the same problems, for it struck even deeper into what the lecture called 'incidents of a kind casting ridicule on our unfortunate nature instead of our conventional life'—exactly the kind to 'provoke derisive laughter, which thwarts the Comic idea'. It was even a question whether they provoked laughter at all, the egoism of Sir Willoughby Patterne was so arch and unrelieved. 'Our unfortunate nature' was the preoccupation, and the egoist as 'our fountain-head, primeval man'. It was in fact more like a moralist's 'excogitation of the comic', Meredith's depreciatory phrase for the comedy of Massinger. To male readers, at any rate, it was too stinging for laughter, until Patterne was aroused to self-defence, when it became farce dependent on misunderstanding. Too much of the discussion of Patterne's egoism was general and theoretical and its fictional embodiment merely repellent, like his treatment of Laetitia Dale whom he knew to be in love with him. His fatuity, theatrical and excessive, and apparent to the reader long before the author pointed it out, diminished the reader's wish to understand; and thoughtful laughter became impossible when thought was dulled by the repetitiveness of the exhibition. Of course what the lecture called 'our conventional life' was important in so far as it was under its forms that Patterne (with his exemplary surname) disguised the drive of the 'primeval man' to dominate; but those forms merely made the disguise too easy for him, with his wealth, position, and entire ability to choose his associates and dependents, keeping at a distance anything from 'the world' which might disturb his self-congratulatory command.

As if realizing the difficulties raised by the canons of the lecture on comedy, Meredith offered them, in the introductory chapter, in the form of parody. He later warned his French translator that this 'pretended testimony to the merits of Comedy is in the vein of Carlyle'. So it was in the extravagant argument of 'an enthusiast' that 'the vigilant Comic' was justified. In a Carlylean fervour of incongruous juxtapositions, comedy became 'the ultimate civilizer, the polisher, a sweet cook', corrector of 'pretentiousness, of inflation, of dulness, and of the vestiges of rawness and grossness to be found among us'. Even if this first chapter was not written last (as Meredith once claimed was his practice), its mocking of the moral urgency of Carlyle suggested that the corrective function of the Comic Spirit might readily lead to didactic excess if it failed to release the 'delicate spirit' of charity; in this the parody seemed either an admission of failure or an appeal to the reader to make up for the novelist's deficiency with charitable resources of his own. For in the treatment of the chief egoist the author's aim to correct was all too apparent and the reader's charity unprovoked.

What kept the reader going was not the spectacle of pretentiousness corrected but the simple doubt as to what a tyrant would do next when he was, superficially, civilized, so that the cruder options were closed to him. Meredith had hold of another strong action which the reader might actually take pleasure in watching, unlike the continually embarrassing behaviour of the Poles, or Rhoda Fleming's hideous mistakes, and it was simpler than Beauchamp's equivocal career. The escape of a girl of 19 from a man who seemed all-powerful, endlessly wily, apparently commanding the fates of her two main confidants, and subduing even her father, was a wonderful development of 'The House on the Beach', beginning where it had ended, with the dropping of the scales from her affianced eyes.

The triumph of the book was to centre its action in the consciousness of the fiancée, Clara Middleton, with her unusual humour and intelligence. Deprived of her natural allies by being motherless, while her father was even more devoted to his library than Elizabeth Bennett's father in *Pride and Prejudice*, she at once aroused interest by seeking for allies in the limited country-house setting where everything was controlled by the fiancé from whom she was trying to free herself. So it became plain just what advantage Meredith had gained from a rekindled interest in comedy: its benefit to writers, he said at the end of his lecture, was that 'they must have a clear scheme . . . a definite plan'. The distinctive thing was not the exposure of egoism, in which Meredith had long been engaged, but the managing at last of a coherent whole sequence, the finding of a form to make dramatic the results of Clara Middleton's awakening to egoism, and her complementary understanding that, however repulsive its complete form in Patterne, she must have a proper proportion of it herself:

'I must be myself to be of any value to you, Willoughby.'

I must be myself or I shall be playing hypocrite to dig my own pitfall,' she said to herself . . .

Her isolation and her longing for 'some little . . . free play of mind in a house that seemed to wear . . . a cap of iron' made the devices of French comedy, in the arrangement of scenes between pairs of actors, into something more than pedantic imitation, for it showed the proliferating effect of Patterne's egoism in the increase of her own: she had no option but to make use of other people, one at a time, in order to prevent the forcible reduction of her own identity. Each of the small number of persons available for the role of confidant represented a different aspect of the pursuit of love seen as, unavoidably, in some degree, the pursuit of power: Laetitia Dale, the devoted and unrewarded woman, who gained, from conversing with Clara Middleton, a 'newly enfranchised individuality', Colonel De Craye, with his experienced adroitness and wit in the traditional amorous game which the action exposed to criticism, Mrs Mountstuart Jenkinson, convinced quite simply that Patterne should be supported in his conceit of having attained perfection, 'for the more men of that class, the greater our influence'. At the opposite extreme were the too-silent self-suppression of Vernon Whitford and the freshly springing individuality of a boy of 12, Crossjay, exposing them all by his freedom not from sexual feeling but from the need to dominate by means of it.

Allusion to the stage—in the subtitle, 'A Comedy in Narrative' and in the chapter titles, or in an entry like that of Mrs Mountstuart 'like a royal barge in festival trim', echoing Millamant's entry quoted from *The Way of the World* in the lecture—heightened the amusement of palpable artifice. The action was delectably managed to imitate the drama, with the evidence of concealed events furnished by those who had suffered, all unwittingly, from their consequences, with characters entering whose presence would be above all others inconvenient at that moment, the anticipation of crucial action 'within an hour', involuntary eavesdropping by the one character, Crossjay, who would not be suspected of malice or falsehood, and the gathering of all on stage for the climactic misunderstandings.

The engaging *coups de théâtre*, however, obscured the fact that the action was much less exploratory of character than would normally be expected in a novel of such length. The surprises were all of situation; the characters continued to behave typically. The exception was the one about whom the reader knew least, Laetitia Dale. The undivulged degree of her transformation might conceivably lead, at one extreme, to radical change in the Egoist himself, at the other to no more than the sharpening of his will to power. The conclusion left him, admittedly, appreciating her possession of brains, as he had failed to do with Clara, but that recognition still implied an unjustified depreciation of Clara's intelligence, in his confidence that he would get the best of any bargain. The author's last word seemed generally grateful, or ominous with personal experience, according to how it was read:

> . . . he had the lady with brains! He had: and he was to learn the nature of that possession in the woman who is our wife.

The outcome, which would have been of first importance in a profounder treatment of egoism, mattered less in a comedy of freedom gained and a tyrant defeated—possibly temporarily and certainly in only one of his aims.

'The Egoist', said the Prelude, 'surely inspires pity.' If he did, it was less for being what he was than for being the subject of the author's inexorable analysis. The discursive subtleties, however, were not matched by the dramatization of Patterne's habit of mind, which was too often cruder than the commentary:

> 'Whenever the little brain is in doubt, perplexed, undecided which course to adopt, she will come to me, will she not? I shall always listen . . .'

Success here appeared to be measured by the power merely to make the reader wince, without anything like the laughter aroused in chapter 23 by the brilliant mixture of dramatization and commentary as Patterne, inspired by his 'favourite readings', imagined a later meeting with an estranged Clara, in the language of popular romance:

'Must I recognize the bitter truth that we two, once nearly one! so nearly one! are eternally separated?'

'... In mercy let it be for ever. Oh, terrible word! Coined by the passions of our youth, ... it ... is the passport given by Abnegation unto Woe that prays to quit this probationary sphere. ... It is better so.'

The generalizations about egoism were often acute, but the figure of the actual egoist remained fixed, his 'inflation' and 'rawness' quite uncorrected, unpitied in his preservation of appearances whatever his inward humiliation. At the conclusion the effect less resembled anything discussed in the Prelude or in the lecture than Wilkie Collins's impudent laughter at the success of the conventional villain. Like the comedies of Beaumarchais this one suggested grounds for contemporary rebellion—against the little tyrant of so many fields and against the male 'aesthetic gluttony, craving ... purity infinite, spotless bloom' in women. But, finally, power remained little diminished.

Egoism was better dramatized in other characters, especially in Clara Middleton. Although Meredith could ironically call 'infantile' the precept she adopted, as if in a George Eliot novel, 'We must try to do good; we must not be thinking of ourselves ...', he showed her following it honestly in a serious matter, the education of the boy, Crossjay. But she did so with a 'remote but pleasurable glimpse of Mr Whitford hearing of it'. George Eliot's subtler forms of egoism were alluded to when Clara was asked 'How of the heart?'

'None,' Clara sighed.

The sigh was partly voluntary, though unforced; as one may with ready sincerity act a character that is our own only through sympathy.

This was a refinement built upon the unreflecting egoism of Rosamond Vincy, who 'acted her own character, and so well, that she did not know it to be precisely her own'. Its general application was again deftly inserted in 'our'. Similarly, the unacknowledged wishes for a deliverer which had been dramatized in Dorothea Brooke's interest in a portrait of Ladislaw's ancestor took a more economical form in Clara Middleton's slips of the tongue (which caused Freud, with his concern for such things, to make his very high estimate of Meredith): the Captain Harry Oxford, with whom Patterne's previous fiancée had escaped, was called 'Harry Whitford' and Whitford 'Mr Ox-Whitford'. The spontaneous springs of the necessary forms of egoism seemed to open without effort.

Equally dexterous, the handling of metaphor was rather more mannered and selfconscious. Where it arose out of an affected strain in the dialogue, as with Mrs Mountstuart Jenkinson's 'rogue in porcelain' figure for Clara, it served a satirical purpose in contrasting with the livelier language and aspirations of the girl herself. Repeated in the porcelain of de Craye's broken wedding present and in Clara's indication of her changed position by rejecting such a present from another, the literal image became metaphorical in its connection with the convention of 'ornamental whiteness' demanded by the delicate sensual glutton. Again the contrast was made with the actual 'captured wild creature' in 'all the health of her nature'. The kind of whiteness more appropriate to her was imagined in terms of naturally growing things, her figure within its summer dress like 'the white stem's line' within 'the silver birch in a breeze', or the wild cherry blossom, contemplation of which took her 'from deep to deeper heavens of white, . . . soaring into homes of angel-crowded space, sweeping through folded and on to folded white fountain-bow of wings, in innumerable columns' until her vision was contracted to Whitford awaking beneath, who then saw her in the dazzling halo. It was all beautifully arranged, but bore the same relationship to the appreciation of the girl as her appreciation of the countryside bore to that of the actuality: 'I can read of this rich kind of English country with pleasure in poetry. But it seems to me to require poetry.' The author's own poetry here seemed to prompt a question like hers: 'What would you say of human beings requiring it?' All was clear, charming, controlled. The author knew exactly what he was doing, so that he felt, as he said in a letter, that the result, for all its 'roundness and finish', came 'mainly from the head'. Yet the artificial scheme made his access to the inner life of his characters more moving than that would suggest—not in the elaborate analysis of his Patterne egoist, but when inner life appeared as a phenomenon to be understood by one character of another only by surmise, and by the respective characters of themselves only as a ringing of alarm bells over the changing unplumbed depths, 'the shifting black and grey of the flood underneath'. Here was the continuity with the previous novel, where that flood, appearing in the literal narrative in the form of the sea, had disturbed the social surface to stranger purposes.

The Tragic Comedians (1880–1) brought into the open how much more Meredith's view of comedy depended on thought than on laughter. The actors in a calamitous love-story could be called comedians because it exposed them, in the terms of the lecture on comedy, as 'out of proportion, overblown, affected, pretentious, bombastical, . . . self-deceived . . . or mined with conceit'. Unfortunately there was in this instance little to separate their bombast and pretentiousness from that of the author. The story, based on that of Lassalle, the German Social Democrat leader, dead only fifteen years, was told in an idiom which concealed his nature and that of his fiancée under the coruscations of exalted metaphor. The one partner, Sigismund Alvan, appeared as eagle, giant, Titan, sun-god; he compared his Clothilde to the city of Paris, 'symbolized goddess of the lightning brain that is quick to conceive, eager to realize ideas . . .', while to her his eyes were 'light-giving' and the emotion of his eloquence 'like the play of sheet lightning'. Nothing that the reader was presented with corroborated these exaggerations; he was expected to be overcome by the sheer force of their accumulation in Alvan's case, but to believe them 'strained' in their application to Clothilde merely because the author said so. She was said to 'set fire to her brain to shine intellectually', empty words without some presentation of the result, and yet it was only by some such presentation that the 'strained' comparison with Paris could be judged. At the same time Clothilde was credited with the ability 'to measure the bombastical and distinguish it from the eloquently lofty' so that the quality of Alvan's affected monologues was apparently to be underwritten by her acceptance. When he on his side, with his 'splendid intelligence', praised light literature which, no doubt, contained this kind of stuff (for 'They were not members of a country where literature is confined to its little paddock, without influence on the . . . social world'), there seemed no possibility of ironic interpretation of it.

'Shun those who cry out against fiction and have no taste for elegant writing.'

'We are the choice public which will have good writing for light reading.'

'Light literature is the garden and the orchard, the fountain, the rainbow, the far view; the view within us as well as without. Our blood runs through it, our history in the quick. The Philistine detests it, because he has no view, out or in ... I have learnt as much from light literature as from heavy—as much, that is, from the pictures of our human blood in motion as from the clever assortment of our forefatherly heaps of bones.'

The hasty excesses of the apologia were matched in the story of Alvan, caught 'in the season before he had subdued his blood' into a death which was 'a derision because the animal in him ran unchained and bounding to it'.

Meredith seemed bent on cheapening the best of his own treatment of the story, for this exaltation of Alvan's vehemence obscured the vanity which had moved him at the turning-point of the action. When he surrendered Clothilde in order to win her back as a 'parent-blest bride',

The compressed energy of the man under his conscious display of a great-minded deference to the claims of family ties and duties intoxicated him. He ... reasoned by the grandeur of his exhibition of generosity ... that the worst was over; he had to deal no more with silly women: now for Clothilde's father!

It was here that, for once, the heroic exaggerations were ironic. Although the irony was unsubtle—especially in the contrast between 'silly women' and Alvan's earlier infatuated praise of Clothilde—there was an action to corroborate it, an action already foreshadowed in her fears that any hint of failure would turn him into 'a vain pretender to the superhuman', prompted 'to fling the gambler's die by the swollen conceit in his own fortune rather than by his desire for the prize'. In measuring the phrases against what the hero did, the reader had at last something to do himself so that the characters could for a moment seem to be more than mere generators of hyperbole and artificial conversation. Her feminine desire to submit her individuality to his led to her suspicions of his love; his over-confidence in himself as a man prevented him from understanding her. They were old themes to Meredith, but given new interest at the centre of the story. (Other old themes went their rounds, the English as a people 'gone to fat', the 'sentimental ... form of reasoning' in 'a lady of romantic notions', even her proviso, like Cornelia Pole's, 'I am his *if* he comes.') Picturesque hyperbole, however, soon forfeited genuine vitality, arising from the reader's participation, to the forced vitalism of the maker of metaphors. Alvan became 'figurable by nothing so much as a wild horse in captivity sniffing the breeze, . . . flame kept under and straining to rise'. So, though a 'contemner' of duels, he issued the challenge which led to his death. Just how this 'last temptation' stamped him as 'a grand pretender, a self-deceiver' was not clear. Nevertheless, like Clothilde's laugh, though 'unaware of her ground for laughing', at the news that he was wounded, his calamity was said to justify the title of a tragic comedian. The attempt to confer 'stature' by means of hyperbole was a reversion to the sensational methods of *Vittoria*. In taking his plot from history Meredith appeared to have committed himself to an action only part of which was susceptible to his more acute methods and then to have filled it out with deplorable materials meant to dazzle.

In the work of his final decade as a novelist, Meredith's main interest was in the reasons women might have for breaking the conventions governing their behaviour in relation to marriage. *Diana of the Crossways* (1884–5) was based, like *The Tragic Comedians*, upon history, and again emphasized the private aspects of a public career. Classical myth once more came into play in the attempt to make this private life of universal importance. The analogues which had inflated the character of Alvan were now limited to those connected with a single goddess (though of double form and triple face) and applied to the heroine's nature and society's misjudgement of it. Believing herself to be Diana goddess of chastity, she recognized that she was in her other form 'Hecate queen of witches', the form public opinion assigned to her after her separation from her husband. She also assigned it to herself because 'I am always at crossways', that is, the places of the meeting of three roads (where in antiquity the shrines of Hecate were erected). In the novel these were the places of cross purposes and divided aims, centred at first upon her house called Crossways. The analogues magnified her stature, though less outrageously than in Alvan's case, in preparation for her discovery that she was 'the Woman of Two Natures'. The story of the lengthening of her adolescence in the disasters following a hasty marriage bore sufficient resemblance, in outline, to the career of Caroline Norton to hold a large contemporary audience with the hope of revelations about it. The more important mysteries of the fictional character were imperfectly resolved—the nature of her relationship, first to older men, then, in turn, to a rising young politician and to a callow youth. Her final acceptance of a man who had concealed his devotion since before her first marriage suggested that his silence was the main cause of the whole imbroglio.

Devoting three volumes to this one career resulted neither in the creation of an individual character nor in the analysis of an interesting case. Her quick mind and a curious countervailing proneness to irrational behaviour were visible, despite the continued use of the impressionistic terms of *The Tragic Comedians*, 'blood', 'brain', 'lightning', 'fire'. The sharpness of wit and range of theoretical knowledge of life shown in her moral and social aphorisms could be seen to contrast strikingly with her relatively late awakening to sexual passion. In her justification the novel even offered a theory of wit resembling the former theory of comedy. The first chapter of *The Egoist* had called comedy 'the laughter of reason', approved of 'wise men' as 'the true diversion' which condensed the 'repleteness' of the Book of Egoism; in the corresponding chapter of *Diana of the Crossways* 'true wit' was praised as 'truth itself, the gathering of the precious drops of right reason, wisdom's lightning'. As Clara Middleton was the agent of comedy in its exposure of the Egoist, so Diana was to exercise wit as the revealer of truth, the gatherer of those precious drops. But the truth with which the novel was concerned was limited to the truth of her own nature and that was concealed from her wit, far beyond the reach of epigram, however much involved in paradox.

Diana, like Mrs Norton, was a novelist, and her novels, if more had been known of them, might have revealed more of that truth. Meredith, however, was interested only in the general features of her work which could be made to resemble his own. Diana was 'not one of the order whose Muse is the Public Taste', but one who 'wrote more and more realistically of the characters and the downright human emotions, less of the wooden supernumeraries of her story, labelled for broad guffaw or deluge tears—the grappling natural links between our public and an author'; she aimed at 'comic analysis that does not tumble into farce', though able if necessary to produce 'any amount of theatrical heroics, pathos, and clown-gabble'. Her work, then, became part of Meredith's continuing battle with popular taste, continuous with his own claims for the novel in hand. It was to be characterized by 'philosophy', to show us that 'we are not so pretty as rose-pink, not so repulsive as dirty drab'—in particular to show Diana as 'the flecked heroine of Reality: . . . a growing soul; but not one whose purity was carved in marble'. She was not to be 'the puppet-woman, mother of Fiction and darling of the multitude'. But that did not stop her from being the puppet of the author's 'philosophy'. At the end it was claimed that only 'those who read her woman's blood and character with the head' would care for her. All too often this meant caring for his somewhat abstract explanations of her inner development. Even the images for it tended to be general rather than specific. To imagine passion as a troubled sea over which her soul, taking 'short flights . . . flew out like lightning', when a more 'remorseless' introspection 'might have . . . shown her to herself even then a tossing vessel as to the spirit, far away from that firm land she trod so bravely', seemed mere picturesque periphrasis for her self-deception, when compared with the analysis of Cecilia Halkett in *Beauchamp's Career*.

When Meredith was forced to go outside the heroine's consciousness for the explanation of her conduct, as in the central precipitating event, her betrayal to a newspaper of her lover's confidence, the sketchiness about the external aspects of her life made the novelist's difficulties almost insurmountable. Labouring over the passage more than over any other part of the book, he still left it indistinct.

For readers in 1884–5 this sketchiness was mitigated by an undercurrent of topicality. An Irish heroine married to an Englishman who wished to end their separation while she did not, held overtones of the state of the Union which the parliamentary tactics of Parnell had been keeping before the public since 1877. The suspicion arising from the connection of Diana's name with that of a prominent politician even had its contemporary parallel in the common talk about Parnell and Mrs O'Shea, an acquaintance of Meredith's. Reflections about democracy, 'a perilous flood' only when 'dammed', were as appropriate in the 1840s, the period of the novel, as in that leading up to the franchise reform of 1884. Even Diana's private position was relatively topical: she was, like Mrs Norton, an estranged wife, of precarious financial position before the Married Women's Property Acts of 1870 and 1882. But all public matters were in the background and of little help to the substantiality of the novel. That no political scandal, for instance, appeared to be caused by the newspaper revelations for which Diana bore responsibility made the climactic episode dream-like. Even the tattlers about her private life became welcome to the reader as evidence that society existed, to dread and detest 'brains in women'.

In *One of our Conquerors* (1890–1) Meredith spoke of himself as concerned with both public and private life, attempting 'both a broad and a close observation of the modern world'. The central character, Radnor, a conqueror in the sense that he had made 'one of the biggest fortunes of modern times', was a topical development of Mr Pole from *Sandra Belloni* in the sixties. Radnor was not a merchant but a financier, not a sporadically active figure, alternately pathetic and farcical, but the book's animating consciousness, seen from within with critical, though curiously sympathetic, understanding.

The phrase which formed the title had first been used by Lady Camper in her 'homeopathic cure' of General Ople's vanity: he had 'nursed the absurd idea of being one of our Conquerors', 'our' referring there primarily to women. Around Radnor, who had conquered fortune as well, was disposed a remarkable group of women, his *de facto* wife Nataly and illegitimate daughter Nesta, to both of whom he was devoted, a casual temptress Lady Grace Halley, who believed that 'he and she together might have the world under their feet', and, in the background, his elderly legal wife, Mrs Burman Radnor, ill and, for most of the action, about to die. With such materials for external conflict of the kind beloved by Radnor in novels—'I can't read dull analytical stuff or "stylists" when I want action—if I'm to give my mind to a story'—Meredith characteristically chose the inner view. External events were disproportionate to the inward conflicts for which they formed the occasion, in a manner much less troubling to the reader in the century of *Mrs Dalloway* or *Under the Volcano* than in 1891. Radnor, after slipping, apparently on a banana skin, became subject to a 'morbid indulgence in reflection: a disease never afflicting him anterior to the stupid fall'. 'Shadowed by doubts of his infallible instinct for success', he found 'he had taken to look behind him', in search of the ghost of an Idea glimpsed in the aftermath of his fall. In this pursuit the mode of the narrative moved both back to the compendious history presented in Carlyle's *Chartism* and forward to a mingling of townscape with curious sequences of mental association such as readers have become familiar with in Joyce and Virginia Woolf.

Radnor's visions in the London streets made pictorial many of the ideas which Meredith had long absorbed from Carlyle. Indeed the very conception of the book seems to arise from Carlyle's speaking in chapter 8 of *Chartism* of 'the Saxons fallen under that fierce-hearted *Conquaestor*, Acquirer or Conqueror', and of 'the fate of the Welsh ... of the Celts generally, whom a fiercer race swept before them'. At any rate the novel's strange apocalyptic tone was such as might have been encouraged by rereading Carlyle. Just what suggestion was to be carried by the Welsh surname of Meredith's 'Acquirer or Conqueror', Radnor, to go with his Christian name Victor, and by the daughter's Welsh Christian name, Nesta, along with Victoria, 'victory', is not clear. But the tendency of the plot is quite plain, to present the decline of the father and the increasing stature of the daughter. Radnor, with his hope that 'a new Æra might begin' if the world would learn 'that Nature—*honest* Nature—is more to be prized than Convention', was obviously moved primarily by concern for his own position as a man whose wife refused to divorce him. But his meditations also set in an equivocal light favourite ideas from Carlyle which were presented as rather beyond his grasp. Radnor's fear of 'the Mammonism we're threatened with' echoed Book III of *Past and Present*; to deal with it he recommended 'dispersion of wealth', echoing Carlyle's contrast, in *Chartism* and later, between 'conquering ... this Planet' and 'sharing ... the fruit of said conquest'; but after claiming that the secret was 'decent poverty', Radnor went on, 'I've a steam yacht in my eye, for next month on the Mediterranean.' His Carlylean belief in 'true aristocracy', the hope he entertained, before his first public appearance in support of his parliamentary candidature, of giving 'the lighted leadership' which would bring 'the English ... into cohesion', formed the culmination of a more egotistical 'resolve to cast off the pestilential cloak of obscurity shortening his days, and emerge before a world he could illumine to give him back splendid reflections'.

His anxieties, his vanity, his volatility were anatomized with humour without being derided. His relations with the women of the novel followed a parallel course of ambitious planning and limited insight. Before the final calamity, after busily nurturing his

Idea 'to have Something inside him, to feel just that sustainment', he was seen inwardly testing his love for Nataly 'by his readiness to die for her: which is heroically easier than the devotedly living, and has a weight of evidence in our internal Courts for [*sic*] surpassing the latter tedious performance'. A more tedious performance was exactly what she would have preferred. 'Deceiving' people with her 'habitual serenity in martyrdom', while suffering from a concealed illness, she was shown placed under strain by Radnor's social ambition when her wish was to 'live obscurely' until they could be legally married. His ambition for an aristocratic marriage for their daughter caused her to turn a 'penetrative hard eye on herself'. Her self-analysis, both simpler and subtler than Diana's, was dramatized in a social situation more fully understood. And there was little of the enthusiasm with which the narrator had tried, in the novels before this, to express his own feeling about the heroine rather than show her to the reader. Figurative language had the function of indicating something done, not merely of evoking emotion in her favour, as so often with Diana Warwick. Nataly had given to Radnor her life, 'morally subdued, physically as well, swept onward'; now, at the prospect of deceiving an eligible suitor for her daughter, 'she was arrested . . . like a waif of the river-floods by the dip of a branch'. The duty she imposed on herself, of telling this very proper young man that his fiancée's parents were unmarried, showed her ability to face the intolerable and even, with her precarious health, to 'meet death, if need be; readily face it as the quietly grey tomorrow', in contrast to Radnor's heroics. The strain of living in defiance of conventions to which she herself deferred was shown in her distress at Nesta's friendship with a woman of tarnished reputation. It made her one of the most moving of the author's heroines, and that because he was commonly reticent about her feelings. In the fine economical scene in which they received the forgiveness of the dying wife, Nataly was seen only silent at prayer, in contrast with the intimate observation of Radnor's wandering thoughts: in the incipient disturbance of his mental balance, 'he beheld his prayer dancing across the furniture; a diminutive black figure, elvish, irreverent, appallingly unlike his proper emotion', until 'He heard the cluck of a horrible sob coming from him. After a repetition of his short form of prayer deeply stressed, he thanked himself with the word "sincere". . .'

Nesta, the daughter, was more elaborately, even allegorically, presented, 'taught to know herself for the weak thing, the gentle parasite, which the fiction of our civilization expects her ... to become in the active' and 'in the passive ... a rock-fortress ... magically encircled'. She was, in the story, somewhat magically encircled, subjected to a degree of mythological idealization exactly recalling Diana—'with the mark in sight, however distant, she struck it, unerring as an Artemis for blood of beasts'. A suitor asking her father's permission to speak was imagined, with an invitation to the reader to 'catch the fine flavour of analogy', as Orpheus charming Cerberus when seeking Eurydice, on 'the dizzy line of division between the living and the dead'. It was a selfconsciously comic imitation, for the suitor was no sweet singer but a clergyman, Barmby, of cavernous voice, a 'blast at the orifice'; situating his request, however, on that 'dizzy line' carried the 'fine flavour' of allusion to Radnor's position in the plot and to Barmby's sermons after the three deaths at the conclusion. He was to lose Nesta, but she was to give him, through 'his uninterred passion', the impetus to 'clutch by the neck' his 'vile self'. She shared her father's proneness to the seeing of visions and they were equally ominous, as in her strange apocalypse when visiting the Mausoleum at Dreux, her hopes for a King under whom 'there would be a union of the old order and the new, cessation to political turmoil: Radicalism, Socialism, all the monster names ... appeased, transfigured'. Her role in the final pages was that of an idealized representative of hope for the future, befitting her second name of Victoria and conveyed in abstractions flushed with intemperate imagery—her 'ruthless penetrativeness' making her intolerant 'of a world whose tolerance of the infinitely evil stamped blotches on its face and shrieked in stains across the skin beneath its gallant garb'. And in a return to Meredith's old polemics, she had, 'strongly backing her, upholding her', men who were 'anything but sentimentalists'. Nevertheless the girl was well defined by her actions. She was shown moved by what she saw of the position of other people, including her father, whom, for instance, she successfully kept from meeting Lady Grace by walking home with him herself in the evenings. Her father's anxiety that she should be out of the way when his wife died (and he could marry Nataly) led to her making a friendship of which he disapproved; but she defended it, and so her independence grew at the same time as her mother was for the first time discovering her own. In the clash of father and daughter her naïve loyalty was indicated by its effect upon her friend, Mrs Marsett, an effect in which the author's irony was no doubt too much in abeyance, but which was presented as affording the girl some preparation for the courage she would have to show in chosing her mate: 'but for love of her father, for love and pity of her mother, she would have ventured the step to make the man who had her whole being accept or reject her.'

At the heart of this strange, over-written book, the observation of the relationships between the three chief characters was as fine as anything in Meredith. To develop from them a 'broad' as well as a 'close' view of 'the modern world', he relied not only on metaphor and myth but on the suggestions, in George Eliot's late manner, of cardinal contemporary examples—Bismarck, 'The Prince of thunder and lightning of his time', now supplanted by 'the Berlinese New Type' who 'put on frankness as an armour over wariness, holding craft in reserve', Japan transformed and considering the adoption of a new language, the 'Kelts' who 'can't and won't forgive injuries'. To H. G. Wells's mouthpiece in *The New Macchiavelli* (1911), 'It discovered Europe to me, as watching and critical', and constituted, in its criticism of an Englishman, 'a supplement and corrective of Kipling'. No doubt the story took too long to develop, clogged by digressions—on old wine, on the Voltairean novel written by one of the characters, or even on the habits of an irrelevant pet dog—and often hindered by the style, unusually cumbersome even for Meredith, of the narrator's colloquy with the reader. But the central action when it came, came with striking impetus and economy.

Before writing *One of our Conquerors* Meredith had been at work, immediately after finishing *The Egoist*, on a version of what was to be his last novel. Even before that, as 'The Amazing Marriage (Gossip as Chorus)' it had appeared in the list of projects he made in the 1860s. Although the surviving manuscripts of part of it show that the novel underwent important changes before it was published in 1895, the main elements of its plan persisted, especially the contrast between the telling of the story by a gossip, as a sequence of scraps of amazing news, and the understanding of it from within the marriage. It gave another example of the contrast, underlying most of his work, between the popular novel of incident written for 'an audience impatient for blood and glory' and the novel written out of what he had called in 1860 'this cursed desire I have haunting me to show the reason for things'. For this kind of novel he worked to create an audience not only interested in seeing the inner 'elementary machinery at work', but ready to be moved by the difference such knowledge of its working made to the moral assessment of the resulting actions.

The preliminary events narrated by Dame Gossip, in a narrative idiom contrasting markedly with the involuted style of *One of our Conquerors*, appears to have shown him their possibilities for independent treatment. So before he could finish the longer novel, he produced, relatively quickly, *Lord Ormont and his Aminta* (1893–4). Interest in a woman's beginning 'to think' about her marriage obviously suited the decade of the 1891 Court of Appeal judgement limiting a husband's authority in the leading case *Reg.* v. *Jackson*, and of the 1893 extension of the Married Women's Property Act. The author himself played Gossip to Aminta's marriage by titillating curiosity with the likelihood that she was mistress not wife at all. He then experimented with the viewpoints of husband and wife. Overmuch attention was given to her commonplace thoughts and feelings, while Lord Ormont was seen, with skill, from the outside, until quite suddenly credited with a 'big heart'. His position as a man caught between the apparently slight tug of his wife and the stronger tug of his sister was developed to evoke the magazine-reader's curiosity. The result was some vigorous characterization of the sister, but a line of interest insufficiently related to the rebellion of the wife. Meredith showed his own hand with some preaching about laws which 'grind the sane human being to dust for their maintenance', in the period of the tale, the 1840s.

The Amazing Marriage (1895) was presented more fully from the opposite side, that of the husband. The period of the story (which could be calculated, from its internal evidence, also to belong to the 1840s) was not stressed; contemporary overtones and analogies were avoided in the interests of the tragedy of Lord Fleetwood's coming to value the wife, Carinthia, whom he had ill treated, and her opposite and growing tendency to refuse value to him, in her ignorance of his nature. The actual affinity between them was emphasized by the romantic device of Fleetwood's attraction to Carinthia from the description of her by an itinerant

Welshman. Elated with an account which 'fused a woman's face and grand scenery, to make them inseparable', sated with beautiful women, so that 'a haggard Venus', likened to a rock, was something he 'should think . . . sacred', Fleetwood was moved by a 'submerged self' which he did not understand. In the brisk, unreliable account by Dame Gossip of the antecedents of the pair (one on each side being Celtic), the reader could glimpse peculiarities such as Meredith experimented with in the marriage of Lord Ormont and Aminta. Like Aminta, Fleetwood's mother was not to be seen in society, while Carinthia's mother, a commoner, having married an Earl before she knew her own mind, left him once she did; like Ormond, Carinthia's father 'nursed a grudge against his country'. Now, however, the whole point of Gossip's narrative was in its thrusting the characters aside with 'the breath . . . out of them', in contrast to the deuteragonist, the second contender for the prize, who 'can at times tell us more of them than circumstances at furious heat will help them to reveal'. The furious heat of the circumstances in which Fleetwood married because he would not withdraw his word, and then subjected his wife to gossip through his melodramatic neglect, was contrasted with the moving development of his 'Cambrian's reverential esteem for high qualities': he found more and more of them in Carinthia, but thrust the knowledge away because of the strange complications of his pride. 'She was thrust away because she had offended: still more because he had offended.'

The reader saw Carinthia almost as he did, defined by her 'bold mountain stride', her 'rocky brows' while standing 'like a lance in the air . . . an Amazon schooled by Athene'. The idealizing techniques which had been used with variable success to present Diana and Nesta Victoria were now vindicated within a subtler structure in which the heroine's qualities could be strikingly glimpsed without needing to be analysed, because the emphasis fell on the tragedy of her husband's delayed appreciation of them. She was shown as he saw her, in action and in brief pregnant sequences of dialogue in which she could, for instance, 'say to her husband, "I guard my rooms", without sign of the stage-face of scorn or defiance or flinging of the glove'. His recognition of her as 'an equal whom he had injured', his 'bowing to the visible equality' while chafing 'at a sense of inferiority following his acknowledgement of it', his ability to be 'plucked out of himself', despite the tightening grip upon him of his past deeds, were marvellously observed. Past habits which prevented him from speaking his entire repentance foreshadowed the subject of Lowry's *Under the Volcano*. There was the same poignant sense of a relationship lost and of the magnitude of the loss for the more flawed of the two parties to it. As repeatedly with Meredith, the predisposing conditions of that loss were in the fat soil of an England grown suddenly wealthy. Fleetwood, like Patterne in his tyranny over his associates, seemed 'the monster a rich country breeds under the blessing of peace'. The story of his discovery of his wife was that of his discovery of himself beneath the habits of superiority. As in many tragedies, the discovery came too late.

It was a striking last throw in the game Meredith had been playing ever since *Feverel*, of depreciating the more popular novels which contained 'traffic of the most animated kind' (the last words of *The Amazing Marriage*) in order to emphasize the equal animation of the unseen play of character and motive. He had finally made his case, and in doing so had charted a course for the psychological novel of the next century.

Chronological Table
1832–80

The major and most of the minor English serial fiction appears under date of commencement, followed (in brackets) by date of volume publication

Date	*General History*	*Literary History*
1832	Grey PM. First Reform Act. Durham University founded. Constable, *Grove, Hampstead.* Turner, *Staffa, Fingal's Cave.* Mendelssohn, *Fingal's Cave,* Chopin's first Paris concert. Wilkins' National Gallery (to 1838).	Bentham d. Crabbe d. Scott d. Goethe d. 'Lewis Carroll' b. Stephen b. *Penny Magazine. Chambers's Journal.* Goethe, *Faust,* ii. Sand, *Valentine.*
1833	Slavery abolished in British colonies (as from Aug. 1834). Factory Act forbids employment of children under 9. First government grant for schools. Oxford Movement starts. Landseer, *Hunted Stag.* Mendelssohn, 'Italian' Symphony.	Hannah More d. A. H. Hallam d. Dixon b. *Penny Cyclopaedia* (to 1844). *Bridgewater Treatises* (to 1836). Balzac, *Le Médecin de campagne.* Béranger, *Chansons nouvelles et dernières.* Laube, *Das neue Europa* (to 1837). Sand, *Lélia.*
1834	Owen's Grand National Trades Union. 'Tolpuddle Martyrs'. Melbourne PM, then Peel. His Tamworth Manifesto commits Tories to reform. Houses of Parliament burnt down. Poor Law Amendment Act starts workhouses. Turner, *Venice.* Berlioz, *Harold en Italie.*	Coleridge d. Lamb d. Malthus d. Seeley b. Thomson b. Balzac, *Eugéne Grandet.* Sand, *Jacques.*
1835	Melbourne PM. Municipal Corporations Act reforms local government. Turner, *Burning of the Houses of Lords and Commons, Keelmen Heaving in Coals by Night.* Donizetti, *Lucia di Lammermoor.*	Cobbett d. Mrs Hemans d. J. Hogg d. Austin b. M. E. Braddon b. Butler b. 'Mark Twain' b. Balzac, *Le Père Goriot.* Gautier, *Mlle de Maupin.* Strauss, *Das Leben Jesu* (to 1836). Tocqueville, *La Démocratie en Amérique.*
1836	Newspaper tax reduced. London University given royal charter as examining body. London Working Men's Association leads towards Chartism. First train in London (to Greenwich). Forms of telegraph being devised in America and England. Pugin's *Contrasts* advocates Gothic architecture. Turner, *Juliet and her Nurse.*	Godwin d. James Mill d. Gilbert b. Balzac, *Le Lys dans la vallée. Dublin Review.* Eckermann, *Gespräche mit Goethe.* Gogol, *Inspector-General.* Heine, *Die Romantische Schule.*

Fiction	*Verse, Prose, Drama*
Bulwer, *Eugene Aram*. Chamier, *Life of a Sailor*. Disraeli, *Contarini Fleming*. Gore, *Fair of May Fair, Opera*. James, *Henry Masterton, String of Pearls*, Marryat, *Newton Foster, Peter Simple* (1834). Sterling, *Fitzgeorge*. F. Trollope, *Refugee in America*.	Tennyson, *Poems*. Lyell, *Principles of Geology, ii*; F. Trollope, *Domestic Manners of the Americans*. Walter, *Factory Lad*.
Blessington, *Repealers*. Bulwer, *Godolphin*. Carleton, *Traits and Stories*, ser. 2. Dickens, first of *Sketches* (1836). Disraeli, *Alroy, Rise of Iskander*. James, *Delaware, Mary of Burgundy*. Marryat, *Jacob Faithful* (1834). Neale, *Port Admiral*. Sterling, *Arthur Coninsgby*. F. Trollope, *Abbess*.	Browning, *Pauline*. Carlyle, *Sartor Resartus* (to 1834); Keble, 'National Apostasy'; Lyell, *Principles of Geology*, iii; Newman *et al.*, *Tracts for the Times* (to 1841).
Ainsworth, *Rookwood*. Bray, *Trials of Domestic Life*. Bulwer, *Last Days of Pompeii, Pilgrims of the Rhine*. Disraeli (in part), *A Year at Hartlebury*. Gore, *Hamiltons*. Grattan, *Heiress of Bruges*. Howard, *Rattlin the Reefer* (1836). James, *John Marston Hall*. Marryat, *Japhet in Search of a Father* (1836). Marsh, *Two Old Men's Tales*. Neale, *Will Watch*. M. Scott, *Cruise of the 'Midge'* (1836).	H. Taylor, *Philip van Artevelde*. Hallam, *Essays and Remains*; Somerville, *On the Connexion of the Physical Sciences*; Thackeray, first articles in *Fraser's Magazine*. Gore, *Modern Honour, Queen's champion*.
Blessington, *Two Friends*. Bulwer, *Rienzi*. Chamier, *Unfortunate Man*. Grattan, *Agnes de Mansfeldt*. James, *Gypsey*. Norton, *Wife, Woman's Reward*. F. Trollope , *Tremordyn Cliff*.	Browning, *Paracelsus*. Macaulay, 'Sir James Mackintosh'; Thirlwall, *History of Greece* (to 1844).
Blessington, *Confessions of an Elderly Gentleman*. Chamier, *Ben Brace*. Dickens, *Pickwick Papers* (1837). Gore, *Mrs Armytage*. Howitt, *Wood Leighton*. James, *Desultory Man*. Marryat, *Mr Midshipman Easy, Snarleyyow* (1837). Marsh, *Tales of the Woods and Fields*. Morier, *Abel Allnut*. Neale, *Priors of Prague*. Stickney, *Home*.	Browning, 'Porphyria's Lover', 'Johannes Agricola in Meditation'. Thackeray, articles in *Constitutional* (to 1837). Dickens, *Strange Gentleman, Village Coquettes*.

Date	*General History*	*Literary History*
1837	William IV d. Accession of Queen Victoria. Paper duty halved. Isaac Pitman, *Stenographic Sound-Hand*. Landseer, *Old Shepherd's Chief Mourner*. Turner, *Interior at Petworth*. Basnevi, Fitzwilliam Museum, Cambridge (to 1847).	Green b. Swinburne b. Leopardie d. Pushkin d. *Bentley's Miscellany*. Balzac, *César Birotteau, Illusions perdues* (to 1843). Bremer, *Grannarne*. Sand, *Mauprat*.
1838	Afghan War. Irish Poor Law. Cobden's Anti-Corn Law League. Working Men's Association drafts People's Charter. Chartist *Northern Star*: circulation 50,000. Public Record Office. Regular steamship service between England and USA. London–Birmingham railway completed. Royal Orthopaedic Hospital. Barry, Reform Club (to 1840). Donizetti's *Lucia* produced in London.	Landon d. Lecky b. Morley b. Sidgwick b. Newman, Keble, Pusey (edd.), *Library of the Fathers of the Holy Catholic Church* (to 1885).
1839	War with China: Hong Kong taken. Chartist riots. Photography invented (Daguerre, Fox Talbot). Maclise, *Charles Dickens*. Turner, *Fighting Téméraire*. Berlioz, *Roméo et Juliette*.	Galt d. Praed d. Pater b. Bremer, *Hemmet*. Stendhal, *La Chartreuse de Parme*.
1840	Queen Victoria marries Prince Albert. Afghans surrender. Maoris yield sovereignty of New Zealand. Rowland Hill starts penny post. Barry and Pugin, Houses of Parliament (to 1865). Etty, *Mars, Venus and Attendant*. Turner, *Slavers*. Schumann, *Dichterliebe*.	Fanny Burney d. Blunt b. Hardy b. Symonds b. *Chronica Jocelini de Brakelonda* published by Camden Society. Lermontov, *Hero of our Time*. Sainte-Beuve, *Port-Royal* (to 1859).

Fiction	*Verse, Prose, Drama*
Ainsworth, *Crichton*. Blessington, *Victims of Society*. Bulwer, *Ernest Maltravers*. Carleton, *Fardorougha the Miser* (1839). Chamier, *Arethusa*. Dickens, *Oliver Twist* (1838). Disraeli, *Henrietta Temple, Venetia*. Gore, *Stokeshill Place*. Howard, *Old Commodore*. James, *Attila*. Lever, *Harry Lorrequer* (1839). Lowndes, *Child of Mystery*. Marryat, *Phantom Ship* (1839). Neale, *Gentleman Jack*. Thackeray, *Yellowplush Correspondence* (1838). F. Trollope, *Vicar of Wrexhill*.	Carlyle, *French Revolution*; Martineau, *Society in America*; Ruskin, *Poetry of Architecture* (to 1838). Browning, *Strafford*; Dickens, *Is She his Wife?*
Bulwer, *Alice*. Chamier, *Jack Adams the Mutineer*. Dickens, *Nicholas Nickleby* (1839). Gore, *Woman of the World*. Howard, *Outward Bound*. James, *Robber*. Manning, *Village Belles*. Surtees, *Handley Cross* (1843). F. Trollope, *Romance of Vienna*.	Dickens (ed.), *Memoirs of Grimaldi*; Martineau, *Retrospect of Western Travel*: Maurice, *Kingdom of Christ*; Mill, 'Bentham'.
Ainsworth, *Jack Sheppard*. Bray, *Trials of the Heart*. Gore, *Cabinet Minister, Courtier Days of Charles II*. James, *Charles Tyrrell, Gentlemen of the Old School, Henry of Guise, Huguenot*. Martineau, *Deerbrook*. Mozley, *Fairy Bower*. Neale, *Flying Dutchman*. M. Taylor, *Confessions of a Thug*. Thackeray, *Catherine* (to 1840). F. Trollope, *Michael Armstrong* (1840), *Widow Barnaby*. Warren, *Adventures of an Attorney, Ten Thousand a Year* (1841).	Sterling, *Poems*. Carlyle, *Chartism*; Darwin, *Journal of Researches into the Geology and Natural History of . . . Countries Visited by HMS Beagle*. Bulwer, *Richelieu*.
Ainsworth, *Tower of London, Guy Fawkes* (1841). Chamier, *Spitfire*. Dickens, *Old Curiosity Shop* (1841). Gore, *Dowager, Preferment*. Howard, *Jack Ashore*. James, *King's Highway, Man at Arms*. Lever, *Charles O'Malley* (1841). Marryat, *Poor Jack, Joseph Rushbrook* (1841) Martineau, *Hour and the Man*. M. Taylor, *Tippoo Sultan*. Thackeray, *A Shabby Genteel Story* (1850). F. Trollope, *One Fault*.	Browning, *Sordello*. Macaulay, 'Lord Clive'; Mill, 'Coleridge'; Thackeray, *Paris Sketch Book*. Bulwer, *Money*.

Date	*General History*	*Literary History*
1841	Hong Kong, New Zealand proclaimed British. Peel PM. Second Afghan War. Arc-lamp street-lighting demonstrated in Paris. Scott, Martyrs' Memorial, Oxford.	Newman's *Tract* 90 condemned by Oxford University. *Punch*. Balzac, *Le Curé de village*. Emerson, *Essays*. Feuerbach, *Das Wesen des Christentums*. Poe, 'Murders in the Rue Morgue'.
1842	End of wars with China and Afghanistan. Mines Act. Chadwick report on 'sanitary condition' of working classes. Chartist riots. Turner, *Snow Storm, Peace*. Wagner, *Rienzi*.	Copyright Act. Gogol, *Dead Souls*, Part 1. Sand, *Consuelo*. Sue, *Les Mystères de Paris* (to 1843). Mallarmé b.
1843	Sind, Natal annexed. Brunel, Thames Tunnel. Landseer, *Queen Victoria, Prince Consort and Princess*. Watts, *Lady Holland*. Wagner, *Fliegende Holländer*.	Southey d. Wordsworth Poet Laureate. Doughty b. Dowden b. *Economist*. Liddell and Scott, *Greek–English Lexicon*. Henry James b.
1844	Factory Act limits working hours for women and children. 'Rochdale Pioneers': first co-operative society. Ragged School Union. Washington–Baltimore telegraph. First public baths (Liverpool). London YMCA. Turner, *Rain, Steam, and Speed*. Joachim plays Beethoven in London, Mendelssohn conducting. Berlioz, *Carnaval Romain*.	Campbell d. Bridges b. Hopkins b. Lang b. *North British Review*. R. Chambers, *Cyclopaedia of English Literature*. Dumas, *Les trois Mousqetaires, Le Comte de Monte-Cristo* (to 1845). Heine, *Neue Gedichte*. Sue, *Le Juif errant* (to 1845). Nietzsche b. Verlaine b.

Fiction	*Verse, Prose, Drama*
Ainsworth, *Old St Paul's*. Bulwer, *Night and Morning*. Chamier, *Tom Bowling*. Crowe, *Susan Hopley*. Dickens, *Barnaby Rudge*. Gore, *Cecil, Cecil: a Peer, Greville*. Gresley, *Charles Lever*. James, *Ancient Regime, Corse de Leon, Jacquerie*. Marryat, *Masterman Ready*. Mozley, *Lost Brooch*. Neale, *Naval Surgeon, Paul Periwinkle*. Paget, *St Antholin's*. Sinclair, *Modern Flirtations*. Thackeray, *Great Hoggarty Diamond* (1848). Tonna, *Helen Fleetwood*. F. Trollope, *Ward*. Whitehead, *Richard Savage* (1842).	Browning, *Pippa Passes*. Carlyle, *On Heroes, Hero-Worship, and the Heroic in History*; Miller, *Old Red Sandstone*; Newman, *Tract 90*. Boucicault, *London Assurance*.
Ainsworth, *Windsor Castle* (1843). Bulwer, *Zanoni*. Chamier, *Passion and Principle*. Gore, *Ambassador's Wife*. James, *Morley Ernstein*. Lever, *Jack Hinton* (1843). Marryat, *Percival Keene*. Mozley, *Louisa*. Neale, *Captain's Wife*. Phillips, *Caleb Stukely* (1844). Smith, *Mr Ledbury* (1844). Thackeray, *Fitz-Boodle's Confessions* (1852). F. Trollope, *Blue Belles of England*.	Browning, *King Victor and King Charles, Dramatic Lyrics*; Macaulay, *Lays of Ancient Rome*; Tennyson, *Poems*. Dickens, *American Notes*.
Blessington, *Meredith*. Bulwer, *Last of the Barons*. Chamier, *Ben Bradshawe*. Crowe, *Men and Women*. Dickens, *Christmas Carol, Martin Chuzzlewit* (1844). Gore, *Banker's Wife*. James, *False Heir, Forest Days*. Lever, *Arthur O'Leary* (1844). Marryat, *Adventures of M. Violet*. Neale, *Lost Ship*. Surtees, *Hillingdon Hall* (1845). F. Trollope, *Barnabys in America, Jessie Phillips* (1843).	Browning, *Return of the Druses*; Horne, *Orion*. Borrow, *Bible in Spain*; Carlyle, *Past and Present*; Macaulay, *Essays*; Mill, *System of Logic*; Ruskin, *Modern Painters* (to 1856); Thackeray, *Irish Sketch Book*. Browning, *A Blot in the 'Scutcheon*.
Ainsworth, *St James's*. Chamier. *Mysterious Man*. Dickens, *Chimes*. Disraeli, *Coningsby*. Fullerton, *Ellen Middleton*. Gore, *Agathonia*. James, *Agincourt, Arabella Stuart, Rose d'Albret*. Lever, *Tom Burke*. Marryat, *Settlers in Canada*. Robinson, *Whitefriars*. Sewell, *Amy Herbert*. Smith, *Scattergood Family* (1845). Thackeray, *Barry Lyndon* (1852). F. Trollope, *Young Love*. Yonge, *Abbeychurch*	Barnes, *Poems*; Browning, *Colombe's Birthday*; Patmore, *Poems*. (R. Chambers), *Vestiges of the Natural History of Creation*; Finlay, *Greece under the Romans*; Kinglake, *Eothen*; Stanley, *Life and Correspondence of Thomas Arnold*.

Date	*General History*	*Literary History*
1845	Expedition against Madagascar. War with Sikhs. Potato famine in Ireland. Maynooth grant for Irish education. Stephenson and Thompson, Britannia Bridge, Menai Strait. Franklin, North-West Passage expedition. Wagner, *Tannhäuser*	Barham d. Hood d. Sydney Smith d. Colvin b. Saintsbury b. Dumas, *La Reine Margot, Vingt ans après.* Engels, *Die Lage der arbeitenden Klassen in England.* Mérimée, *Carmen.*
1846	Sikhs defeated. Louis Napoleon escapes from prison to England. Corn Laws repealed. Russell PM. Planet Neptune, protoplasm discovered. Rawlinson deciphers cuneiform. Ether used as anaesthetic in operation. Evangelical Alliance to stop 'encroachments of Popery and Puseyism'. Turner, *Angel Standing in the Sun.* Mendelssohn conducts *Elijah* in Birmingham.	Frere d. F. H. Bradley b. Hakluyt Society (to publish travel-books). Bohn's *Standard Library. Daily News.* George Eliot's translation of *Das Leben Jesu.* Sand, *La Mare au diable.* Senancour d.
1847	Factory Act: ten-hour day for women and young persons. Simpson uses chloroform as anaesthetic. Jenny Lind sings in England. Smirke, British Museum south front completed. Cruickshank, *The Bottle.* Verdi, *Macbeth.*	Alice Meynell b. Balzac, *La Cousine Bette.* Emerson, *Poems.* Longfellow, *Evangeline.*
1848	Revolutions in Europe. Louis Napoleon President of France. Second Sikh War. Cholera in England. Public Health Act. Marx–Engels *Communist Manifesto* (first printed in London). Chartist demonstration ends in fiasco. Rossetti, Millais, and Hunt form Pre-Raphaelite Brotherhood. Queen's College for women founded.	E. Brontë d. I. D'Israeli d. Marryat d. Bohn's *Classical Library.* Milnes (ed.), *Life, Letters and Literary Remains of John Keats.* Grillparzer, *Bruderzwist in Habsburg.* Sand, *La Petite Fadette.* Chateaubriand d.

Fiction	*Verse, Prose, Drama*
Carleton, *Art Maguire, Valentine McClutchy, Rody the Rover, Tales and Sketches*, Chamier, *Count Königsmark*. Cobbold, *Margaret Catchpole*. Crowe, *Martha Ginnis*. Dickens, *Cricket on the Hearth*. Disraeli, *Sybil*. Gore, *Self*. James, *Arrah Neil, Beauchamp* (1848), *Smuggler*. Jewsbury, *Zoe*. Le Fanu, *Cock and the Anchor*. Lever, *O'Donoghue*. Marryat, *Mission, Privateersman* (1846). Marsh, *Mt Sorel*. Robinson, *Whitehall, Caesar Borgia* (1846). Savage, *Falcon Family*. Sewell, *Gertrude*. Thackeray, *Jeames's Diary* (1846).	Browning, *Dramatic Romances and Lyrics*. Carlyle, *Cromwell*; Martineau, *Letters on Mesmerism*; Newman, *Development of Christian Doctrine*.
Blessington, *Memoirs of a Femme de Chambre*. Bulwer, *Lucretia*. Carleton, *Black Prophet* (1847). Dickens, *Dombey and Son* (1848). Gore, *Debutante, Men of Capital, Peers and Parvenus*. Gresley, *Colton Green*. James, *Heidelberg*. Le Fanu, *Fortunes of Col Torlogh O'Brien* (1847). Lever, *Knight of Gwynne* (1847). Marryat, *Valerie* (1849). Marsh, *Emilia Wyndham*. Paget. *Milford Malvoisin*. Sinclair, *Jane Bouverie*. Surtees, *Hawbuck Grange* (1847). Thackeray, *Book of Snobs* (1848). F. Trollope, *Robertses on their Travels*.	*Poems by Currer, Ellis and Acton Bell*. Dickens, *Pictures from Italy*; Grote, *History of Greece* (to 1856); Thackeray, *Journey from Cornhill to Grand Cairo*.
Ainsworth, *James II* (1848). Blessington, *Marmaduke Herbert*. A. Brontë, *Agnes Grey*, C. Brontë, *Jane Eyre*. E. Brontë, *Wuthering Heights*, Crowe, *Lily Dawson*. Disraeli, *Tancred*, Fullerton, *Grantley Manor*. Froude, *Shadows of the Clouds*. Gore, *Castles in the Air*. James, *Castle of Ehrenstein, Convict, Margaret Graham* (1848), *Russell*. Linton, *Azeth the Egyptian*. Marryat, *Children of The New Forest*. Sewell, *Margaret Percival*. Thackeray, *Vanity Fair* (1848), *Mr Punch's Prize Novelists* (1853). A. Trollope, *Macdermots of Ballycloran*. F. Trollope, *Father Eustace*. Yonge, *Scenes and Characters*.	Tennyson, *Princess*. Coyne, *How to Settle Accounts with your Laundress*; Morton, *Box and Cox*.
Ainsworth, *Lancashire Witches* (1849). A. Brontë, *Tenant of Wildfell Hall*. Bulwer, *Harold, Caxtons* (1849). Carleton, *Emigrants of Ahadarra*. Gaskell, *Mary Barton*. James, *Last of the Fairies, Sir Theodore Broughton*. Jewsbury, *Half Sisters*. C. Kingsley, *Yeast* (1851). Lever, *Roland Cashel* (1850). Linton, *Amymone*. Marryat, *Little Savage*. Marsh, *Angela*. Newman, *Loss and Gain*. Savage, *Bachelor of the Albany*. Smith, *Christopher Tadpole*. Thackeray, *Pendennis* (1850). A. Trollope, *Kellys and the O'Kellys*. F. Trollope, *Town and Country*. Warren, *Now and Then*.	Clough, *Bothie of Toper-na-fuosich*; Kingsley, *Saint's Tragedy*. Forster, *Oliver Goldsmith*; Martineau, *Eastern Life*; Mill, *Political Economy*.

Date	*General History*	*Literary History*
1849	Sikhs surrender, Punjab annexed. Rome proclaimed republic under Mazzini, then taken by French. Communist riots suppressed in Paris. Revolt against British in Montreal. Disraeli becomes Conserative leader. Bedford College for women. Maurice, Kingsley, Hughes preach Christian Socialism. Millais, *Isabella*. Rossetti, *Girlhood of Mary Virgin, Dante drawing an Angel.* Hunt, *Rienzi*. Liszt, *Tasso*.	Beddoes d. A. Brontë d. Henley b. *Notes and Querries*. Poe d. Strindberg b.
1850	Navy blockades Piraeus for Gibraltarian Don Pacifico. Anglo-Kaffir War. Forts on Gold Coast bought from Denmark. Public Libraries Act. Factory Act: sixty-hour week for women and young persons. Miss Buss starts North London Collegiate School. Re-establishment of RC hierarchy: 'Papal Aggression'. First Dover–Calais telegraph-cable. Butterfield, All Saints', Margaret Street (to 1859). Millais, *Christ in the House of his Parents*. Rossetti, *Ecce Ancilla Domini*. Hunt, *Claudio and Isabella*. Collins, *Convent Thoughts*. Wagner, *Lohengrin*.	Bowles d. Jeffrey d. Wordsworth d. Stevenson b. Tennyson Poet Laureate. *Harper's Magazine* (to republish English authors in America). *Household Words*. *The Germ*. *Christian Socialist*. Hebbel, *Herodes und Marianne*. Turgenev, *A Month in the Country*.
1851	Louis Napoleon's *coup d'état*. Great Exhibition in Paxton's Crystal Palace (Hyde Park). Mayhew, *London Labour and the London Poor*. Tenniel starts cartoons in *Punch*. Cubitt, King's Cross Station (to 1852). Millais, *Mariana*. Hunt, *Hireling Shepherd*.	Lingard d. A. C. Bradley b. H. A. Jones b. M. A. Arnold (Mrs Humphrey Ward) b. *New York Times*. H. B. Stowe, *Uncle Tom's Cabin* (to 1852).

Fiction	*Verse, Prose, Drama*
C. Brontë, *Shirley*. Dickens, *David Copperfield* (1850). Froude, *Nemesis of Faith*. Gore, *Diamond and the Pearl*. James, *Forgery, Woodman*. Lever, *Con Cregan*. Manning, *Mary Powell*. Marsh, *Mordaunt Hall*. Mulock, *Ogilvies*. M. Oliphant, *Margaret Maitland*. Robinson, *Owen Tudor*. Savage, *My Uncle the Curate*. Sinclair, *Edward Graham*. Smith, *Pottleton Legacy*. Surtees, *Mr Sponge's Sporting Tour* (1853). F. Trollope, *Lottery of Marriage*. 'Waters', *Recollections of a Police Officer* (1853).	(Arnold), *Strayed Reveller*. Layard, *Nineveh*; Macaulay, *History of England* (to 1861); Martineau, *Thirty Years' Peace* (to 1850); Ruskin, *Seven Lamps of Architecture*.
Blessington, *Country Quarters*. Bulwer, *My Novel* (1853). W. Collins, *Antonina*. James, *Old Oak Chest*. Jewsbury, *Marian Withers* (1851). C. Kingsley, *Alton Locke*. Lever, *Maurice Tiernay*. Manning, *Household of Sir Thomas More*. Marsh, *Wilmingtons*. Mulock, *Olive*. Thackeray, *Rebecca and Rowena*. A. Trollope, *La Vendée*. F. Trollope, *Petticoat Government*. Yonge, *Kenneth*.	E. B. Browning, *Poems*; Browning, *Christmas-Eve and Easter-Day*; Tennyson, *In Memoriam A. H. H.*; Wordsworth, *Prelude*. Carlyle, *Latter-Day Pamphlets*.
Ainsworth, *Mervyn Clitheroe* (1858). Borrow, *Lavengro*. Drury, *Eastbury*. Gaskell, *Cranford* (1853). James, *Fate, Henry Smeaton*. Le Fanu, *Ghost Stories and Tales of Mystery*. Lever, *Daltons* (1852). Linton, *Realities*. Marsh, *Ravenscliffe, Time the Avenger*, Norton, *Stuart of Dunleath*. M. Oliphant, *Caleb Field, Merkland*. Robinson, *Gold Worshippers*. Surtees, *Young Tom Hall* (to 1853). F. Trollope, *Second Love*. Yonge, *Castle Builders* (1854), *Little Duke* (1854), *Pigeon Pie* (1860).	E. B. Browning, *Casa Guidi Windows*; Meredith, *Poems*. Carlyle, *Life of John Sterling*; Ruskin, *Stones of Venice* (to 1853). Simpson, *Captain Cutter, Only a Clod*.

Date	General History	Literary History
1852	Louis Napoleon proclaimed Emperor Napoleon III. Derby PM, then Aberdeen. Burmese War. Duke of Wellington d. Brunel and Wyatt, Paddington Station (to 1854). Millais, *Ophelia*. Hunt, *Awakening Conscience*. Brown, *Work* (to 1865). Schumann, *Manfred*	T. Moore d. Lady Gregory b. G. Moore b. Roget, *Thesaurus*. Gautier, *Émaux et camées*. Leconte de Lisle, *Poèmes antiques*. Sand, *Les Maîtres sonneurs*.
1853	Turkey rejects Russian ultimatum, declares war on Russia. Turkish fleet destroyed, Anglo-French fleet off Dardanelles. Hunt, *Light of the World*. Verdi's *Rigoletto* produced in London.	*Encyclopaedia Britannica*, 8th edn. (to 1860), Heine, *Die Götter im Exil*.
1854	Crimean War: Alma, siege of Sebastopol, Balaclava (Charge of Light Brigade), Inkerman. Maurice founds Working Men's College. University College, Dublin. Pius IX makes Immaculate Conception article of faith. Florence Nightingale at Scutari. Frith, *Ramsgate Sands*. Millais, *John Ruskin*. Hunt, *Scapegoat*. Brown, *English Autumn Afternoon*. Rossetti begins *Found*.	S. Ferrier d. Lockhart d. Wilson d. Wilde b. George Eliot, translation of Feuerbach. Thoreau, *Walden*. G. de Nerval, *Les Chimères*. Rimbaud b.
1855	Palmerston PM. Fall of Sebastopol. London–Balaclava telegraph. Deane and Woodward, Oxford University Museum (to 1859). Brown, *Last of England*. Hughes, *April Love* (to 1856). Rossetti, *Paolo and Francesca*. London production of Verdi's *Il Trovatore*. Wagner conducts Philharmonic Society concerts.	C. Brontë d. Mary Mitford d. Rogers d. Pinero b. Smith, *Latin–English Dictionary*. Newspaper tax abolished. First Hoe rotary press in England. *Daily Telegraph*. *Saturday Review*. Longfellow, *Song of Hiawatha*. Stendhal, *Lucien Leuwen*. Whitman, *Leaves of Grass*. Nerval d.

Fiction	*Verse, Prose, Drama*
Carleton, *Real Hall, Squanders of Castle Squander*. W. Collins, *Basil*. Crowe, *Adventures of a Beauty*. Dickens, *Bleak House* (1853). Fulleron, *Lady-bird*. James, *Adrian, Pequinillo, Revenge*. C. Kingsley, *Hypatia* (1853). Lever, *Dodd Family Abroad* (1854), *Sir Jasper Carew* (1855). Marsh, *Castle Avon*. Mullock, *Head of the Family*. M. Oliphant, *Adam Graeme of Mossgray*. Savage, *Reuben Medlicott*. Sinclair, *Beatrice*. Thackeray, *Esmond*. F. Trollope, *Uncle Walter*.	*Arnold, Empedocles on Etna*. Newman, *University Education*. Boucicault, *Corsican Brothers*; T. Taylor and Reade, *Masks and Faces*.
Ainsworth, *Star Chamber* (1854). C. Brontë, *Villette*. Gaskell, *Ruth*. Gore, *Dean's Daughter*. James, *Agnes Sorrel, Vicissitudes of a Life*. Manning, *Cherry and Violet, Provocations of Mme Palissy*. Marsh, *Longwoods of the Grange*. Mulock, *Agatha's Husband*. Reade, *Christie Johnstone, Peg Woffington*. Sewell, *Experience of Life*. Thackeray, *Newcomes* (1854–5). F. Trollope, *Young Heiress*, Yonge, *Daisy Chain* (1856), *Heir of Redclyffe, Lances of Lynwood* (1855).	Arnold, *Poems*; Patmore, *Tamerton Church-Tower*. Wallace, *Travels on the Amazon*. Browning, *Colombe's Birthday*; Reade, *Gold*.
W. Collins, *Hide and Seek*. Crowe, *Linny Lockwood*. Dickens, *Hard Times*. Gaskell, *North and South* (1855). Gore, *Progress and Prejudice*. James, *Ticonderoga*. Lever, *Martins of Cro'Martin* (1856). Manning, *Claude the Colporteur, Hill Side, Jack and the Tanner of Wymondham*. Marsh, *Aubrey*. M. Oliphant, *Magdalen Hepburn*. Robinson, *Westminster Abbey*. Sewell, *Katherine Ashton*. Yonge, *Heartease*.	Arnold, 'Balder dead'; Patmore, *Angel in the House* (to 1862); Tennyson, 'Charge of the Light Brigade'. Huxley, 'Educational Value of the Natural History Sciences'.
Carleton, *Willy Reilly*. Clive, *Paul Ferroll*. Dickens, *Little Dorrit* (1857). Gore, *Mammon*. Jewsbury, *Constance Herbert*. C. Kingsley, *Westward Ho!* Lever, *Fortunes of Glencore* (1857). Manning, *Old Chelsea Bun-House*. Marsh, *Heiress of Houghton*. M. Oliphant, *Lilliesleaf*. Ogle, *A Lost Love*. Sewell, *Cleve Hall*. Sinclair, *Cross Purposes*. Thackeray, *Rose and the Ring*. A. Trollope, *Warden*. F. Trollope, *Gertrude*.	Arnold, 'Stanzas from the Grande Chartreuse'; Browning, *Men and Women*; MacDonald, *Within and Without*; Tennyson, *Maud and Other Poems*. W. Collins, *Lighthouse*.

Date	*General History*	*Literary History*
1856	Crimean War ends. Queen institutes VC. Second Chinese War. War with Persia. Bessemer patents steel-making process. Oudh annexed. First stage appearance of Henry Irving. Millais, *Blind Girl, Autumn Leaves*. Wallis, *Death of Chatterton*. London production of Verdi's *La Traviata*.	Miller d. Rider Haggard b. Shaw b. Bremer, *Hertha*. Emerson, *English Traits*. Hugo, *Les Contemplations*. Ludwig, *Zwischen Himmel und Erde*. Heine d.
1857	Persian War ends. Chinese fleet destroyed. Indian Mutiny. Translantic cable laid (to 1865). Divorce Courts established. Queen opens Museum of Ornamental Art (Victoria and Albert Museum, 1899). Millais, *Sir Isumbras at the Ford*. Rossetti, *Tune of Seven Towers, The Wedding of St George*. Wallis, *The Stonebreaker*.	Conrad b. Gissing b. Jerrold d. Davidson b. Arnold Professor of Poetry at Oxford. Oxford Union frescos: Rossetti, Morris, Burne-Jones. Swinburne, 'Jovial Campaign', Baudelaire, *Les Fleurs du mal*. Flaubert, *Madame Bovary*. Scheffel, *Ekkehard*. Musset d.
1858	Derby PM. Chinese War ends. Indian Mutiny suppressed. Powers of East India Company transferred to Crown. Jewish Disabilities Act. Property qualification for MPs abolished. Fenian Brotherhood in Ireland. I. K. Brunel's *Great Eastern* (biggest ship ever built). Darwin and Wallace give joint paper on evolution, Frith, *Derby Day*. Millais, *William Gladstone*. Morris, *Queen Guenevere*. E. M. Barry, Covent Garden Opera House. Offenbach, *Orphée aux enfers*.	R. Owen d. Watson b. Ruskin's 'unconversion' in Turin. Gautier, *Le Roman de la momie*.

Fiction	*Verse, Prose, Drama*
W. Collins, *After Dark*. James, *Old Dominion*. Jewsbury, *Angelo, Sorrows of Gentility*. Manning, *Tasso and Leonora*. Meredith, *Shaving of Shagpat*. Mulock, *John Halifax Gentleman*. M. Oliphant, *Zaidee*. Newman, *Callista*. Reade, *It Is Never too Late to Mend*. Savage, *Clover Cottage*. Sewell, *Ivors*. F. Trollope, *Fashionable Life*. Yonge, *Young Stepmother* (1861).	Burton, *First Footsteps in East Africa*; Froude, *History of England* (to 1870); C. Kingsley, *Heroes*.
Borrow, *Romany Rye*. C. Brontë, *Professor*. Bulwer, *What Will He Do with It?* (1859). W. Collins, *Dead Secret*. Eliot, *Scenes of Clerical Life* (1858). Gore, *Two Aristocracies*. Hughes, *Tom Brown's Schooldays*. James, *Leonora d'Orco*. C. Kingsley, *Two Years Ago*. Lawrence, *Guy Livingstone*. Lever, *Davenport Dunn* (1859). Manning, *Good Old Times, Helen and Olga*. Marsh, *Rose of Ashhurst*. Meredith, *Farina*. M. Oliphant, *Athelings*. Reade, *White Lies*. Surtees, *Ask Mama* (1858). Thackeray, *Virginians* (1858–9). Trollope, *Barchester Towers*. Yonge, *Dynevor Terrace*.	E. B. Browning, *Aurora Leigh*; MacDonald, *Poems*. Buckle, *History of Civilization in England* (to 1861); Gaskell, *Life of Charlotte Brontë*; Miller, *Testimony of the Rocks*; Ruskin, *Political Economy of Art*. W. Collins, *Frozen Deep*; Yates, *After the Ball*.
Clive, *Year after Year*. Farrar, *Eric*. Gaskell, *My Lady Ludlow* (1859). Gore, *Heckington*. James, *Lord Montagu's Page*. MacDonald, *Phantastes*. Manning, *Ladies of Bever Hollow, Town and Forest* (1860), *Year Nine*. M. Oliphant, *Laird of Norlaw*. Reade, *Autobiography of a Thief*. Robinson, *Mauleverer's Divorce*. Sewell, *Ursula*. A. Trollope, *Dr Thorne, Three Clerks*.	Clough, *Amours de Voyage*; C. Kingsley, *Andromeda and Other Poems*; Morris, *Defence of Guenevere and Other Poems*. Carlyle, *Frederick the Great* (to 1865). W. Collins, *Red Vial*.

Date	General History	Literary History
1859	Palmerston PM, Gladstone Chancellor of Exchequer. Franco-Austrian War. Austrian defeat at Solferino. Peace of Villafranca. Napoleon III negotiates to form Italian Confederation. Nightingale, *Notes on Hospitals, Notes on Nursing*. Webb, Red House for Morris (to 1860). Millet, *L'Angélus*. Gounod, *Faust*.	Conan Doyle b. De Quincy d. H. Hallam d. Leigh Hunt d. Macaulay d. K. Grahame b. Housman, b. F. Thompson b. *Chambers's Encyclopaedia* (to 1868), *Macmillan's Magazine*. Goncharov, *Oblomov*. Hugo, *La Légende des siècles* (to 1883).
1860	Maori War. First Italian Parliament. Lincoln President of USA. Huxley–Wilberforce debate at Oxford. Scott, St Pancras Station. Hunt, *Finding of the Saviour in the Temple*. Millais, *Black Brunswicker*. W. B. Scott, *A. C. Swimburne*.	G. P. R. James d. Barrie b. Inge b. *Cornhill Magazine*, ed. Thackeray: 120,000 copies sold. Bradlaugh's *National Reformer*. Burckhard, *Die Kultur der Renaissance in Italien*. Turgenev, *On the Eve*. Schopenhauer d.
1861	Prince Consort d. Victor Emmanuel King of Italy. Maori War ends. American Civil War starts. Lowe's Revised code: payment by results in grants to schools. Post Office Savings Bank. HMS *Warrior* (all-iron warship). Mrs Beeton, *Houshold Management*. Hughes, *Home from Work*. Brahms, first piano concerto.	E. B. Browning d. Clough d. Hewlett b. Raleigh b. Palgrave's *Golden Treasury of Songs and Lyrics*. Baker's *Hymns Ancient and Modern*. Morris and Co. founded. Max Müller, *Science of Languages* (to 1864). Paper tax abolished.
1862	*Alabama*, built at Birkenhead for Confederates, allowed to sail. Bismarck Prussian premier. Lancashire cotton famine. Companies Act: Limited Liability. H. Dunant, *Souvenir de Solferino* leads to international Red Cross. Frith, *Railway Station*. Verdi, *Forza del Destino*.	Buckle d. Knowles d. Elizabeth (Siddal) Rossetti d. Newbolt b. Hugo, *Les Misérables*. Turgenev, *Fathers and Sons*. Thoreau d. Maeterlinck b.

Fiction	*Verse, Prose, Drama*
W. Collins, *Queen of Hearts, Woman in White* (1860). Dickens, *A Tale of Two Cities.*Eden, *Semi-Detached House*. Eliot, *Adam Bede*. Gaskell, *Lois the Witch*. Hughes, *Scouring of the White Horse, Tom Brown at Oxford* (1861). James, *Cavalier*. Jewsbury, *Right or Wrong*. H. Kingsley, *Geoffry Hamlyn*. Lawrence, *Sword and Gown*. Lever, *One of Them* (1860). Manning, *Poplar House Academy*. Meredith, *Ordeal of Richard Feverel*. Mulock, *A Life for a Life*. Payn, *Bateman Household* (1860), *Foster Brothers*. Reade, *A Good Fight, Love Me Little Love Me Long*. Surtees, *Plain or Ringlets* (1860). A. Trollope, *Bertrams*. Wills, *Life's Foreshadowings*. Wood, *East Lynne* (1861).	Fitzgerald, *Rubáiyát of Omar Khayyám*; Tennyson, *Idylls of the King* ('Enid', 'Vivien', 'Elaine', 'Guinevere'). Darwin, *Origin of Species*; Mill, *On Liberty*; Ruskin, *Two Paths*,
Carleton, *Evil Eye*. Clive, *Why Paul Ferrol Killed his Wife*. Dickens, *Great Expectations* (1861). Eden, *Semi-Attached Couple*. Eliot, *Mill on the Floss*. James, *Man in Black*. Lever, *Day's Ride* (1863). Manning, *Valentin Duval*. Meredith, *Evan Harrington*. Payn, *Richard Arbour* (1861). Peacock, *Gryll Grange*. Thackeray, *Lovel the Widower* (1861). A. Trollope, *Castle Richmond, Framley Parsonage* (1861). Yonge, *Hopes and Fears, Stokesley Secret* (1861).	Faraday, *Various Forces of Matter*; Wilson, Temple, Williams, Powell, Pattison, Jowett, Goodwin, *Essays and Reviews*. Boucicault, *Colleen Bawn*; Braddon, *Loves of Arcadia*.
Ainsworth, *Constable of the Tower*. Braddon, *Lady Audley's Secret* (1862), *Trail of the Serpent*. Bulwer, *A Strange Story* (1862). Eliot, *Silas Marner*. Fullerton, *Laurentia*. H. Kingsley, *Ravenshoe* (1862). Lawrence, *Barren Honour* (1862). Le Fanu, *House by the Churchyard* (1863). Manning, *Chronicle of Ethelfled*. Reade, *Cloister and the Hearth*. Thackeray, *Philip* (1862). A. Trollope, *Orley Farm* (1862). T. Trollope, *La Beata, Marietta*. Wood, *Shadow of Ashlydyat* (1863). Yonge, *Countess Kate* (1862).	Rossetti (tr.), *Early Italian Poets*. Faraday, *Chemical History of a Candle*; Mill, *Utilitarianism*; Spencer, *Education*; Spedding, *Francis Bacon*.
Ainsworth, *Lord Mayor of London, Cardinal Pole* (1863). Braddon, *Aurora Floyd* (1863), *John Marchmont's Legacy* (1863). Carleton, *Redmond*. W. Collins, *No Name*. Eliot, *Romola* (1863). C. Kingsley, *Water Babies* (1863). Lever, *Barrington*. Mulock, *Mistress and Maid*. M. Oliphant, *Last of the Mortimers*. Robinson, *Cynthia Thorold*. A. Trollope, *Small House at Allington* (1864). Wood, *Channings, Mrs Halliburton's Troubles*. Yonge, *Trial* (1864).	Meredith, *Modern Love*. Ruskin, *Unto this Last, Munera Pulveris* (to 1863).

Date	General History	Literary History
1863	Palmerston cedes Ionian Islands to Greece. Metropolitan line opened. Scott, Albert Memorial (to 1872). Hughes, *Home from Sea*. Gounod's *Faust* produced in London. Bizet, *Pêcheurs de perles*. Stanislavsky b.	F. Trollope d. Thackeray d. Quiller-Couch b. Rossetti (ed.), 'Selections' of Blake (in Gilchrist's *Life*). Flaubert, *Salammbô*. Gautier, *Le capitaine Fracasse*. Laube, *Der deutsche Krieg* (to 1866). Renan, *Vie de Jésus*. Sainte-Beuve, *Nouveaux Lundis* (to 1870). Taine, *Histoire de la littérature anglaise*. Tolstoy, *War and Peace* (to 1869). Verne, *Cinq semaines en ballon*. Vigny d.
1864	Sherman marches through Georgia. Lincoln re-elected President. Geneva Convention. International Working Men's Association founded in London. Octavia Hill starts work on slums. Pius IX condemns liberalism, socialism, rationalism. Clerk-Maxwell predicts electromagnetic radiation. Butler, *Family Prayers*. Landseer, *Man Proposes, God Disposes*. Rossetti, *Beata Beatrix* (to 1870). Whistler, *Symphony in White No 2*.	Clare d. Landor d. Phillips b. *The Month*. Raabe, *Der Hungerpastor*. Verne, *Voyage au centre de la terre*.
1865	Lincoln assassinated. Civil War ends. Slavery abolished in USA. Palmerston d. Russell PM. Transatlantic cable completed. Eyre suppresses Negro revolt in Jamaica. Wagner, *Tristan und Isolde*.	Aytoun d. Gaskell d. Surtees d. Kipling b. Yeats b. *Argosy*. *Fortnightly Review*. *Pall Mall Gazette*. Whitman, *Drum-Taps*. Raabe, *Ferne Stimmen*. Verne, *De la terre à la lune*.

Fiction	*Verse, Prose, Drama*
Ainsworth, *John Law* (1864). Braddon, *Eleanor's Victory*. Gaskell, *A Dark Night's Work, Cousin Phillis* (1865), *Sylvia's Lovers*. H. Kingsley, *Austin Elliot, Hillyars and the Burtons* (1865). Le Fanu, *Wylder's Hand* (1864). Lever, *Luttrell of Arran* (1865), *Tony Butler* (1865). MacDonald, *David Elginbrod*. Manning, *Bessy's Money, Duchess of Trajetto, Meadowleigh*. Norton, *Lost and Saved*. M. Oliphant, *Salem Chapel*. Reade, *Hard Cash*. M. Taylor, *Tara*. A. Trollope, *Rachel Ray*. T. Trollope, *Beppo the Conscript* (1864). *Giulio Malatesta, Lindisfarne Chase* (1864). Wood, *Verner's Pride*,	Butler, *A First Year in Canterbury Settlement*; Gardiner, *History of England from the Accession of James I* (to 1882); Gilchrist, *Life of William Blake*; Huxley, *Man's Place in Nature*; Kinglake, *Invasion of the Crimea* (to 1887). Three stage versions of *Lady Audley's Secret* and five of *Aurora Floyd*.
Ainsworth, *Spanish Match* (1865). Blackmore, *Clara Vaughan*. Braddon, *Doctor's Wife, Henry Dunbar, Only a Clod* (1866). 'Carroll', *Alice's Adventures in Wonderland*. M. Collins, *Who Is the Heir?* (1685). W. Collins, *Armadale* (1866). Clive, *Greswold*. Dickens, *Our Mutual Friend* (1865). Gaskell, *Wives and Daughters* (1866). Lawrence, *Maurice Dering*. Le Fanu, *Uncle Silas*. Lever, *A Rent in a Cloud* (1865). MacDonald, *Adela Cathcart*. Manning, *Interrupted Wedding*. Meredith, *Emilia in England*. M. Oliphant, *Perpetual Curate*. Payn, *Lost Sir Massingberd*. Robinson, *Christmas at Old Court, Madeleine Graham*. Surtees, *Mr Romford's Hounds* (1865). Thackeray, *Denis Duval*. A. Trollope, *Can You Forgive Her?* (1865). Wills, *Wife's Evidence*. Wood, *Lord Oakburn's Daughters, Oswald Cray, Trevlyn Hold*. Yates, *Broken to Harness*. Yonge, *Clever Woman of the Family* (1865).	Browning, *Dramatis Personae*; Tennyson, *Enoch Arden and Other Poems*. Newman, *Apologia pro Vita sua*. Brough and Halliday, *Area Belle*.
Ainsworth, *Constable de Bourbon* (1866). Blackmore, *Cradock Nowell* (1866). Braddon, *Lady's Mile* (1866), *Sir Jasper's Tenant*. Hatton, *Bitter Sweets*. C. Kingsley, *Hereward the Wake*. Lawrence, *Sans Merci* (1866). Le Fanu, *Guy Deverell*. Lever, *Sir Brook Fossbrooke* (1866). Linton, *Grasp your Nettle*. MacDonald, *Alec Forbes of Howglen*. Manning, *Belforest, Selvaggio*. Meredith, *Rhoda Fleming*. L. Oliphant, *Piccadilly* (1870). Payn, *Clyffards of Clyffe* (1866), *Married beneath Him*. Reade, *Griffith Gaunt* (1866). Robinson, *Dorothy Firebrace*. M. Taylor, *Ralph Darnell*. A. Trollope, *Belton Estate* (1866), *Miss Mackenzie*. Wood, *Lady Adelaide's Oath* (1867), *Mildred Arkell*. Yates, *Land at Last*. (1866), *Running the Gauntlet*. Yonge, *Dove in the Eagle's Nest* (1866), *Prince and the Page* (1866).	Newman, *Dream of Gerontius*; Swinburne, *Atalanta in Calydon*. Arnold, *Essays in Criticism*; Lecky, *History of the Rise and Influence of the Spirit of Rationalism in Europe*; Livingstone, *Zambesi*; Ruskin, *Sesame and Lillies*. Robertson, *Society*.

Date	*General History*	*Literary History*
1866	Fenians active in Ireland. Habaes Corpus suspended. Derby PM. Elizabeth Garrett opens dispensary for women. Dr Barnardo starts children's homes. Pasteur, *Études sur le vin* (pasteurization). Work starts on London underground railway. Rossetti, *Monna Vanna*. Smetana, *Bartered Bride*. Thomas, *Mignon*.	Jane Welsh Carlyle d. Keble d. Peacock d. Gilbert Murray b. Wells b. Hopkins received by Newman into RC Church. Stubbs Professor of History at Oxford. *Contemporary Review*. Dostoevsky, *Crime and Punishment*. Gaboriau, *L'Affaire Lerouge*. *Le Parnasse contemporain*. Verlaine, *Poèmes saturniens*.
1867	Fenian outrages in England. Canada becomes Dominion. Second Reform Act. Factory Act. Lister describes antiseptic surgery in *Lancet*. Albert Hall (to 1871). Marx, *Das Kapital*, i. Bizet, *Jolie Fille de Perth*. Strauss, 'Blue Danube' waltz. Verdi's *Forza* produced in London.	Faraday d. Arnold Bennett b. Dowson b. Galsworthy b. Johnson b. Ibsen, *Peer Gynt*. Baudelaire d. Pirandello b.
1868	Disraeli PM. Basutoland annexed. Gladstone PM. Keble College, Oxford, founded. Expedition against Abyssinia. Maori War. Hodgson, *European Curiosities*. Watts, *Thomas Carlyle*. Wagner, *Die Meistersinger von Nürnberg*.	Brougham d. Milman d. Gertrude Bell b. Hopkins becomes Jesuit. Dostoevsky, *Idiot* (to 1869). Raabe, *Abu Telfan*.

Fiction	*Verse, Prose, Drama*
Ainsworth, *Old Court* (1867). Braddon, *Birds of Prey* (1867). Eliot, *Felix Holt the Radical.* Hatton, *Against the Stream.* H. Kingsley, *Leighton Court, Silcote of Silcotes* (1867). Le Fanu, *All in the Dark.* Linton, *Lizzie Norton of Greyrigg.* MacDonald, *Robert Falconer* (1868). Manning, *Lincolnshire Tragedy, Miss Biddy Frobisher.* Meredith, *Vittoria* (1867). M. Oliphant, *Miss Marjoribanks.* Payn, *Mirk Abbey.* A. Trollope, *Claverings* (1867), *Last Chronicle of Barset* (1867). T. Trollope, *Gemma* (1867). Yates, *Black Sheep.* Wood, *Elster's Folly, St Martin's Eve.*	Arnold, 'Thyrsis'; Swinburne, *Poems and Ballads*; Hopkins writes 'Habit of Perfection'. Hopkins starts *Journal*; Pater, 'Coleridge'. Robertson, *Ours.*
Ainsworth, *Myddleton Pomfret* (1868). Braddon, *Dead Sea Fruit* (1868), *Rupert Godwin.* Broughton, *Cometh up as a Flower, Not Wisely but too Well.* M. Collins, *Sweet Anne Page* (1868). Fullerton, *Stormy Life.* Hatton, *Talants of Barton.* H. Kingsley, *Mlle Mathilde* (1868). Le Fanu, *Lost Name* (1868), *Tenants of Malory.* Lever, *Bramleighs of Bishop's Folly* (1868). Linton, *Sowing the Wind.* MacDonald, *Guild Court* (1868). Norton, *Old Sir Douglas.* M. Oliphant, *Madonna Mary.* Payn, *Carlyon's Year* (1868). A. Trollope, *Phineas Finn* (1869). T. Trollope, *Artingale Castle.* Wood, *A Life's Secret, Anne Hereford* (1868), *Orville College.* Yonge, *Danvers Papers, Chaplet of Pearls* (1868).	Arnold, *New Poems*; MacDonald, *Disciple and Other Poems*; Morris, *Life and Death of Jason*; Swinburne, *Song of Italy.* Bagehot, *English Constitution*; Carlyle, 'Shooting Niagara'; Ruskin, *Time and Tide.* Dickens and W. Collins, *No Thoroughfare*; Robertson, *Caste.*
Braddon, *Charlotte's Inheritance, Run to Earth.* W. Collins, *Moonstone.* Hatton, *Christopher Kenrick* (1869). H. Kingsley, *Stretton* (1869). Lawrence, *Brakespeare, Breaking a Butterfly* (1869). Le Fanu, *Haunted Lives.* Lever, *Paul Gosslet, That Boy of Norcott's* (1869). Manning, *Diana's Crescent, Jacques Bonneval.* M. Oliphant, *Brownlows.* Payn, *Bentinck's Tutor, Blondel Parva.* Reade, *Foul Play.* Robinson, *Matrimonial Vanity Fair.* A. Trollope, *He Knew He Was Right* (1869). T. Trollope, *Leonora Casaloni.* Wood, *Red Court Farm, Roland Yorke* (1869). Yonge, *Caged Lion* (1870).	Browning, *Ring and the Book* (to 1869); Morris, *Earthly Paradise* (to 1870); Newman, *Verses on Various Occasions.* Huxley, 'Physical Basis of Life'. Yates, *Tame Cats,* (with Simpson) *Black Sheep.*

Date	*General History*	*Literary History*
1869	Irish Church disestablished. Suez Canal opened. Emily Davies starts Girton College at Hitchin (moved to Cambridge 1873). Sophia Jex-Blake begins medical training at Edinburgh. Wagner, *Rheingold*.	Binyon b. Carleton d. *Graphic*. 'Mark Twain', *Innocents Abroad*. Lamartine d. Sainte-Beuve d.
1870	Franco-Prussian War. Irish Land Act: loans to peasants. Married Women's Property Act gives wives rights over own earnings. Forster's Education Act: board schools funded by local rates. Siege of Paris. Vatican asserts Papal Infallibility. Civil Service opened to competitive examination. Butterfield, Keble College, Oxford. Pearson, St. Augustine's, Kilburn (to 1880). Burne-Jones, *The Mill*. Delibes, *Coppélia*.	Dickens d. Belloc b. Sturge Moore b. Brewer's *Dictionary of Phrase and Fable*. W. M. Rossetti's edn. of Shelley. English Literature made a subject in elementary schools. Flaubert, *L'Education sentimentale*. Raabe, *Der Schüdderump*.
1871	Fall of Paris ends war. Paris Commune set up and suppressed. Bismarck's *Kulturkampf* against Catholics. Kimberley diamond fields annexed. Stanley finds Livingstone. Purchase of army commissions stopped. Religious tests abolished at Oxford, Cambridge, and Durham universities. Anne Clough starts house of residence at Cambridge (Newham College in 1876). Starley's 'penny-farthing' bicycle. Rossetti, *Dante's Dream*. Leighton, *Hercules Wrestling with Death for the Body of Alcestis*.	Grote d. Robertson d. W. H. Davies b. Synge b. Valéry b. Dostoevsky, *Devils* (to 1872). Zola, *La Fortune des Rougon*.

Fiction	*Verse, Prose, Drama*
Ainsworth, *Hilary St Ives* (1870). Blackmore, *Lorna Doone*. M. Collins, *Ivory Gate*. W. Collins, *Man and Wife* (1870). Fullerton, *Mrs Gerald's Niece*. Le Fanu, *Wyvern Mystery*. Manning, *Spanish Barber*. Payn, *A County Family, A Perfect Treasure*. Reade, *Put Yourself in his Place* (1870). A. Trollope, *Vicar of Bullhampton* (1870). T. Trollope, *Garstangs of Garstang Grange*. Wood, *Bessie Rane* (1870), *George Canterbury's Will* (1870).	Clough, *Poems and Prose Remains*; Gilbert, *Bab Ballads*; Tennyson, *Holy Grail and Other Poems*. Arnold, *Culture and Anarchy*; Galton, *Hereditary Genius*; Lecky, *History of European Morals*; Mill, *On the Subjection of Women*. Robertson, *School, Progress*.
Braddon, *Fenton's Quest* (1871). Broughton, *Red as a Rose Is She*. M. Collins, *Vivian Romance*. Dickens, *Mystery of Edwin Drood*. Disraeli, *Lothair*. Hatton, *Behind the Mask*. Lawrence, *Anteros* (1871). Lever, *Lord Kilgobbin* (1872). MacDonald, *Wilfred Cumbermede* (1872). Meredith, *Harry Richmond* (1871). M. Oliphant, *John, Three Brothers*. Payn, *Gwendoline's Harvest* (1872). A. Trollope, *Sir Harry Hotspur* (1871), *Ralph the Heir* (1871). Wood, *Dene Hollow* (1871). Yonge, *Pillars of the House* (1873).	Rossetti, *Poems*. Arnold, *St Paul and Protestantism*; Newman, *A Grammar of Assent*. Gilbert, *Palace of Truth*; Reade, *Put Yourself in his Place, Robust Invalid*; Robertson, *Birth*.
Blackmore, *Maid of Sker* (1872). Braddon, *Lovels of Arden*. Bulwer, *Coming Race*. M. Collins, *Marquis and Merchant, Two Plunges for a Pearl* (1872). W. Collins, *Poor Miss Finch* (1872). Eliot, *Middlemarch* (1872). Hardy, *Desperate Remedies*. Hatton, *In the Lap of Fortune* (1873). H. Kingsley, *Old Margaret*. Le Fanu, *Checkmate, Rose and Key, Chronicles of Golden Friars*. MacDonald, *At the Back of the North Wind, Vicar's Daughter* (1872). M. Oliphant, *Squire Arden*. Payn, *Cecil's Tryst* (1872), *Like Father like Son, Not Wooed but Won*. Reade, *A Terrible Temptation*. A. Trollope, *Eustace Diamonds* (1873). T. Trollope, *Durnton Abbey*. Wood, *Within the Maze* (1872).	Browning, *Balaustion's Adventure*; Swinburne, *Songs before Sunrise*. Arnold, *Friendship's Garland*; Darwin, *Descent of Man*; Ruskin, *Fors Clavigera*. Wills, *Hinko*.

Date	General History	Literary History
1872	Trading posts on Gold Coast bought from Holland. Viceroy of India murdered. Jesuits expelled from Germany. Conscription in France. Ballot Act in England. Rossetti, *Bower Meadow*. Bizet, *L'Arlésienne*.	Maurice d. Beerbohm b. Russell b. Lever d. Gautier d. Nietzsche, *Geburt der Tragödie*. Zola, *La Curée*.
1873	Napoleon III d. at Chislehurst. Ashanti War. Falk Laws in Germany. Russians annexe Khiva. Remington typewriter. Moody and Sankey evangelize England. Cambridge University starts extension lectures. Butterfield, Keble College Chapel (to 1876). Rossetti, *La Ghirlandata*. Brahms, *German Requiem* performed in London.	Bulwer-Lytton d. Le Fanu d. Mill d. Walter de la Mare b. Rimbaud, *Une Saison en enfer*. Galdós, *Trafalgar*. Tolstoy, *Anna Karenina* (to 1877). Verne, *Le Tour du monde en 80 Jours*. Zola, *Le Ventre de Paris*.
1874	Duke of Edinburgh marries Grand Duchess Marie Alexandrovna. Disraeli PM. LSE opened to women. Strike of agricultural workers. Fiji Islands annexed. Tichborne claimant convicted of perjury. Impressionist exhibition in Paris. Butler, *Mr Heatherley's Holiday*. Burne-Jones, *Beguiling of Merlin*. Strauss, *Fledermaus*. Smetana, *Má Vlast* (to 1879).	Dobell d. Baring b. Chesterton b. Winston Churchill b. Hugo, *Quatre-vingt-treize*.
1875	France becomes Republic. Wreck of *Deutschland*. Disraeli buys Suez Canal shares. Public Health Act. Prince of Wales visits India. Madame Blavatsky founds Theosophical Society. Mary Baker Eddy, *Science and Health*. Rossetti, *Blessed Damozel* (to 1878). Bizet , *Carmen*. Verdi conducts his *Requiem* in London.	Finlay d. C. Kingsley d. Lyell d. Thirlwall d. *Encyclopaedia Britannica*, 9th edn. (to 1889). Henry James, *A Passionate Pilgrim*.

Fiction	*Verse, Prose, Drama*
Ainsworth, *Boscobel*. Braddon, *Robert Ainsleigh, Strangers and Pilgrims* (1873), *To the Bitter End*. Bulwer, *Parisians* (1873). Butler, *Erewhon*. 'Carroll', *Through the Looking Glass*. M. Collins, *Princess Clarice*. W. Collins, *New Magdalen* (1873). Hardy, *Under the Greenwood Tree*. Hatton, *Clytie* (1874). H. Kingsley, *Valentin, Harveys*. Le Fanu, *In a Glass Darkly*. Linton, *Joshua Davidson*. M. Oliphant, *At his Gates, Ombra*. Payn, *Woman's Vengeance*. Reade, *Wandering Heir*. M. Taylor, *Seeta*. T. Trollope, *Stillwinches of Combe Mavis*. Wood, *Master of Greylands* (1873).	Browning, *Fifine at the Fair*; Morris, *Love is Enough*; Tennyson, *Gareth and Lynette*. Bagehot, *Physics and Politics*; Forster, *Life of Dickens* (to 1874). Wills, *Charles I*.
Ainsworth, *Good Old Times*. Braddon, *Lost for Love* (1874), *Lucius Davoren*. Broughton, *Nancy*. Bulwer, *Kenelm Chillingly*. M. Collins, *Mr Carrington, Squire Silchester's Whim*. H. Kingsley, *Oakshott Castle*. Lawrence, *Silverland*. Le Fanu, *Willing to Die*. M. Oliphant, *May*. Payn, *At her Mercy* (1874). A. Trollope, *Lady Anna* (1874), *Phineas Redux* (1874).	Browning, *Red Cotton Night-Cap Country*. Arnold, *Literature and Dogma*; Mill, *Autiobiography*; Newman, *Idea of a University*; Pater, *Studies in the Renaissance*; Stephen, *Freethinking and Plainspeaking*.
Ainsworth, *Merry England*. Blackmore, *Alice Lorraine* (1875). Braddon, *Hostages to Fortune* (1875), *Taken at the Flood*. M. Collins, *Frances, Transmigration*. Hardy, *Far from the Madding Crowd*. H. Kingsley, *Reginald Hetherege*. Lawrence, *Hagarene*. Linton, *Patricia Kemball* (1875). Manning, *Lord Harry Bellair, Monk's Norton*. Meredith, *Beauchamp's Career* (1876). M. Oliphant, *For Love and Life, A Rose in June*. Payn, *Best of Husbands*. A. Trollope, *Way We Live Now* (1875). Yonge, *Lady Hester, Three Brides* (1876). Wood, *Johnny Ludlow* (6 series, to 1889).	Thomson, *City of Dreadful Night*. Green, *Short History of the English People*; Sidgwick, *Methods of Ethics*; Stephen, *Hours in a Library* (to 1879); Stubbs, *Constitutional History* (to 1878). Gilbert, *Sweethearts*; Simpson, *Lady Dedlock's Secret*.
Ainsworth, *Goldsmith's Wife, Preston Fight*. Braddon, *A Strange World, Joshua Haggard's Daughter* (1876). M. Collins, *Sweet and Twenty*. W. Collins, *Law and the Lady*. H. Kingsley, *Number Seventeen*. Linton, *Atonement of Leam Dundas* (1876). MacDonald, *Malcom*, M. Oliphant, *Valentine and his Brother, Whiteladies*. Payn, *Halves* (1876), *Walter's Word*. A. Trollope, *Prime Minister* (1876). Wood, *Bessie Wells, Edina* (1876).	Browning, *Inn Album*; Hopkins writes 'Wreck of the Deutschland'. Pattison, *Isaac Casaubon*; Ruskin, *Proserpina* (to 1886). Gilbert, *Tom Cobb*, (with Sullivan) *Trial by Jury*; Hatton, *Clytie*.

Date	General History	Literary History
1876	Turks massacre Bulgarians. Disraeli becomes Earl of Beaconsfield. Bell patents telephone. Otto's four-stroke internal-combustion engine. Rossetti, *Spirit of the Rainbow*. Brahms, first symphony. Wagner, *Ring des Nibelungen*. Tchaikovsky, *Swan Lake*.	Forster d. H. Kingsley d. Harriet Martineau d. G. M. Trevelyan b. Daudet, *Jack*. Galdós, *Doña Perfeta*. Henry James, *Roderick Hudson*. George Sand d.
1877	Queen proclaimed Empress of India. Transvaal annexed. Praxiteles' *Hermes* found at Olympia. Edison invents phonograph. Pasteur starts work on causes of infectious diseases. Burne-Jones, *Perseus Slaying the Sea-Serpent* (*c.*1875–7). Rossetti, *Astarte Syriaca*. Whistler, *Nocturne in Black and Gold*. Brahms, second symphony.	Bagehot d. Granville-Barker b. *Nineteenth Century*. Morris, Webb, Faulkner found Society for Protection of Ancient Buildings. Ibsen, *Pillars of Society*. Henry James, *American*. Zola, *L'Assommoir*.
1878	'Jingoism' against Russia. Beaconsfield gets Cyprus from Turkey in return for promise of protection. Afghan War. First electric street-lighting in London. Booth founds Salvation Army. 'Cleopatra's Needle' brought to London. London University opens all degrees to women. Woolner, *Lady Godiva Unrobing*.	G. H. Lewes d. Masefield b. Thomas b. Whistler gets farthing damages against Ruskin for libel. *English Men of Letters* series. Fontane, *Vor dem Sturm*. Henry James, *Europeans*.
1879	Zulu War. British legation massacred at Kabul: Afghanistan invaded. Electric light bulb invented. Ayrton addresses British Association on 'Electricity as a Motive Power'. Somerville Hall (later College) founded at Oxford. Pearson, Truro Cathedral (to 1910). Tchaikovsky, *Eugene Onegin*. Joachim plays Brahms, violin concerto.	E. M. Forster b. Monro b. *Cambridge Review*. Skeat's *Etymological Dictionary of the English Language*. Lewis and Short's *Latin Dictionary*. Grove's *Dictionary of Music and Musicians* (to 1889). Daudet, *Les Rois en exil*. Dostoevsky, *Brothers Karamazov* (to 1880). Henry George, *Progress and Poverty*. Ibsen, *A Doll's House*. Henry James, *Daisy Miller*.

Fiction	*Verse, Prose, Drama*
Ainsworth, *Chetwynd Calverley, Leaguer of Lathom*. Blackmore, *Cripps the Carrier, Erema* (1877). Broughton, *Joan*. Bulwer, *Pausanias the Spartan*. W. Collins, *Two Destinies*. Eliot, *Daniel Deronda*. H. Kingsley, *Grange Garden*. MacDonald, *Marquis of Lossie* (1877), *St George and St Michael, Thomas Wingfold Curate*. M. Oliphant, *Curate in Charge, Phoebe Junior*. Payn, *Fallen Fortunes, What he cost her* (1877). A. Trollope, *American Senator* (1877). Wood, *Adam Grainger*.	Browning, *Pacchiarotto*; 'Carroll', *Hunting of the Snark*; Tennyson, *Harold*. Bradley, *Ethical Studies*; Stephen, *History of English Thought in the Eighteenth Century*. Tennyson, *Queen Mary*.
Ainsworth, *Fall of Somerset*. Hatton, *Gay World, Queen of Bohemia*. H. Kingsley, *Mystery of the Island*. M. Oliphant, *Carita, Mrs Arthur, Young Musgrave*. Payn, *By Proxy* (1878). A. Trollope, *Is He Popenjoy?* (1878). Wood, *Pomeroy Abbey* (1878). Yonge, *Magnum Bonum* (1879).	Hopkins writes 'God's Grandeur', 'Windover', etc.; Patmore, *Unknown Eros*. Martineau, *Autobiography*; Meredith, *Idea of Comedy*. Gilbert, *Engaged*, (with Sullivan) *Sorcerer*; Pinero, *Two Hundred a Year*.
Ainsworth, *Beatrice Tyldesley*. Braddon, *An Open Verdict, Vixen*. W. Collins, *Fallen Leaves* (1879), *Haunted Hotel*. M. Collins, *You Play Me False*. Hardy, *Return of the Native*. Hatton, *Cruel London*. MacDonald, *Sir Gibbie*. Payn, *Less Black than We're Painted, Under One Roof* (1879). M. Taylor, *A Noble Queen*. A. Trollope, *An Eye for an Eye*. (1879). *John Caldigate* (1879).	Browning, *La Saisiaz*; Patmore, *Amelia*. Lecky, *History of England in the Eighteenth Century*; Stevenson, *An Inland Voyage*. Gilbert and Sullivan, *HMS Pinafore*; Pinero, *La Comète, Two can Play at that Game*.
Ainsworth, *Beau Nash*. Blackmore, *Mary Anerley* (1880). Braddon, *Cloven Foot*. Linton, *Under which Lord?* MacDonald, *Paul Faber Surgeon*. Meredith, *Egoist*. M. Oliphant, *Greatest Heiress in England, Within the Precincts*. Payn, *Confidential Agent*. A. Trollope, *Cousin Henry, Duke's Children* (1880).	Browning, *Dramatic Idyls* (to 1880); Hopkins writes 'Binsey Poplars', etc.; Morris, *Gwen*. Arnold, *Mixed Essays*; Mill, 'Chapters on Socialism'; Morris, 'Art of the People'; Stevenson, *Travels with a Donkey*. Pinero, *Daisy's Escape*; Tennyson, *Falcon*.

Date	*General History*	*Literary History*
1880	Gladstone PM. Bradlaugh, MP, refuses to swear on Bible. Attendance at elementary schools made compulsory. Manchester University founded. Burne-Jones, *Golden Stairs*. Dvořák, symphony in D.	George Eliot d. H. Taylor d. Noyes b. Lytton Strachey b. Ward's *English Poets* (to 1881). Ibsen's *Pillars of Society* (tr. Archer, as *Quicksands*) staged in London. Zola, *Nana*. Apollinaire b.

Fiction	*Verse, Prose, Drama*
Braddon, *Asphodel* (1881), *Just as I Am, Story of Barbara*. Broughton, *Second Thoughts*. W. Collins, *Black Robe* (1881), *Jezebel's Daughter*. Disraeli, *Endymion*. Le Fanu, *Purcell Papers*. Linton, *Rebel of the Family*. Meredith, *Tragic Comedians*. M. Oliphant, *He that Will not when He May*. Payn, *Confidential Agent*. A. Trollope, *Dr Wortle's School* (1881). Wood, *Court Netherleigh* (1881).	Hopkins writes 'Felix Randal', 'Brothers', etc. Arnold, 'Study of Poetry'; Huxley, 'Science and Culture'. Gilbert and Sullivan, *Pirates of Penzance*; Pinero, *Bygones*, *Hester's Mystery*, *Moneyspinner*.

Select Bibliography

Note. Selection is subject to present editorial policy, which excludes journal articles and items in the bibliographies listed, and in that in Vol. XIV (with a few exceptions for the reader's convenience).

1. Bibliographical Guides

Ford, G. H. (ed.), *Victorian Fiction: A Second Guide to Research* (New York, 1978).

Freeman, R. E. (ed.), *Bibliographies of Studies in Victorian Literature for the Ten Years, 1965–1974* (New York, 1981).

Harris, M., *A Checklist of the 'Three Decker' Collection in the Fisher Library, University of Sydney* (Sydney, 1980).

Rosenbaum, B., and White, P., *Index of English Literary Manuscripts, 4. 1, 1800–1900, Arnold–Gissing* (London, 1982).

Sadleir, M., *XIX Century Fiction: A Bibliographical Record Based on his own Collection* (Cambridge, 1951).

Stevenson, L. (ed.), *Victorian Fiction: A Guide to Research* (Cambridge, Mass., 1964).

Vann, J. D., *Victorian Novels in Serial* (New York, 1985).

Watson, G. (ed.), *The New Cambridge Bibliography of English Literature, 3, 1800–1900* (Cambridge, 1969).

Wolff, R. L., *Nineteenth-Century Fiction: A Bibliographical Catalogue Based on the Collection Formed by Robert Lee Wolff* (New York, 1986).

2. Background

Barker, F. *et al.* (edd.), *1848: The Sociology of Literature: Proceedings of the Essex Conference . . .* (Colchester, 1978).

Bentley, M., *Politics without Democracy, 1815–1914: Perception and Preoccupation in British Government* (London, 1984).

Bourne, J. M., *Patronage and Society in Nineteenth-Century England* (Baltimore, Md., 1986).

Brown, L., *Victorian News and Newspapers* (Oxford, 1985).

Burnett, J., *Destiny Obscure: Autobiographies and Childhood, Education and Family from the 1820s to the 1920s* (London, 1982).

Chaloner, W. H., and Richardson, R. C., *Bibliography of British Economic and Social History* (Manchester, 1984).

Chandos, J., *Boys Together: The English Public School, 1800–1864* (London, 1983).

Chapple, J. A. V., *Science and Literature in the Nineteenth Century* (London, 1986).

Church, R. A., *The Great Victorian Boom, 1850–1873* (London, 1975).
Colloms, B., *Victorian Visionaries* (London, 1982).
Crowther, M. A., *The Workhouse System, 1834–1929* (London, 1981).
Dennis, B., and Skilton, D. (edd.), *Reform and Intellectual Debate in Victorian England* (London, 1987).
Faber, R., *Young England* (London, 1987).
Flint, K. (ed.), *The Victorian Novelist: Social Problems and Social Change* (London, 1987).
Gallagher, C., *The Industrial Reformation of English Fiction: Social Discourse and Narrative Form, 1832–1867* (Chicago, 1985).
Girouard, M., *The Return to Camelot: Chivalry and the English Gentleman* (New Haven, Conn., 1981).
Golby, J. M. (ed.), *Culture and Society in Britain, 1850–1890: A Source Book of Contemporary Writings* (Oxford, 1986).
Goodway, D., *London Chartism, 1838–1848* (Cambridge, 1982).
Gourvish, T. R., *Railways and the British Economy, 1830–1914* (London, 1980).
Helmstadter, R. J., and Phillips, P. T. (edd.), *Religion in Victorian Society: A Sourcebook of Documents* (Lanham, Md., 1985).
Jay, E. (ed.), *Faith and Doubt in Victorian Britain* (London, 1986).
Jordanova, L. (ed.), *The Languages of Nature: Critical Essays on Science and Literature* (New Brunswick, NJ, 1986).
Kindleberger, C. P., *Manias, Panics, and Crashes: A History of Financial Crises* (London, 1978).
Lindberg, D. C., and Numbers, R. L. (edd.), *God and Nature: Historical Essays on the Encounter between Christianity and Science* (Berkeley, Calif., 1986).
McLevy, J., *The Casebook of a Victorian Detective* (Edinburgh, 1975).
Mason, P., *The English Gentleman: The Rise and Fall of an Ideal* (London, 1982).
Mingay, G. E., *The Transformation of Britain, 1830–1939* (London, 1986).
Norman, E., *The Victorian Christian Socialists* (Cambridge, 1987).
Phillipps, K. C., *Language and Class in Victorian England* (Oxford, 1985).
Rose, M. E. (ed.), *The Poor and the City: The English Poor Law in its Urban Context, 1834–1914* (New York, 1985).
Scull, A. (ed.), *Madhouses, Mad-Doctors and Madmen: The Social History of Psychiatry in the Victorian Era* (London, 1982).
Stein, R. L., *Victoria's Year: English Literature and Culture, 1837–1838* (Oxford, 1987).
Strange, K. H., *The Climbing Boys: A Study of Sweeps' Apprentices* (London, 1981).
Valenze, D. M., *Prophetic Sons and Daughters: Female Preaching and Popular Religion in Industrial England* (Princeton, NJ, 1985).
Vance, N., *The Sinews of the Spirit: The Ideal of Christian Manliness in Victorian Literature and Religious Thought* (Cambridge, 1985).
Vicinus, M., *Independent Women: Work and Community for Single Women, 1850–1920* (Chicago, 1985).
Walton, J. K., *A Social History of Lancashire, 1558–1939* (Manchester, 1986).
Wright, T. R., *The Religion of Humanity: The Impact of Comtean Positivism on Victorian Britain* (Cambridge, 1986).

3. General Studies

Armstrong, N., *Desire and Domestic Fiction: A Political History of the Novel* (Oxford, 1987).

Bann, S., *The Clothing of Clio: A Study of the Representation of History in Nineteenth-Century Britain and France* (Cambridge, 1984).

Kaplan, F., *Sacred Tears: Sentimentality in Victorian Literature* (Princeton, NJ, 1987).

Levine, G., *The Realistic Imagination* (Chicago, 1981).

Thesing, W. B., *Dictionary of Literary Biography:* 55, *Victorian Prose Writers before 1867* (Detroit, 1987); 57, *Victorian Prose Writers after 1867* (Detroit, 1988).

Wheeler, M., *English Fiction of the Victorian Period* (London, 1985).

4. Special Studies

Alexander, J. H., and Hewitt, D. (edd.), *Scott and his Influence* (Aberdeen, 1983).

Argyle, G., *German Elements in the Fiction of George Eliot, Gissing, and Meredith* (Frankfurt am Main, 1979).

Ashton, R., *The German Idea: Four English Writers* [Coleridge, Carlyle, G. H. Lewes, George Eliot] *and the Reception of German Thought, 1800–1860* (Cambridge, 1980).

Carlisle, J., *The Sense of an Audience: Dickens, Thackeray, and George Eliot at Mid-Century* (Brighton, 1981).

Chapman, R., *The Sense of the Past in Victorian Literature* (London, 1986).

Cross, N., *The Common Writer: Life in Nineteenth-Century Grub Street* (Cambridge, 1985).

David, D., *Fictions of Resolution in . . . 'North & South', 'Our Mutual Friend', 'Daniel Deronda'* (London, 1981).

—— *Intellectual Women and Victorian Patriarchy: Harriet Martineau, Elizabeth Barrett Browning and George Eliot* (Ithaca, NY, 1987).

Edwards, P. D., *Some Mid-Victorian Thrillers: The Sensation Novel, its Friends and its Foes* (St Lucia, Queensland, 1971).

Eigner, E., and Worth, G. (edd.), *Victorian Criticism of the Novel* (Cambridge, 1985).

Engel, E., and King, M. F., *The Victorian Novel before Victoria* (London, 1984).

Feltes, N. N., *Modes of Production of Victorian Novels* (Chicago, 1986).

Foster, S., *Victorian Women's Fiction: Marriage, Freedom, and the Individual* (London, 1985).

Garrett, P. K., *The Victorian Multiplot Novel: Studies in Dialogical Form* (New Haven, Conn., 1980).

Gilmour, R. *The Idea of the Gentleman in the Victorian Novel* (London, 1981).

Glynn, J., *The Prince of Publishers: A Biography of George Smith* (London, 1986).

Gregor, I. (ed.), *Reading the Victorian Novel: Detail into Form* (London, 1980).

Hardy, B., *Forms of Feeling in Victorian Fiction* (London, 1985).

Hawthorn, J. (ed.), *The Nineteenth-Century British Novel* (London, 1986).

Houfe, S., *The Dictionary of British Book Illustrators and Caricaturists, 1800–1914* (London, 1978).
Hughes, W., *The Maniac in the Cellar: Sensation Novels of the 1860s* (Princeton, NJ, 1980).
Jay, E., *The Religion of the Heart: Anglican Evangelicalism and the Nineteenth-Century Novel* (Oxford, 1978).
Kearns, M. S., *Metaphors of Mind in Fiction and Psychology* (Lexington, Ky., 1987).
Kettle, A., *The Nineteenth-Century Novel: Critical Essays and Documents* (London, 1981).
Klaus, H. G., *The Literature of Labour: 200 Years of Working-Class Literature* (Brighton, 1985).
Lascelles, M., *The Story-Teller Retrieves the Past: Historical Fiction and Fictitious History in the Art of Scott, Stevenson, Kipling and Some Others* (Oxford, 1980).
Lowenthal, D., *The Past is a Foreign Country* (Cambridge, 1985).
Muresianu, S. A., *The History of the Victorian Christmas Book* (New York, 1987).
Nestor, P., *Female Friendships and Communities: Charlotte Brontë, George Eliot, Elizabeth Gaskell* (Oxford, 1985).
Olmsted, J. C. (ed.), *A Victorian Art of Fiction: Essays on the Novel in British Periodicals, 1830–1850* (New York, 1979); *1851–1869* (New York, 1979); *1870–1900* (New York, 1980).
Ousby, I., *Bloodhounds of Heaven: The Detective in English Fiction from Godwin to Doyle* (Cambridge, Mass., 1976).
Postlethwaite, D., *Making it Whole: A Victorian Circle and the Shape of their World* [George Eliot, G. H. Lewes, Harriet Martineau, Herbert Spencer, Robert Chambers] (Columbus, Ohio, 1984).
Qualls, B. V., *The Secular Pilgrims of Victorian Fiction: The Novel as Book of Life* (Cambridge, 1982).
Ross, A. M., *The Imprint of the Picturesque on Nineteenth-Century British Fiction* (Waterloo, Ont., 1986).
Sanders, A., *The Victorian Historical Novel, 1840–1880* (London, 1978).
Smith, P., *Public and Private Value: Studies in the Nineteenth-Century Novel* (Cambridge, 1984).
Smith, S. M., *The Other Nation: The Poor in English Novels of the 1840s and 1850s* (Oxford, 1980).
Srebrnik, P. T., *Alexander Strahan: Victorian Publisher* (Ann Arbor, Mich., 1986).
Stewart, G., *Death Sentences: Styles of Dying in British Fiction* (Cambridge, Mass., 1984).
Stone, D. D., *The Romantic Impulse in Victorian Fiction* (Cambridge, Mass., 1980).
Sullivan, A. (ed.), *British Literary Magazines, 3, The Victorian and Edwardian Age, 1837–1913* (Westport, Conn., 1984).
Sutherland, J. A., *Victorian Novelists and Publishers* (London, 1977).
Terry, R. C., *Victorian Popular Fiction, 1860–1880* (London, 1983).
Thomson, P., *George Sand and the Victorians: Her Influence and Reputation in Nineteenth-Century England* (London, 1977).

Weiner, J. H. (ed.), *Innovators and Preachers: The Role of the Editor in Victorian England* (Westport, Conn., 1985).
Wheeler, M., *The Art of Allusion in Victorian Fiction* (London, 1979).
Wilson, C., *First with the News: The History of W. H. Smith* (London, 1985).
Wolff, R. L., *Gains and Losses: Novels of Faith and Doubt in Victorian England* (New York, 1977).
Yeazell, R. B. (ed.), *Sex, Politics, and Science in the Nineteenth-Century Novel: Selected Papers from the English Institute, 1983–84* (Baltimore, Md., 1986).

5. Individual Authors

Ainsworth, William Harrison

Worth, G. J., *William Harrison Ainsworth* (New York, 1972).

Blackmore, Richard Doddridge

Sutton, M. K., *R. D. Blackmore* (Boston, Mass., 1979).

Borrow, George

Collie, M., and Fraser, A., *George Borrow: A Bibliographical Study* (Winchester, 1984).

Braddon, Mary Elizabeth

Wolff, R. L., *Sensational Victorian: The Life and Fiction of M. E. Braddon* (New York, 1979).

The Brontës

Berg, M., *Jane Eyre: Portrait of a Life* (Boston, Mass., 1987).
Chitham. E., *The Brontës' Irish Background* (London, 1986).
—— *A Life of Emily Brontë* (Oxford, 1987).
Crump, R. W., *Charlotte and Emily Brontë, 1846–1915: A Reference Guide* (Boston, Mass., 1982); *1916–1954* (1985); *1955–1983* (1986).

Bulwer, Edward George Earle Lytton

Campbell, J. L., *Edward Bulwer-Lytton* (Boston, Mass., 1986).

Carleton, William

Boué, A., *William Carleton, 1794–1869: Romancier Irlandais* (Paris, 1973).
Chesnutt, M., *Studies in the Short Stories of William Carleton* (Göteborg, 1976).
Hayley, B., *A Bibliography of the Writings of William Carleton* (Gerrard's Cross, 1985).
—— *Carleton's 'Traits and Stories' and the Nineteenth-Century Anglo-Irish Tradition* (Gerrard's Cross, 1983).
Sullivan, E. A., *William Carleton* (Boston, Mass., 1983).
Wolff, R. L., *William Carleton, Irish Peasant Novelist: A Preface to his Fiction* (New York, 1979).

Chamier, Frederick

Van der Voort, P. J., *The Pen and the Quarter-Deck: a Study of the Life and Works of Captain Frederick Chamier, R.N.* (Leiden, 1972).

Collins, Wilkie

Andrew, R. V., *Wilkie Collins* (New York, 1979).

Beetz, K. H., *Wilkie Collins: An Annotated Bibliography, 1889–1976* (London, 1978).

Lonoff, S., *Wilkie Collins and his Victorian Readers* (New York, 1982).

Dickens, Charles

Andrews, M., *Dickens on England and the English* (Hassocks, Sussex, 1979).

Bolton, H. P., *Dickens Dramatized* (London, 1987).

Brown, J. M., *Dickens: Novelist in the Market-Place* (London, 1982).

Cohen, J. R., *Charles Dickens and his Original Illustrators* (Columbus, Ohio, 1980).

Collins, P. (ed.), *Dickens: Interviews and Recollections* (London, 1981).

Daldry, G., *Charles Dickens and the Form of the Novel* (London, 1987).

Dunn, R. J., *David Copperfield: An Annotated Bibliography* (and subsequent Garland Bibliographies of individual novels, New York, 1981–)

Fenstermaker, J. J., *Charles Dickens, 1940–1975: An Analytical Subject Index to Periodical Criticism of the Novels and Christmas Books* (Boston, Mass., 1979).

Giddings, R. (ed.), *The Changing World of Charles Dickens* (London, 1983).

Hollington, M., *Dickens and the Grotesque* (London, 1984).

Horton, S. R., *The Reader in the Dickens World: Style and Response* (London, 1981).

Kaplan, F. (ed.), *Charles Dickens' Book of Memoranda* (New York, 1981).

McMaster, J., *Dickens the Designer* (Totowa, NJ, 1987).

MacPike, L., *Dostoevsky's Dickens: A Study of Literary Influence* (London, 1981).

Moss, S. P., *Charles Dickens' Quarrel with America* (New York, 1984).

Oppenlander, E. A., *Dickens' 'All the Year Round': Descriptive Index and Contributor List* (New York, 1984).

Patten, R. L., *Charles Dickens and his Publishers* (Oxford, 1978).

Schlicke, P., *Dickens and Popular Entertainment* (London, 1985).

Schwarzbach, F. S., *Dickens and the City* (London, 1979).

Slater, M., *Dickens and Women* (London, 1983).

—— (ed.), *Dickens on America and Americans* (Hassocks, Sussex, 1979).

Stokes, E., *Hawthorne's Influence on Dickens and George Eliot* (St Lucia, Queensland, 1985).

Stone, H., *Dickens and the Invisible World: Fairy Tales, Fantasy, and Novel-Making* (London, 1980).

Thurley, G., *The Dickens Myth: Its Genesis and Structure* (London, 1976).

Walder, D., *Dickens and Religion* (London, 1981).

Welsh, A., *From Copyright to Copperfield: The Identity of Dickens* (London, 1987).

Disraeli, Benjamin

Braun, T., *Disraeli the Novelist* (London, 1981).

Gunn, J. A. W., Matthews, J., Schurman, D. M., Weibe, M. G. (edd.), *Benjamin Disraeli: Letters*, 1, *1815–1834* (Toronto, 1982); 2, *1835–1837* (Toronto, 1982); 3, *1838–1841* (Toronto, 1987).

Hibbert, C., *Disraeli and his World* (London, 1979).

Horsman, E. A., *On the Side of the Angels? Disraeli and the Nineteenth-Century Novel* (Dunedin, NZ, 1974).

Schwarz, D. R., *Disraeli's Fiction* (London, 1979).

George Eliot

Atkins, D., *George Eliot and Spinoza* (Salzburg, 1978).
Beer, G., *George Eliot* (Brighton, 1986).
Carpenter, M. W., *George Eliot and the Landscape of Time: Narrative Form and Protestant Apocalyptic History* (Chapel Hill, NC, 1986).
Cottom, D. *Social Figures: George Eliot, Social History, and Literary Representation* (Minneapolis, 1987).
Ermarth, E. D., *George Eliot* (Boston, 1985).
Graver, S., *George Eliot and Community: A Study in Social Theory and Fictional Form* (Berkeley, Calif., 1984).
Levine, G., *An Annotated Critical Bibliography of George Eliot* (Brighton, 1988).
Myers, W., *The Teaching of George Eliot* (Leicester, 1984).
Shuttleworth, S., *George Eliot and Nineteenth-Century Science* (Cambridge, 1984).
Smith, A. (ed.), *George Eliot: Centenary Essays and an Unpublished Fragment* (London, 1980).
Uglow, J., *George Eliot* (London, 1987).
Welsh, A., *George Eliot and Blackmail* (Cambridge, Mass, 1985).
Witemeyer, H., *George Eliot and the Visual Arts* (London, 1979).

Gaskell, Elizabeth

Easson, A., *Elizabeth Gaskell* (London, 1979).
Fryckstedt, M. C., *Elizabeth Gaskell's 'Mary Barton' and 'Ruth': A Challenge to Christian England* (Stockholm, 1982).
Gérin W., *Elizabeth Gaskell: A Biography* (Oxford, 1976).
Lansbury, C., *Elizabeth Gaskell* (Boston, Mass., 1984).
Selig, R. L., *Elizabeth Gaskell: A Reference Guide* (Boston, Mass., 1977).
Stoneman, P., *Elizabeth Gaskell* (Brighton, 1987).
Welch, J., *Elizabeth Gaskell: An Annotated Bibliography, 1929–1975* (London, 1977).

Hughes, Thomas

Worth, G. J., *Thomas Hughes* (Boston, Mass., 1984).

Jewsbury, Geraldine

Fryckstedt, M. C., *Geraldine Jewsbury's 'Athenaeum' Reviews: A Mirror of Mid-Victorian Attitudes to Fiction* (Stockholm, 1986).

Kingsley, Charles

Harris, S., *Charles Kingsley: A Reference Guide* (Boston, Mass., 1981).

Kingsley, Henry

Mellick, J. S. D. (ed.), *Henry Kingsley* (St Lucia, Queensland, 1982).
—— *The Passing Guest: A Life of Henry Kingsley* (St Lucia, Queensland, 1983).
Scheuerle, W. H., *The Neglected Brother: A Study of Henry Kingsley* (Tallahassee, Fl., 1971).

Le Fanu, Sheridan

McCormack, W. J., *Sheridan Le Fanu and Victorian Ireland* (Oxford, 1980).
Melada, I., *Sheridan Le Fanu* (Boston, Mass., 1987).

Linton, Eliza Lynn

Anderson, N. F., *Woman against Women in Victorian England: A Life of Eliza Lynn Linton* (Bloomington, Ind., 1987).

Van Thal, H. M., *Eliza Lynn Linton, the Girl of the Period: A Biography* (London, 1979).

MacDonald, George

Raeper, W., *George MacDonald* (Tring, 1987).

Triggs, K., *The Stars and the Stillness: A Portrait of George MacDonald* (Cambridge, 1986).

Marryat, Frederick

Gautier, M. P., *Captain Frederick Marryat: L'Homme et l'œuvre* (Paris, 1973).

Martineau, Harriet

Pichanick, V. K., *Harriet Martineau: The Woman and her Work* (Ann Arbor, Mich., 1980).

Sanders, V., *Reason over Passion: Harriet Martineau and the Victorian Novel* (Brighton, 1986).

Thomas, G., *Harriet Martineau* (Boston, Mass., 1985).

Meredith, George

Beer, G., and Harris, M., *The Notebooks of George Meredith* (Salzburg, 1983).

Collie, M., *George Meredith: A Bibliography* (London, 1974).

Moses, J., *The Novelist as Comedian: George Meredith and the Ironic Sensibility* (New York, 1983).

Olmsted, J. C., *George Meredith: An Annotated Bibliography of Criticism, 1925–1975* (New York, 1978).

Stone, J. S., *George Meredith's Politics as Seen in his Life, Friendships, and Works* (Port Credit, Ont., 1986).

Norton, Caroline

Hoge, J. O., and Olney, C., *The Letters of Caroline Norton to Lord Melbourne* (Columbus, Ohio, 1975).

Oliphant, Laurence

Taylor, A., *Laurence Oliphant* (Oxford, 1982).

Oliphant, Margaret

Clarke, J. S., *Margaret Oliphant (1828–1897): A Bibliography* (St Lucia, Queensland, 1986).

Reade, Charles

Smith, E. E., *Charles Reade* (Boston, Mass., 1976).

Ritchie, Anne Thackeray

Gérin, W., *Anne Thackeray Ritchie: A Biography* (Oxford, 1981).

Surtees, Robert Smith

Johnston-Jones, D. R., *The Deathless Train: The Life and Work of Robert Smith Surtees* (Salzburg, 1975).

Welcome, J., *The Sporting World of R. S. Surtees* (Oxford, 1982).

Thackeray, William Makepeace

Collins, P. (ed.), *Thackeray: Interviews and Recollections* (London, 1982).

Colby, R. A., *Thackeray's Canvas of Humanity: An Author and his Public* (Columbus, Ohio, 1979).

Harden, E. F., *The Emergence of Thackeray's Serial Fiction* (London, 1979).

Olmsted, J. C., *Thackeray and his Twentieth-Century Critics: An Annotated Bibliography, 1900–1975* (New York, 1977).

Peters, C., *Thackeray's Universe: Shifting Worlds of Imagination and Reality* (London, 1987).

Phillipps, K. C., *The Language of Thackeray* (London, 1978).

Trollope, Anthony

Bareham, T. (ed.), *Anthony Trollope* (London, 1980).

Edwards, P. D., *Anthony Trollope: His Art and Scope* (Hassocks, Sussex, 1978).

Hall, N. J., *Trollope and his Illustrators* (London, 1980).

—— (ed.), *The Trollope Critics* (London, 1981).

—— (ed.), *the Letters of Anthony Trollope* (Palo Alto, Calif., 1983).

Halperin, J., *Trollope and Politics: A Study of the Pallisers and Others* (London, 1977).

—— (ed.), *Trollope Centenary Essays* (London, 1982).

Hamer, M., *Writing by Numbers: Trollope's Serial Fiction* (Cambridge, 1987).

Harvey, G., *The Art of Anthony Trollope* (London, 1980).

Kendrick, W. M., *The Novel Machine: The Theory and Fiction of Anthony Trollope* (London, 1980).

Kincaid, J. R., *The Novels of Anthony Trollope* (Oxford, 1977).

Lansbury, C., *The Reasonable Man: Trollope's Legal Fiction* (Princeton, NJ, 1981).

Letwin, S. R., *The Gentleman in Trollope: Individuality and Moral Conduct* (London, 1982).

MacDonald, S. P., *Anthony Trollope* (Boston, Mass., 1987).

McMaster, J., *Trollope's Palliser Novels: Theme and Pattern* (London, 1978).

McMaster, R. D., *Trollope and the Law* (London, 1986).

Morse, D. D., *Women in Trollope's Palliser Novels* (Ann Arbor, Mich., 1987).

Olmsted, J. C., and Welch, J. E., *The Reputation of Trollope: An Annotated Bibliography, 1925–1975* (New York, 1978).

Overton, B., *The Unoffical Trollope* (Hassocks, Sussex, 1982).

Pollard, A., *Anthony Trollope* (London, 1978).

Super, R. H., *Trollope in the Post Office* (Ann Arbor, Mich., 1981).

Terry, R. C., *Anthony Trollope: The Artist in Hiding* (London, 1977).

—— (ed.), *Trollope: Interviews and Recollections* (Basingstoke, 1987).

Tracy, R., *Trollope's Later Novels* (London, 1978).

Wright, A., *Anthony Trollope: Dream and Art* (London, 1983).

Trollope, Frances

Heineman, H., *Mrs Trollope, the Triumphant Feminine in the Nineteenth Century* (Athens, Ohio, 1979).

—— *Frances Trollope* (Boston, Mass., 1984).

Johnston, J., *The Life, Manners, and Travels of Fanny Trollope: A Biography* (New York, 1978).

Yates, Edmund

Edwards, P. D., *Edmund Yates, 1831–1894: A Bibliography* (St Lucia, Queensland, 1980).

Index

Main entries are in bold figures. An asterisk indicates a biographical footnote. Novels by Victorians are separately indexed under short titles. The Chronological Table and the Bibliography are not included.